The
Regional
Geography
of Canada

The Regional Geography of Canada

Robert M. Bone

OXFORD
UNIVERSITY PRESS

ORD
TY PRESS

ord Drive, Don Mills, Ontario M3C 1J9
can.com

New York

Auckland Bangkok Bogotá Buenos Aires Calcutta
Cape Town Chennai Dar es Salaam Delhi Florence Hong Kong Istanbul
Karachi Kuala Lumpur Madrid Melbourne Mexico City Mumbai
Nairobi Paris São Paulo Singapore Taipei Tokyo Toronto Warsaw

with associated companies in Berlin Ibadan

Oxford is a trade mark of Oxford University Press
in the UK and in certain other countries

Published in Canada
By Oxford University Press

Canadian Cataloguing in Publication Data

Bone, Robert M.
  The regional geography of Canada

Includes bibliographical references and index.
ISBN 0-19-541095-5

1. Canada – Geography. I. Title.

FC75.B66 1999   917.1   C99-931336-3
F1011.3.B66 1999

Statistics Canada information is used with the permission of the Minister of Industry, as Minister
responsible for Statistics Canada. Information on the availability of the wide range of data from Statistics
Canada can be obtained from Statistics Canada's Regional Offices, its World Wide Web site at
http://www.statcan.ca, and its toll-free access number 1-800-263-1136.

Cover and Text Design: Brett J. Miller
Cover Images: Corel Photos
Maps: Paul Sneath free&Creative

2 3 4 — 02 01 00

This book is printed on permanent (acid-free) paper ∞.

Printed in Canada

Contents Overview

Detailed Contents

List of Figures

List of Tables

The purpose of this book is to introduce university students to Canada's regional geography. In studying the regional geography of Canada, the student not only gains an appreciation of the country's amazing diversity but also learns how its regions interact with one another. By developing the central theme that Canada is a country of regions, this text presents a number of images of Canada, revealing its physical, cultural, and economic diversity as well as its regional complexity. A number of features such as photos, key terms, maps, vignettes, tables, graphs, and references and further readings are designed to facilitate and enrich student learning.

The Regional Geography of Canada divides Canada into six geographic regions: Ontario, Québec, British Columbia, Western Canada, Atlantic Canada, and the Territorial North. Each region has a particular regional geography, history, population, and a unique location. These factors have determined each region's character, set the direction for its development, and created a sense of place. In examining these themes, this book underscores the dynamic nature of Canada's regional geography. Part of the dynamism of Canada's regional geography is the changing relationship between Canada's six regions. Trade liberalization provides one element of change while regional tensions provide another element of change.

Following the Second World War, the liberalization of world trade and the creation of the North American trading bloc exposed Canada's economy to new challenges and dangers. The consequences of trade liberalization were fivefold:

- Canada's old spatial economic structure was transformed from one that is national to one that is continental.
- Canada's economy was integrated into the North American market.
- Canada's manufacturing sector was restructured.
- Rates of unemployment increased.
- Canada's regional economies were reoriented to adjust to the larger North American market.

This book explores both the national and regional implications of these economic changes and proposes scenarios for greater stability in the new global economic order.

Regional tensions have always existed in Canada. These tensions often place pressure on the existing social and political systems. This pressure for change reflects the 'potential' for a redress of the division of power within Canada's federation. From this perspective, regional tensions can lead to a shift of power within a region and may even alter relations between regions and/or the federal government. In this book, three tensions or faultlines are examined. These tensions occur between Aboriginal and non-Aboriginal Canadians, French and English Canadians, and centralist and decentralist forces. Each of these tensions exists in all parts of Canada but, because of spatial variations in Canada's human and/or physical geography, each appears more prominently in a particular region or regions. For instance the Aboriginal/non-Aboriginal faultline is most deeply felt in the Territorial North, where a larger majority of the population is Aboriginal. The French/English faultline manifests itself most commonly in Québec. The centralist/decentralist faultline is frequently played out between the federal government and Western Canada, often over resources. This book explores the nature of these faultlines, the need to reach compromises, and the fact that these faultlines provide the country with its greatest strength—diversity. Ultimately, these faultlines are shown to be not divisive forces but forces of change that ensure Canada's existence as a country of regions.

Organization of the Text

This book consists of eleven chapters. Chapters 1 through 4 deal with general topics related to Canada's national and regional geographies—Canada's physical, historical, and

human geography—thereby setting the stage for a discussion of the six main geographic regions of Canada. Chapters 5 through 10 focus on these six geographic regions, and Chapter 11 provides a conclusion.

Chapter 1 discusses the nature of regions and regional geography, including the core/periphery model and its applications. Chapter 2 introduces the major physiographic regions of Canada and other elements of physical geography that affect Canada and its regions. Chapter 3 is devoted to Canada's historical geography, such as its territorial evolution and the emergence of regional tensions and regionalism. This discussion is followed, in Chapter 4, by an examination of the basic demographic, economic, and social factors that influence both Canada and its regions. This chapter explores the national and global forces that have shaped Canada's regions as well as the features of its population (size, urbanization, etc.). To sharpen our awareness of how economic forces affect local and regional developments, three major themes running throughout this text are introduced in these first four chapters. The primary theme is that Canada is a country of regions. Two secondary themes reflect the recent shift in economic circumstances and its effects on regional geography. The first theme is the integration of the North American economy, and the second is the changing world economy. These two economies, described as continentalism and globalization, exert both positive and negative impacts on Canada and its regions. These two economic forces are explored through the core/periphery model.

In Chapters 5 to 10, the text moves from a broad, national overview to a more regional focus. Each of these six chapters profiles one of Canada's large geographic regions: Ontario, Québec, British Columbia, Western Canada, Atlantic Canada, and the Territorial North. These regional chapters explore the physical and human characteristics that distinguish each region from the others and that give each region its special sense of place. To emphasize the economic specialty of each region, a pre-dominant economic activity is identified and explored through a *Key Topic*. They are: the Automobile Industry in Ontario; Hydro-Québec in Québec; Forestry in British Columbia; (Agricultural) Transition in Western Canada; the Fishing Industry in Atlantic Canada; and Megaprojects in the Territorial North. From this presentation, the unique character of each region emerges. The book concludes with Chapter 11, which discusses the future of Canada as a country of regions.

Acknowledgements

For over five years, I have relied heavily on the help of colleagues across Canada while writing this text. During that period, I have benefited from the constructive comments of anonymous reviewers selected by Oxford University Press at various stages of the preparation of this manuscript. I especially owe a debt of thanks to one of those anonymous reviewers who spiced his critical comments with words of encouragement that kept me going.

Professors at the University of Saskatchewan were particularly helpful—several provided me with their comments on specific matters that fall under their areas of specialization while some were pressed into reviewing draft versions of regional chapters. Most contributed photographs. These colleagues and friends include Alec Aitken, Bill Archibold, William Barr, Dirk de Boer, K.I. Fung, Walter Kupsch, Lawrence Martz, John McConnell, John Newell, Jim Randall, Jack Stabler, and Mike Wilson. Professors from other institutions that played a role in shaping this book include Jim Miller, Robert MacKinnon, and Gilles Viaud from Cariboo College; Hugh Gayler and Dan McCarthy from Brock University; Ben Moffat from Medicine Hat College; and Keith Storey from Memorial University.

Thanks to a sabbatical leave granted to me by the University of Saskatchewan, I was able to spend the fall term of 1996 at the Département de géographie at the Université de Laval. My generous hosts not only put up with my limited command of French but more importantly, contributed significantly to the design and content of my chapter on Québec. Among those contributing to my education were Louis-Edmond Hamelin, who must be considered the Dean of Canadian geographers; Jean-Jacques Simard, a well-known sociologist who has written extensively on Québec's north; and Benoit Robitaille, the Chair of the Département de géographie at the Université de Laval. Discussions of critical issues with Jacques Bernier, Marc St-Hilaire, and Eric Waddell proved extremely fruitful. For me, one of the highlights at Laval was the opportunity to present my ideas on Québec at one of the Friday afternoon seminars organized by Dean Louder. The constructive and stimulating responses from the cultural geographers attending that seminar were especially helpful to me.

Maps are a critical element in geography. I must thank the cartographer in the Department of Geography, Keith Bigelow, for the help he provided on the maps. The library staff in Government Documents at the University of Saskatchewan was particularly helpful in my search for detailed information of Canada's regions.

Last but not least, the staff at Oxford University Press, including Andréa Vanasse, who undertook the final editing, made the preparation of this book a pleasant experience. Jane McNulty and Valerie Ahwee, both Oxford University Press editors, deserve special mention for guiding me through some rough patches in an early version of the manuscript. Finally, a special note of appreciation to my wife, Karen.

The
Regional
Geography
of Canada

Chapter 1

Overview

Canada is a country of regions. In this chapter the concepts of geographic regions and a sense of place are introduced. In shaping these two concepts, history and geography have played a key role. Canada has six geographic regions, each with its own sense of place. In a series of regional profiles, each region's major characteristics are presented and form the basis of more detailed discussion later in this book. These regions interact with each other and with the outside world. To understand this interaction, a conceptual framework based on the core/periphery model is presented and then discussed in terms of a shift from a national version of this model to a global one. The implications for Canada and its geographic regions are then outlined.

Objectives

- Define regional geography.
- Discuss the notion of a sense of place.
- Present the geographic regions of Canada.
- Describe the dynamic nature of regions.
- Examine the core/periphery model.
- Apply that conceptual framework to Canada.
- Outline a series of regional profiles.

Regions of Canada

Introduction

A fundamental goal of geography is to gain an understanding of our changing world by accurately describing, analysing, and interpreting its physical and human characteristics. Geography's broad array of subject matter and its integrating approach originated in classical Greek times. The ancient Greek civilization stimulated the travels, writings, and map-making of scholars such as Herodotus (484–*c*.425 BC), Aristotle (384–332 BC), Thales (*c*.625–*c*.547 BC), and Ptolemy (AD 90–168). Eratosthenes (*c*.273–*c*.192 BC) coined the word 'geography' and mapped the known world of ancient Greece. By considering both human and physical aspects of a region, geographers have developed an integrative approach to the study of our world. This approach, which is the essence of geography, separates geography from other disciplines. The richness of geography is revealed in Vignette 1.1, drawn from

---

Vignette 1.1

The Nature of Geography

[Geography is] the study of the earth's surface as the space within which the human population lives. The word comes from the Greek *geo,* the earth, and *graphein,* to write. Perhaps the best-known formal definition of the field was provided by the American geographer, Richard Hartshorne, in his *Perspective on the Nature of Geography* (1959): 'geography is concerned to provide accurate, orderly, and rational description and interpretation of the variable character of the Earth surface'. The last two terms in this definition need some elaboration. By 'variable character' geographers mean the spatial variation that can occur between the character of the earth's surface at one location and another. This variation may occur at all map scales, from the globe itself, say between continent and continent, down to a local level, say between one district and another within an urban area. By 'Earth surface' is meant that rather thin shell, only one-thousandth of the planet's circumference thick, that forms the habitat or environment within which the human population is able to survive.

Source: Peter Haggett, 'Geography', in *The Dictionary of Human Geography*, edited by R.J. Johnston, Derek Gregory, and David M. Smith (Oxford: Blackwell, 1986):175. Reprinted by permission of Blackwell Publishers.

The Dictionary of Human Geography, in which Peter Haggett's incisive definition of geography is provided.

Regional geography simplifies the fundamental goal of geography by dividing the complex world into smaller, more manageable spatial units called regions. A **region** is an area of the earth's surface that has distinctive characteristics. These characteristics vary according to the criteria geographers use.[1] A region may be defined in terms of its physical or human characteristics (Vignette 1.2). The Rocky Mountains is an example of a physical region, while Acadia is a cultural region. Physical characteristics of a region include:

- geographic location (latitude and longitude)
- landforms (mountains, plateaux, and plains)
- climate
- soil
- natural vegetation

Human or cultural characteristics include:

- culture (history, language, ethnicity, and religion)
- economy (resources, industries, and transportation)
- political identity (boundaries, political structure, and international relations)
- demographics (population and migration)
- urbanization
- a sense of place

A **sense of place** is a powerful psychological bond between people and a region. The physical and human characteristics of a region combine to form this psychological bond. Canada, with its vast spaces and magnificent diversity of physical settings, lends itself to strong regional bonds. The sea, for instance, exerts a profound influence on Maritimers who live along its coastline. The land-locked Prairie region also exerts a powerful influence

Vignette 1.2

Cultural Regions

Cultural regions are classified into three types: formal, functional, and vernacular. A formal cultural region is an area where people have one or more cultural traits in common. The Inuit, for example, inhabit the Canadian Arctic. They have a number of common cultural traits, including language (Inuktitut). A functional cultural region is distinguished by its political, social, or economic functions. Provinces and territories represent functional administrative regions. The Québec govern-ment has also engaged in a number of social functions designed to promote the franco-phone language and culture. Vernacular cultural regions are a third type. Such regions are held together by a sense of belonging. People living in vernacular regions have had a number of historical experiences that combine to provide them with a sense of regional identity. The Canadian Prairies, Central Canada, and the Maritimes are examples of vernacular regions.

Low tide in Peggy's Cove, Nova Scotia. (*Courtesy Walter Kupsch*)

on its inhabitants. The North has a cold climate to which people have had to adapt in order to survive. Residents living and working in any area are exposed to contemporary issues and memories of past events that continue to shape their lives. Not all these memories are pleasant ones. For example, the Dust Bowl of the 1930s deeply marked Prairie farmers, forcing many to abandon their homesteads. While the Dust Bowl has faded into past collective memory, this semi-arid environment continues to influence life on the Prairies, making drought a real threat to Prairie agriculture. The concept of a sense of place, then, recognizes that people living in a region have undergone a collective experience

that leads to shared aspirations, concerns, goals, and values. Over time, such experiences develop into a social cohesiveness among those people living within a spatial unit. Often these experiences are coloured by extreme weather events in a region. The rainstorm and subsequent flooding in the Saguenay Valley of Québec in July 1996 was such an experience.

A sense of place can also evolve from a region's history and geography. Early in its history, British Columbia was isolated from the rest of Canada by the Rocky Mountains, which form the largest mountain system in Canada, extending 1200 km from the US–Canada border northward towards 60° N. While this formidable physical barrier, considered by many

to be the backbone of Canada, has been sur-
mounted by modern transportation systems, it
remains an important physical and cultural
feature unique to British Columbia. In his
book, *The Unknown Country,* Hutchison
(1942) described the Rocky Mountains' effect
on people as he observed changes in the landscape from a passenger car on the Canadian
Pacific train: 'Unlike the life of the Prairies or
the East—so unlike it that, crossing the Rockies, you are in a new country, as if you had
crossed a national frontier. Everyone feels it,
even the stranger feels the change of outlook,
tempo and attitude. What makes it so, I do not
know.' An equally powerful expression of
place exists in Québec, where history and
geography have had four centuries to nurture
a strong sense of place. New France was born
and grew within the confines of the St
Lawrence Lowlands. Since Québec's early
days, the St Lawrence River has played a key
role in its settlement and economic development. Prior to the British conquest of New
France in 1760, this mighty river promoted
the interests of this French colony by:

- providing a supply route to France, the
  mother country
- facilitating internal movement within the
  colony, particularly to and from Québec
  City
- encouraging each tenant farmer
  (*habitant*) to build his farmhouse near the
  river, which resulted in landholdings that
  took the form of long lots
- allowing the fur traders (*coureurs du bois*)
  to penetrate far inland to trade with distant Indian tribes

- giving French explorers access into the
  heart of North America, while the British
  settlers along the Atlantic Coast had no
  similar river system and had to contend
  with a formidable physical barrier, the
  Appalachian Mountains.

While Québec's territory greatly expanded
after Confederation, the St Lawrence Lowlands
region remains the core of the North American
francophone community.

A region, then, is a synthesis of physical
and human characteristics which, combined
with its distinctiveness from surrounding
regions, produce a regional character and a
sense of place. People living and working in a
region are conscious of belonging to that place
and frequently demonstrate an attachment
and commitment to their 'home' region. Such
a sense of belonging shows that people are
aware of common issues and challenges confronting their region and are collectively seeking solutions to them. Indeed, the theme of
this book is that Canada is a country of
regions, each of which has a strong sense of
place.

The Nature of Regions

Geographers conceptualize space in terms of
regions. Regions are a critical intellectual concept and represent a framework for geographic
studies. The geographer's challenge is to divide
a large spatial unit like Canada into a series of
'like places'. To do so, a regional geographer
selects the critical physical and human characteristics that logically divide a large spatial unit
into a series of regions and that distinguish

Table 1.1

Regions of the World

Geographic Realms After De Blij and Muller	World Regions After Clawson and Fisher
North America (includes Canada)	Anglo America (includes Canada)
Europe	Western Europe
Russia	Eastern Europe, Russia, and the Eurasian states of the former Soviet Union
Australia and New Zealand	Australia, New Zealand, and Japan
South Asia	Monsoon Asia
North Africa/Southwest Asia	The Middle East and North Africa
Sub-Saharan Africa	Africa south of the Sahara
South America	Latin America
Middle America	
The Chinese Realm	
Japan	
The Pacific Realm (a subregion of Southeast Asia)	

Sources: H.J. De Blij and Peter O. Muller, *Geography: Realms, Regions and Concepts* (Toronto: John Wiley, 1994); David L. Clawson and James S. Fisher, *World Regional Geography: A Development Approach* (Toronto: Prentice-Hall, 1998).

each region from adjacent ones. Towards the margins of a region, its core characteristics become less distinct and merge with those characteristics of a neighbouring region. In that sense, boundaries separating regions are best considered transition zones rather than finite limits.

Regions vary in many ways, one of which is size. At the global scale, the continents and major cultural groupings constitute a dozen or so world regions. One popular world regional geography text divides the world into twelve world geographic realms (De Blij and Muller 1994), while another has eight world regions (Clawson and Fisher 1998). Each of these world divisions has a region that includes Canada.

At the scale of a nation-state, geographers organize space into much smaller regions. In this text, Canada is divided into six geographic regions such as Atlantic Canada and the Territorial North. Smaller regions can also be conceived. For example, most urban geographers and planners conceive of a city as consisting of two basic parts: its central business district and outlying areas or suburbs. In summary, geographic regions vary in size from **mega-regions** (world regions) to **mesoregions** (segments of a nation-state) to **microregions** (areas within a locale). They may also vary from **natural regions** (for example, climatic regions) to **cultural regions** (for example, political regions).

Geographers sometimes divide regions further into subregions, which are simply a finer spatial presentation of a region. For example, the Cordillera is a major landform in North America. The Cordillera (as a region), however, contains a number of distinct landforms that form subregions. Three examples are the Rocky Mountains, the Interior Plateau, and the Coast Mountains. A more detailed account of Canada's landforms (including the Cordillera and its subregions) is presented in Chapter 2 under 'Physiographic Regions'. **Physiography** is the study of landforms, their underlying geology, and the processes that shape these landforms.

The Nature of Regional Geography

Once regions have been identified, the geographer's focus in describing and analysing these can vary—this is the nature of regional geography. Often a regional geography approach can begin by presenting a region's geographic situation and physical nature, even though the main discussion focuses on a region's human geography, particularly its economic activities and social characteristics. A theoretical framework is sometimes employed to provide an abstract image of that space and the linkages between its various parts. This helps in understanding the main elements underlying a region's basic geographic structure and the processes operating within it. The nature of regional geography, however, goes well beyond theory and the host of facts associated with regional characteristics. It is concerned with a spatial synthesis that depicts a region's distinctive character.

There have been several approaches to regional geography. Traditionally, geographers perceive people and nature as interacting in a cause-and-effect interplay—an interplay between culture and the physical environment. Geographers may emphasize physical geography's influence upon human activities; that is, how each region's physical base provides a number of opportunities and limitations for the settlement of the land, including a set of resources that provides the basis of economic development. Geographers may also recognize that people, acting through their culture, have an impact on the physical environment. In this text, physical and historical geography continue to play a role, but more emphasis is given to contemporary economic and social events that, in today's world, are transforming relationships between Canada's regions and its place in the North American and world economies. The most critical events driving this transformation are new trading arrangements. NAFTA (the North American Free Trade Agreement) and GATT (the General Agreement on Tariffs and Trade) have unleashed powerful forces that are recasting Canada's economy into a continental and global context.

A new economic reality for Canada and its regions is emerging. As the trade barriers protecting the old national economic system are eliminated, the volume of trade with the United States and, to a lesser degree, Mexico is increasing by leaps and bounds, thereby stimulating north-south trade patterns in North America. As well, Canada's global trade is increasing, particularly with the Far East, the emerging industrial giant. As part of the Pacific Rim, Canada's West Coast province has benefited greatly from the expansion of trade with Japan, China, and other countries in the

Far East, and from direct investment by companies based in the Far East. All these changes will be explored in this book, as well as their effect on the regions of Canada.

Canada's Geographic Regions

In geography, regions vary and there are different perspectives as to what constitutes a region. This text conceptualizes Canada in terms of six large-scale geographic regions (Figure 1.1): Ontario, Québec, British Columbia, Western Canada, Atlantic Canada, and the

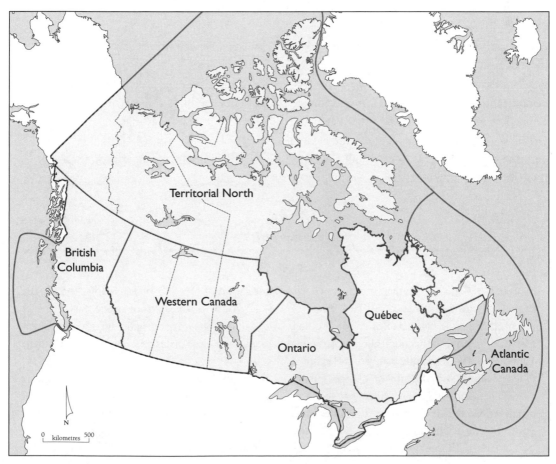

Figure 1.1: The six geographic regions of Canada.

Territorial North. Why have these six regions been selected as a basis of study rather than eight or twelve, for example? What is the logic underlying this selection? First of all, these six major regions represent the division of a huge country into manageable segments. Second, these particular segments (regions) are clearly and obviously distinguished from each other by means of both political boundaries and physical features such as topography, climate, and vegetation. For example, the physical landscapes of the Atlantic provinces are dominated by the Appalachian Uplands. While the Canadian Shield covers most of Ontario and Québec, it reaches much higher elevations in Québec. The Rocky Mountains not only mark the beginning of the Cordillera but also provide a clear physical divide between Western Canada and British Columbia. Furthermore, as Canada acquired new lands, provincial and territorial boundaries changed or were created and the six geographic regions reflect those territorial adjustments. (These adjustments relating to territorial expansion will be discussed in greater detail in Chapter 3.) Finally, these six regions have become firmly fixed in Canadians' popular imagination as constituting logical spatial groupings that make sense historically, politically, and geographically. Geographers and other scholars have long used this regional structure in their scientific studies.[2] Equally important, this set of regions is commonly recognized by the media, thus firmly implanting this spatial pattern of Canada in the Canadian psyche. In sum, then, there are several good reasons for using these six geographic regions:

- They are readily understood by Canadians.
- They reflect the political nature of Canada.
- They facilitate the use of statistical data that are readily available at the provincial level.
- They include the notion of a sense of place.

Vignette 1.3

Nordicity

Nordicity is a quantitative measure of the degree of 'northernness' of a place based on ten natural and social variables. Each variable is assigned a range of values called polar units. Vancouver, for example, has 35 polar units, while Canada's northernmost centre, Eureka, has 857 polar units. The southern edge of northern Canada is defined as 200 polar units. South of this boundary lies the populated area of Canada. The maximum number of polar units (1,000) can only be attained at the North Pole. These polar units measure variations in Canada's cold environment, state of development, and its distance from the Canadian ecumene; that is, its densely inhabited core. Nordicity has accomplished three goals: (1) it provides a numerical definition of each place in Canada; (2) it can accommodate change over time; and (3) it allows for a regionalization of Canada.

As is mentioned earlier, there are different conceptualizations of global division into regions. Geographers classify regions according to a combination of physical and human characteristics. A geographer's selection of one or more of these characteristics depends upon the nature of the investigation. De Blij and Muller, for example, divided the world into twelve regions (geographic realms) based on culture; that is, the way society organizes itself. Once selected, the characteristics serve as the criterion for defining the type and spatial extent of a particular region. The range in the type and extent of regions can be considerable. For instance, Canadian geographer Louis-Edmond Hamelin, in seeking to differentiate areas in Canada's North, conceived the notion of nordicity based on a composite of ten natural and social factors (Vignette 1.3). The result was a vision of nordicity dividing Canada into five zones, two representing southern Canada and three, northern Canada (Figure 1.2). Compare Hamelin's regional division of Canada with that shown in Figure 1.1.

Given the dominant demographic and economic position of Ontario, Québec, and British Columbia, they are each designated a

Cape Dorset, situated on Dorset Island southwest of Baffin Island, has a traditional economy based on hunting and sealing. (*Courtesy R.M. Bone*)

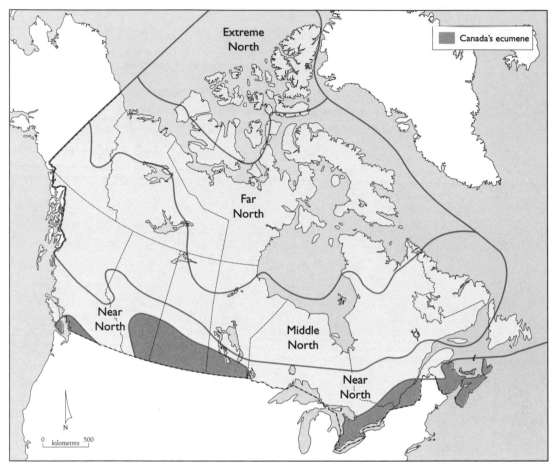

Figure 1.2 Canadian nordicity. Nordicity, as expressed by the Near North, Middle North, Far North, and Extreme North, covers nearly 90 per cent of Canada.
Source: Adapted from Louis-Edmond Hamelin, *Canadian Nordicity: It's Your North Too*, translated by W. Barr (Montréal: Harvest House Ltd, 1979). Reprinted by permission of the authors.

geographic region in this book. Ontario and Québec have traditionally been the more powerful provinces in Canada. Known as Central Canada, they are the country's centre of economic and political power. Joined now by British Columbia, which has undergone unprecedented demographic and economic growth, these three provinces have 75 per cent of the Canadian population and account for over 75 per cent of Canada's gross domestic product (Table 1.2). Economic output is measured by **gross domestic product** (GDP), the monetary value of all goods and services produced by an economy over a specified period. The remaining seven provinces, which account for roughly 24 per cent of Canada's population and economic output, are combined into two geographic regions. Atlantic Canada includes

Table 1.2
General Characteristics of the Six Canadian Regions, 1996

Geographic Region	Area* (000km²)	Area (per cent)	Population (000)	Population (per cent)	GDP (per cent)
Ontario	1 068.6	10.8	10,753.6	37.3	40.5
Québec	1 540.7	15.5	7,138.8	24.8	22.0
British Columbia	947.8	9.5	3,724.5	12.9	13.0
Western Canada	1 963.5	19.8	4,801.0	16.6	18.1
Atlantic Canada	540.3	5.4	2,333.8	8.1	6.0
Territorial North	3 909.8	39.0	95.2	0.3	0.4
Canada	9 970.7	100.0	28,846.9	100.0	100.0

Note: *Includes freshwater bodies such as the Canadian portion of the Great Lakes.

Sources: Statistics Canada, 1997c; 1998c:Table 39; Stanford, 1998:185.

Newfoundland, New Brunswick, Nova Scotia, and Prince Edward Island, while Alberta, Saskatchewan, and Manitoba comprise Western Canada. By creating these two regions, the five geographic regions become more similar in territorial size, population levels, and economic output. The remaining geographic region, the Territorial North, extends over one-third of Canada's territory. However, few people live in this region, which accounts for less than 1 per cent of Canada's economic output. It was composed of the Yukon Territory and the Northwest Territories until 1999, when the Northwest Territories was divided into two political units: the territory of Nunavut and the yet-to-be-named western part of the Northwest Territories.

The Dynamic Nature of Regions

Regions are not static entities. They and their characteristics change over time, as do the relationships between regions. Human beings, whose activities affect both the physical and human elements in a region, are the principal agents of regional change. This dynamism is revealed in a variety of ways. Some changes involve humans' alteration of the physical landscape. In the late-nineteenth century, for example, the building of the Canadian Pacific Railway opened the West to settlement, thereby dramatically changing its land use. More recently, the construction of the Confederation Bridge from mainland Canada to Prince Edward Island established a direct highway link. Unfortunately, not all economic

activities have had a singularly positive outcome. Three regional examples resulted in
serious damage to the physical environment
and negative social implications for local
inhabitants. These regional examples are clear-
cut logging practices in the mountainous
areas of British Columbia; overfishing of the
northern cod stocks in the Grand Banks off
Newfoundland; and diversion of the Churchill
River into the Nelson River in northern
Manitoba.

Social changes too have had a profound
impact on the state of Canadian regions. Social
change is sometimes manifested in the courts
or through elected governments' enactment
of legislation. For example, Aboriginal rights
were recognized by the courts and governments, which in turn led to the first comprehensive land-claim agreement in 1984, followed by four others in the early 1990s. Also,
a new territory, Nunavut, has been created.
The net result is a new political geography in
the Territorial North.[3]

Another example of social change is the
empowerment of the French language in Québec. Between 1969 and 1977, three different
provincial governments passed language laws
in order to have the language of the French
majority fully recognized as the province's primary language. The third of these laws, the so-
called Charter of the French Language (Bill
101), was enacted in 1977. The purpose of Bill
101 was to make all of Québec's society function in French—in its various levels of governments, in corporations both large and
small, and in service industries. Even public
advertisements were restricted to the French
language. By 1980, Québec society had been
transformed, with French the dominant language in the province. However, anglophones
retained their own schools, as guaranteed in
the British North America Act and, since
1982, as guaranteed by the Constitution Act of
Canada.

Often economic change also has a regional
impact. The cause of change can originate either
within or outside the country. For Canada's
resource hinterland, sudden changes in world
prices for its primary products often create a
boom or a bust in the local economy. In such
cases, economic change is highly concentrated
in one region. For instance, the sharp rise
in world oil prices in the 1970s suddenly
transformed Alberta, Canada's principal oil-
producing province, into an economically
prosperous 'have' province. In other cases,
economic change can affect more than one
region. This is particularly true when Ottawa
attempts to ameliorate regional disparities. In
1967 Ottawa initiated the fiscal equalization
payments program, which transfers wealth
earned by 'have' provinces (Ontario, Alberta,
and British Columbia) to economically disadvantaged 'have-not' provinces (Newfoundland, New Brunswick, Nova Scotia, Prince
Edward Island, Manitoba, Saskatchewan, and
Québec). Ottawa's goal is to ensure that all
Canadians have access to a basic level of public services.

Over time, the balance of power between
the regions changes, affecting the relationships
between regions. In many ways, changes in
the population size of regions provide a measure of power shifts. In 1871, 97 per cent of
Canada's population lived in Atlantic Canada,
Québec, and Ontario. One hundred twenty-

five years later, the figure had dropped to 70 per cent (Figure 1.3). This remarkable demographic change illustrates the dynamic nature of regions and provides a key to understanding the regional tensions that are discussed later in this text.

Economic Integration

A particularly important aspect of change for Canada's regions has been, and continues to be economic integration. For a long time, Canada was a high-tariff

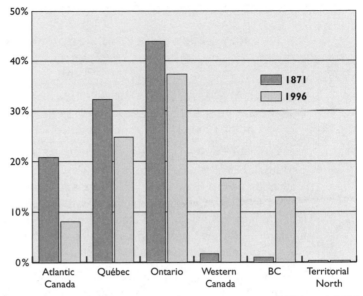

Figure 1.3: Regional populations by percentage, 1871 and 1996.

Source: Adapted from Statistics Canada, 1997c:11.

country.[4] In 1879, John A. Macdonald introduced his economic policy of high tariffs known as the National Policy. His immediate goal was to stimulate the Dominion's faltering economy by encouraging Canadians to buy goods made by other Canadians. This required that tariffs be raised to keep out lower-priced American goods. One consequence of Macdonald's policy was the creation of an industrial core in southern Ontario and Québec. Another consequence was that the rest of the country remained a resource hinterland. Many would argue that this policy served Canada well, while others claim that it favoured Central Canada at the expense of the rest of Canada. Whatever the truth, Canada has abandoned this protectionist policy, turning instead to the North American and global economies (Table 1.3). By unleashing such powerful economic forces, Canada's economy

and society are changing. Canadians, now more outward looking, are seeking their place in the unfolding global economy. At the same time, the implications for Canada's domestic economy and internal trade arrangements are both far-reaching and uncertain. Canada's trade with the United States and, to a lesser degree, other foreign countries has already increased sharply. At the same time, shifts are underway between regions within Canada. Left unchecked, this trend is likely to continue into the next century, creating a new set of international and domestic trading relations.

A Spatial Framework: The Core/Periphery Model

A spatial framework provides an essential foundation for the study of regional geography. A spatial vision of the world enables a geogra-

Table 1.3
Key Steps Altering Canada's Economy

Date	Event
1947	The first GATT agreement was signed by Ottawa and twenty-two other governments. These countries agreed to reduce tariffs and eliminate import quotas among the members of GATT through multilateral negotiations and agreements.
1965	The Auto Pact between Canada and the United States created a continental market for automobiles. An important provision for Canada was that its share of automobile production be maintained. The three large American car manufacturers (General Motors, Ford, and Chrysler) could now expand their production to a continental scale. In this process, Canadian branch plants ceased to produce a wide variety of automobiles for the Canadian market and began to produce only a few models (but more of them) for the continental market.
1985	The Royal Commission on the Economic Union and Development Prospects for Canada examined the question of Canada–United States free trade and concluded that such an agreement would greatly benefit Canada.
1989	The Free Trade Agreement between Canada and the United States was approved. The aim of FTA was to integrate the two economies. Unlike the Auto Pact, however, there were no assurances that Canada's share of economic activities would be maintained. Instead, the North American marketplace would determine the type and amount of economic activity in Canada.
1994	Superseding FTA, NAFTA broadened the geographic area of FTA to include Mexico. The North American economy now had a member with much lower wage rates, causing many labour-intensive industries to close their operations in Canada and reopen in Mexico.
1995	The final GATT treaty (the so-called Uruguay Round) was ratified by 124 governments on 1 January 1995. This agreement is the most far-reaching trade pact in world history and moves closer to the idea of world free trade. Unlike the arrangement under the seven previous GATT negotiating rounds, the World Trade Organization, a permanent institution with powers to enforce trade rules and assess penalties against members, was created.
1995	The proposed Multilateral Agreement on Investment would have augmented the concept of world free trade by requiring governments to treat foreign and domestic companies equally. Canada declared its willingness to participate in these negotiations, conducted by the Organization for Economic Co-operation and Development (OECD), but some Canadians expressed strong concern that such an agreement would preclude Ottawa from acting in the best interests of its own citizens. By 1998, strong opposition caused the OECD to abandon this proposal.

phy student to comprehend more readily the processes affecting different regions of our constantly changing world and to grasp the significance of numerous facts by associating them with a conceptual model. By their very nature, however, abstract models simplify the real world. In doing so, they often lose touch with reality. For that reason, the spatial framework used in this book only serves as a backdrop to the regional geography of Canada. This text emphasizes the economic disparities between, and the interdependency among, regions, as well as the processes that shape and reshape regions. To assist the reader in comprehending this information, a conceptual framework based on the core/periphery model is employed.

The **core/periphery model** is an abstract theory devised and refined by scholars that explains how the capitalist economic system evolved into distinct spatial units. In the most simplistic terms, industrialization takes place in a few favoured areas known as cores. As these industrial cores expand, so too does their need for raw materials, energy, and foodstuffs, so they must turn to their peripheries for sufficient supplies. In this way, the world is divided into two interrelated functional units: industrial cores and resource peripheries. The **core** controls the direction and rate of development of its **periphery**. A core region is also known as a **heartland**, and a periphery is sometimes referred to as a **hinterland**. As a core has virtually total economic and political power, a core and a periphery have a dominant/dependent relationship expressed by the characteristics shown in Table 1.4.

The core/periphery model describes an asymmetrical economic relationship between regions that functions at different geographic levels. One classification identifies three levels, namely global, regional, and local levels. These different levels are interrelated, but the highest-order core (the global level) has ultimate control of the system. Under these conditions, the core/periphery concept may be

Table 1.4
Basic Characteristics of a Core and a Periphery

	Core	Periphery
Geographic:	Space and physical barriers overcome by a modern transportation system	Space and physical barriers continue to hinder economic development
Economic:	Industrial production emphasizes manufacturing	Primary production dominates
Cultural:	Attitudes, language, social customs, and values prevail	Forced to accept the core's culture
Political:	Controls the periphery	Is subservient to the core

Source: Adapted from Edward J. Malecki, *Technology and Economic Development* (New York: Wiley, 1991). Reprinted by permission of Pearson Education Limited.

regarded as a multiple system of 'nested' cores and peripheries, each operating at particular spatial levels but with all levels interconnected and dominated by the highest-order core. This notion of different geographic levels enriches our understanding of Canada's regional nature. Using Ontario as an example of a particular spatial level, we can argue that a core and a periphery exist within that province. Southern Ontario, where the vast majority of the province's population, industry, and political power is found, acts as the core. In sharp contrast, northern Ontario, with its tiny population and primary economy based on forestry, hunting, mining, and trapping, serves as a periphery.

In general, core regions tend to be relatively powerful and wealthy, while peripheral regions tend to be relatively weak and poor. In terms of trade, the core exchanges manufactured goods for primary products from its periphery. While this economic arrangement generates great wealth, the distribution of that wealth is often questioned. Some hold that this trade enriches the core and impoverishes the periphery. Others refute this claim, arguing that the market economy benefits all regions. Pushing the argument of the core/periphery model too strongly, however, leads to a philosophical cul-de-sac known as environmental determinism.[5] Geographers generally take the position that physical geography affects the range of developmental opportunities facing people living in a region. Nevertheless there are several interpretations of the core/periphery model and its ability to distribute wealth. One interpretation is by Neo-

Marxist scholars, while the other is by conventional economists who subscribe to free-market principles.

The Neo-Marxist version of the core/periphery model substitutes regions for classes. The process driving this model is regional exploitation rather than class exploitation. Sometimes this left-wing view takes the form of the **dependency theory** (Frank 1969). According to this theory, rich industrial countries or regions (the so-called core) extract the wealth generated by resource industries in developing countries or regions (the so-called periphery). This exploitive process eventually strips the developing country or region of its natural wealth, leaving it underdeveloped and impoverished. As a form of neo-colonial domination, cities within the peripheral region serve the interests of the core and thereby facilitate the development process unleashed by capitalist firms. Since branch operations of foreign companies are located in these cities, the primary role of these cities is to function as a node for the exploitation of the surrounding hinterland. From an ideological point of view, the functions of these centres are controlled by decisions made in the industrial core where the main headquarters of the foreign companies are located.

Conventional economists have interpreted the core/periphery model differently. While they agree that the core region is a dominant economic force, they believe that the core, by investing in the periphery, is not exploiting but developing that region. This process of regional development is often called 'modernization'. From the perspective of **modernization theory**, regional differences are only temporary and will disappear with time. The mechanism for such change lies in a competitive marketplace where regional inequalities are corrected by the mobility of the factors of production, particularly labour and capital. At the global level, capital and technology are likely to move to developing countries where the rate of return is higher because of lower labour costs. Within a nation-state, workers are attracted to areas of high wages, thereby creating a shortage of workers in the low-wage areas. At that point, wage levels rise, reaching an equilibrium with wages in other regions.

Wallerstein's Global Version

Immanuel Wallerstein (1979) devised a world system analysis to study social change from a Neo-Marxist perspective. He applied the notion of class exploitation to countries of the world by dividing the capitalist world economy into three spatial units. These are the core, semiperipheral, and peripheral units (Table 1.5). Wallerstein echoed the Marxist argument that the core appropriates the wealth generated by the periphery for its own enrichment. Since industrial cores operate within a global environment, a world system approach based on a core/periphery model is necessary to understand social and economic changes. Wallerstein argues that his world system analysis provides the key to understanding how the world core maintains its power by dominating its periphery.

Friedmann's Regional Version

In the 1960s, John Friedmann, an American scholar, was working as a regional planner in Venezuela where he devised a regional version of the core/periphery model (Table 1.5). He applied this model to regions within a single country. He viewed core regions as territorially organized subsystems of society that have a high capacity for generating and absorbing innovative change, and peripheral regions as subsystems whose development is determined chiefly by core region institutions on which they depend. From a modernization perspective, Friedmann conceived four types of regions: he divided a country into three varieties of peripheries, all dominated in different ways by a core region. The three peripheral regions are the upward transitional region, the downward transitional region, and a resource frontier. This division of the periphery into three subregions demonstrates the existence of important variations in peripheral resource bases and their consequent development.

Canada and the Core/Periphery Model

Because Friedman's regional classification closely approximates the economic characteristics of the six geographic regions used in this book, his regional classification is used in this text. The application of Friedmann's version of the core/periphery model that underlies this text must be accompanied by three caveats:

Table 1.5
Two Versions of the Core/Periphery Model

Type of Spatial Unit	Global Version
Core Regions	Highly industrialized regions that dominate trade, control the most advanced technologies, and have high levels of productivity within diversified economies. Western Europe, North America, and Japan form the principal core regions at the world scale.
Semiperipheral Regions	Falling between core and peripheral regions, semiperipheral regions are able to exploit nearby peripheral areas, but are themselves exploited and dominated by the core regions. They are countries that were once peripheral, indicating that peripheral (and possibly core) status are not necessarily permanent. Brazil, India, and South Korea, having developed an industrial base, have moved into the semiperipheral category.
Peripheral Regions	Peripheral regions have undeveloped or very specialized economies characterized by old technologies and low levels of productivity. Egypt, Indonesia, and Zambia fall into the peripheral group.
Type of Spatial Unit	Regional Version
Core Region	The core region is the focus of economic, political, and social activity. Most people live in the core, which is highly urbanized and industrialized. The core has a high capacity for innovation and economic change. Innovations and economic advances are disseminated downward through the national urban hierarchical system to the periphery. Central Canada is Canada's core region.
Upward Transitional Region	The upward transitional region's economy and population are growing as both capital and labour flow into this rapidly developing area. While initial development occurred in the primary sector, there is now a greater emphasis on manufacturing and service activities. British Columbia is an example of an upward transitional region.
Downward Transitional Region	In a downward transitional region, the economy is declining, unemployment is rising, and out-migration is occurring. Often this is an 'old' region that is dependent on resource development for its economic growth. Now that these resources have passed their prime or have been exhausted, the regional economy has stalled. Atlantic Canada is an example of a downward transitional region.

Table 1.5 (continued)	
Resource Frontier	Located far from the core region, few people live in this frontier and little development has taken place. Resource companies are just beginning to penetrate into this remote area. As energy and mineral deposits are discovered, the prospects for economic growth are enhanced. The Arctic represents Canada's last resource frontier.

Sources: John Friedmann, *Regional Development Policy: A Case Study of Venezuela* (Cambridge: MIT Press, 1966); Immanuel Wallerstein, *The Capitalist World-Economy* (Cambridge: Cambridge University Press, 1979).

1. This is only one of several well-known models designed by geographers to conceptualize spatial organization (see the references for this chapter) and it is in no way definitive.

2. Dramatic shifts in interregional relationships can take place over time, prompted by any number of factors, both internal and external.

3. State intervention in the marketplace can either redistribute national wealth (thus counteracting the tendency of market forces to concentrate wealth in a few favoured regions) or it can accelerate this concentration of wealth by using political or military power (thus protecting the wealthy regions).

Despite these caveats, Friedmann's core/ periphery model establishes a reasonable context to analyse the striking regional economic disparities evident across Canada. By combining the abstract notion of the core/periphery model with geographic reality, the model attempts to shed some light on the power imbalances—economic, political, and social— that have existed between the major regions of Canada since Confederation and the effect of differences in resources on the balance of power. As mentioned earlier, in Canada the centre of power has resided in Central Canada, especially in the Windsor–Québec City axis. From the perspective of the core/ periphery model, all the other regions of Canada function as the periphery (hinterlands); they supply the natural resources (raw materials and foodstuffs) that fuel manufacturing activity, employment, and urban growth in the dominant core region. For the first 122 years of Confederation, the natural advantages enjoyed by Central Canada were re-enforced by high tariffs. In this way, Confederation favoured Central Canada, making it dominant and the peripheral areas dependent. Since 1967, Ottawa has directed federal transfers (also known as equalization payments) to economically disadvantaged regions of the country. Ottawa's capacity to redress regional disparities has diminished with economic integration with the United States. In the

1970s, Ottawa eagerly intervened in regional economic matters. The primary federal agency was the Department of Regional Economic Expansion, commonly referred to as DREE (Savoie 1986). From 1967 to 1982, DREE served as a super ministry of regional development. Often its efforts had little lasting impact on reducing the gap between have and have-not regions. At best, these efforts prevented the economic gap from widening.

In terms of application to Canada's regional geography, two major attractions of the core/periphery model are:

• Canada's regional disparities lend themselves to such a spatial framework.
• Relations between Canada's regions, whether they be economic or political, also reflect the core/periphery model.

The core/periphery model has been employed by a number of Canadian scholars over the years. The **staple thesis**, devised by Harold Innis (1930), is a Canadian variant of the core/periphery model.[6] Within this theoretical context, Innis conceived of Canada's early history in terms of its regional geography but with the country functioning as a resource economy driven by foreign demands. From this core/periphery perspective, Innis projected a new vision of Canada's economic history, asserting that the export of Canada's natural resources to Great Britain, the United States, and other advanced industrial countries affected the national and regional economies of Canada and their social and political systems. Innis insisted that Canada is a nation because of its geography; that is,

Canada's history was triggered by the exploitation of a major resource (staple) found in a region and shaped by its trade relation with external markets. In some cases, resource development led to economic diversification and, in other cases, it did not. In several books, Innis provided detailed historical accounts of staple development in several regions of Canada. For instance, the fur trade in northern Canada did not result in the diversification of the northern economy but, rather, in the enrichment of England and France (Innis 1930). The wheat trade, on the other hand, led to the diversification of the economy of Western Canada. The staple thesis, then, provided Innis with a Canadian version of the core/periphery model.

In recent years, many Canadian scholars, particularly geographers, have employed the core/periphery model in their examination of Canada and its regions. Their reasons are straightforward: the physical geography of Canada has divided Canada into favoured and unfavoured regions, and Ottawa's economic policy stimulated economic growth in Central Canada, the most physically favoured region. History also favoured Central Canada. In 1867, Central Canada was already the agricultural, industrial, and population heartland of Canada. Important works by McCann (1982, 1987), McCann and Gunn (1998), and Matthews (1983) exemplify two interpretations of this model. McCann emphasizes a modernization perspective in which economic growth occurs in varying degrees in each region. Much of this growth occurs in cities and is then dispersed throughout the urban hierarchy to smaller and smaller urban centres and eventually to the rural countryside. Matthews has taken a dependency perspective common to many Neo-Marxists. He sees the forces of the market economy leading to greater regional differences, culminating in the impoverishment of the have-not regions. In the Neo-Marxist model, economic growth does take place, but the cities often extract the wealth generated in the smaller centres and rural communities, thus reversing the flow of wealth and concentrating it in the larger cities.

This book attempts to strike a balance between the modernization and dependency perspectives. While it is recognized that the market economy generates unprecedented economic growth, its negative impacts on the environment and society cannot be ignored. Without state intervention, these negative impacts may well lead to what dependency scholars call a state of underdevelopment. While a market economy produces economic growth, undesirable by-products are produced, including a spatial imbalance in economic growth creating have and have-not regions, and hidden environmental and social costs.

How closely does the theoretical model fit the realities of Canada? The answer to this question changes with time. From the days of the National Policy to the Auto Pact, Canada's economy was closed, with high tariffs preventing foreign goods from reaching our domestic market. At that time, the combination of a nationalist economic policy and the regional nature of Canada produced a near mirror image of the core component of the model. Highly industrialized areas in southern Ontario and southern Québec functioned as the industrial core, supplying manufactured products to the rest of Canada. The role of the Canadian resource hinterland, however, was linked much more closely to global markets, so its image as a periphery was somewhat blurred. Why did Canada's resource hinterland sell so much of its primary products at world prices to external markets and yet buy Canadian manufactured goods at artificially high domestic prices? This peculiar situation was created by Ottawa's trade policy. The federal government placed high tariffs on imported foreign manufactured goods, but rarely took similar action to ensure high prices for raw materials and foodstuffs within Canada. Given the small size of the domestic market, many **primary products**, such as grain, lumber, and minerals, became necessarily dependent on world markets.

After the Auto Pact, the economic integration of Canada's economy into a North American one began. The national version of the core/periphery model began to crumble when the Free Trade Agreement was signed in 1989. The FTA forced manufacturing firms based in Canada to compete within the much larger continental market. During this restructuring process, some manufacturing companies survived while others failed. Not surprisingly, this economic integration caused many American-owned branch plants to close their Canadian plants, thereby allowing their larger and more efficient plants in the United States to serve the entire North American market. Canadian-based firms, including a few branch plants, turned to a more specialized production line suited to the North American market. Such an approach allowed these firms to achieve

economies of scale. That is, by increasing their production, these firms reduce the average cost of their product. During this period, three economic events took place within the Canadian hinterland:

- The price of manufactured goods available in the Canadian hinterland declined because less-expensive foreign imports became available, thereby reducing costs to consumers in the hinterland.
- The removal of trade barriers permitted more Canadian primary products into American domestic markets (although in several cases, the Americans found ways to curtail these exports, such as durum wheat and lumber), thereby stimulating the hinterland economy.
- The elimination of the rail transportation subsidy for grain in 1995 made the export of bulky products like spring wheat more expensive for Prairie farmers, forcing them to seek alternative agricultural crops that are either sold locally (canola) or have a high enough value to permit shipment by rail (durum wheat).

Therefore the economic integration of Canada's economy has impacted the state of the core/periphery relationship and expanded it to a continental and global scale.

In the next and final section of this chapter, a brief profile of each of Canada's six geographic regions provides a preview of each region's predominant characteristics. The order of these regional profiles reflects the core/periphery model used in this text. Each profile also touches on some of the major themes to be developed in later chapters and defines the *Key Topics* in Chapters 5 to 10.

Regional Profiles

The statistical data on the area, population, and economic strength of Canada's regions (Table 1.2) and its social characteristics (Table 1.6) underscore the following key factors in studying Canada's regional geography:

- Of the nearly 30 million Canadians, more than 60 per cent live and work in Ontario and Québec. This demographic fact illustrates the dominant position of these provinces in the nation's economic and political matters and supports the notion of a core region.
- With over 80 per cent of the Québec population declaring French as their mother tongue, that province's position as the centre of French language and culture in Canada is indisputable. From a linguistic perspective, Québec is a distinct part of Canada; from a federalist point of view, Québec's geopolitical situation makes it an indispensable part of Canada; but from a separatist perspective, this cultural/linguistic difference provides the *raison d'être* for an independent country.
- While Aboriginal Canadians reside in all provinces, their proportion in the Territorial North is nearly half of the total population. With this demographic power, Aboriginal peoples have a more secure position in the affairs of the Territorial North than in the southern

Table 1.6
Social Characteristics of the Six Canadian Regions, 1996

Geographic Region	French* (000)	French (per cent)	Aboriginal Peoples** (000)	Aboriginal Peoples (per cent)
Ontario	520,860	4.9	141,525	1.3
Québec	5,784,635	82.1	71,415	1.0
British Columbia	60,675	1.6	139,655	3.8
Western Canada	129,615	2.7	362,770	7.6
Atlantic Canada	291,170	12.6	37,785	1.6
Territorial North	2,715	2.9	45,865	48.4
Canada	6,789,665	23.8	799,010	2.8

Notes: *French includes those who speak only French, as well as those who speak French and another language.

**In 1996 the Census asked two questions to determine the number of Aboriginal peoples. The 1996 Aboriginal population, as determined by a question about ancestry, was 799,010. This 'ancestry' question did not appear in past censuses. The 1996 Aboriginal population, as determined by a question based on ethnic origin, was 1,101,995. The explanation for the difference in the number of Aboriginal peoples may be one of 'multiple identity'. For example, Statistics Canada determined that only 625,710 out of the 1,002,670 Canadians who reported some Aboriginal ancestry in the 1991 census considered themselves to be North American Indian, Métis, or Inuit.

Sources: Statistics Canada, 1997a:3 pp.; 1997b:4 and 5 of 15; 1998a:1 and 2 of 10; 1998b:15 to 20 of 21.

geographic regions. In 1999, the territory of Nunavut provided the Inuit with political expression of their dominance in the Canadian Arctic.

Ontario

Ontario is the economic powerhouse of Canada. Its historic development goes back to the eighteenth century. Shortly after the American War of Independence began in 1775, British settlers fled from New England and resettled in the remaining British colonies, including the area now known as southern Ontario. Although Ontario was settled much later than Atlantic Canada and Québec, it soon became the dominant economic and political power in Canada. Much of the credit for its industrial development goes to the protective trade barriers first imposed under the National Policy in 1879, which continued unabated until 1989 when FTA was signed, and the favourable physical features of the Great Lakes–St Lawrence Lowlands. The fertile soils and mild climate of the Great Lakes–St Lawrence Lowlands attracted many settlers. As the number of settlers increased, a large

Rooftop view of downtown Ottawa, which has been redeveloped over several decades with the addition of parks and federal buildings. (*Corel Photos*)

domestic market emerged. In addition, this area's close proximity to the American manufacturing belt enabled it to develop close ties with American manufacturing companies. By 1996, nearly 11 million residents claimed Ontario as their home, making this region the largest in Canada. Toronto, the country's largest city, is located in southern Ontario. As the financial capital of Canada, this megacity houses most of the corporate headquarters of major Canadian and foreign firms. Altogether, southern Ontario, along with southern Québec, form a powerful economic core. Northern Ontario, on the other hand, is a sparsely populated area that functions as a resource hinterland. Ontario is noted for six other key characteristics:

- There are three physiographic regions in Ontario: the Hudson Bay Lowlands, the Canadian Shield, and the Great Lakes–St Lawrence Lowlands.
- The population is heavily concentrated in the Great Lakes–St Lawrence Lowlands with a secondary cluster in the Canadian Shield. Few people live in the Hudson Bay Lowlands.
- Manufacturing is strong, particularly the automobile assembly and parts sector.
- Its economy and population are growing with both capital and people coming to Ontario.
- Outside of Québec, the largest number of francophones reside in Ontario. In 1996, French-speaking Ontarians comprised

4.9 per cent of the total population, or 521,000 people.

- *Key Topic*: Given the effect of the Auto Pact on Ontario's economy, the automobile industry is examined in more detail in Chapter 5.

Québec

Québec is the only province in Canada whose sole official language is French, which reflects its historic origins and francophone majority. Like Atlantic Canada, settlement in Québec began in the seventeenth century, thus giving it a long and rich history. The fertile lowlands of the St Lawrence Valley represent the heartland of Québec and the cultural homeland of francophones across North America. In the nineteenth century, Québec's settlers spread southward into the Appalachian region and the Canadian Shield, but limited agricultural land kept their population numbers low. Currently, agricultural and manufacturing activities remain concentrated in the St Lawrence Lowlands. Montréal, the principal city of Québec, is situated on an island in the St Lawrence River. Because of its language, civil laws, and history, Québec has a culture that is strikingly different from that of other regions of Canada, yet because of its geographic position and economic linkages with the rest of

Montréal has long been a major seaport in eatern North America. (*Corel Photos*)

Canada and the United States, Québec is an integral part of both the Canadian and the larger North American economy. According to the Friedmann classification, Québec consists of both a core and a northern frontier hinterland. Québec is noted for many cultural, economic, and physical features:

- Québec straddles three physiographic units: the Canadian Shield, Appalachia, and the St Lawrence Lowlands.
- With a varied agricultural and resource base, Québec has a highly diversified economy. Agriculture is mostly dairy, fruit, and vegetable production, while the resource base includes forests, minerals, and hydroelectricity.
- Most manufacturing is concentrated in the Montréal region. Shifts are occurring with labour-intensive firms declining and high-technology firms (particularly in the field of transportation) increasing.
- Québec's population reached 7.1 million by 1996. However, slow population growth is weakening the province's political position within Canada. From 1871 to 1996, Québec's share of Canada's population fell from 34.2 per cent to 24.8 per cent.
- Most Quebeckers subscribe to the 'two founding nations' perception of Canada.
- *Key Topic*: The development of hydro-electricity in northern Québec to supply southern consumers and factories demonstrates the heartland/hinterland structure of this province. These key elements are discussed at greater length in Chapter 6.

British Columbia

The settlement of British Columbia began in the mid-nineteenth century when Victoria was established in 1843 as the headquarters of the Hudson's Bay Company. In the past, British Columbia was isolated from the more populated areas of Canada because of the long distance to eastern Canada and because of the barrier to transportation posed by the Cordillera. While modern transportation systems have overcome these two natural barriers, British Columbians still find themselves looking southwards to the United States and across the Pacific to Japan and China for business. For these two reasons, this geographic region is more closely linked to Asian markets than are the other five geographic regions. Of all the geographic regions, British Columbia has the smallest proportion (1.6 per cent) of francophones in its population. The total number of French-speaking Canadians in this Pacific region was approximately 60,000 in 1996.

With its mild climate, vast forests, and varied mineral wealth, British Columbia is one of the most richly endowed geographic regions. Like Western Canada, British Columbia is an upward transitional region. If its rapid economic and population growth continues, this Pacific region is destined to become a core region in the next century. While staple production played a key role in its early economic development, service activities based on high-technology and tourist-oriented businesses have now become its strongest economic sector. Even so, forestry and, to a lesser degree, mining and fishing, remain crucial economic activities. A current economic slump in British Columbia was caused by a drop in Asian customers' demand for primary products. British Columbia is distinguished by six additional characteristics:

- Settlement of this British colony began in the mid-nineteenth century well before 1871 when British Columbia joined Canada.
- In terms of physiography, British Columbia lies within the Cordillera, except for its northeastern sector where the Interior Plains are found.
- Because of the Pacific Ocean's influence, British Columbia's West Coast region has the mildest climate in Canada.
- Most of the 3.7 million British Columbians reside along the Pacific Coast.
- Because of its robust economy and mild climate, British Columbia attracts many newcomers from other provinces and foreign countries, particularly the Far East.
- *Key Topic*: The extensive forests of British Columbia are a renewable resource that has reached its natural limit of sustainability.

Western Canada

The settlement of Western Canada was relatively recent. After the Canadian Pacific Railway was completed in 1885, thousands of homesteaders moved to the Prairies. Attracted by the promise of fertile land, this flood of newcomers continued to arrive until the outbreak of the First World War. Today, nearly 5 million Canadians live in this western hinterland. While agriculture remains a key to the region's prosperity, the western economy

Vancouver's Canada Place, which was the Canada Pavilion during Expo '86, is now a convention centre. (*Corel Photos*)

Western Canada's economy is characterized by agriculture, including livestock production. (*Corel Photos*)

has diversified with most growth occurring in its service and resource industries. This trend towards diversification is most noticeable in Alberta where oil and gas developments propelled the province's economy into high gear, making it wealthy. Saskatchewan and Manitoba, with their fewer energy resources, still depend heavily on their agricultural sectors. For these reasons, Alberta is the leading economic force within Western Canada. With the end of a rail transportation subsidy for grain farmers, major changes are anticipated in agriculture, including a shift away from wheat to other types of crops and livestock. Western Canada is clearly an upward transitional region, according to Friedmann's classification system. These are some other characteristics of Western Canada:

- The Interior Plains and the Canadian Shield are the main physiographic regions.
- Most of the population is in the Interior Plains.
- Agriculture is limited by a dry continental climate that can cause crop failures.
- Distance from major markets remains a limiting factor for agricultural and industrial exports.
- Approximately 150,000 francophones live in this region, comprising 3.3 per cent of the total population.
- *Key Topic*: Agriculture remains an essential element in Western Canada and is the focus for this region.

Atlantic Canada

Atlantic Canada is the smallest geographic

region in Canada. Except for the Territorial North, it has the smallest population (2.3 million in 1996). From a geographic and historical perspective, Atlantic Canada consists of two parts: the Maritimes and Newfoundland/Labrador. Unlike other regions, the sea has played a strong, even dominant, role in the lives of Atlantic Canada's inhabitants. Atlantic Canada has not fared well in Confederation, perhaps because of its small resource base and a geographic location that limited the region's access to its natural market in New England with high tariffs, and to Central Canada with high shipping costs. As well, the region's fish, forests, and minerals have been exploited for a long time, which has sometimes resulted in depreciation of the resource base, notably the mineral resources and cod stocks. The eco-nomy and population of Atlantic Canada are growing more slowly than those of other regions. Atlantic Canada is a disadvantaged region of Canada. Its heavy dependency on equalization payments, unemployment insurance, and other transfer payments is a measure of its weak economic position within Confederation. Other indicators are high unemployment rates and a steady out-migration of its residents. In terms of Friedmann's core/periphery model, Atlantic Canada most closely approximates his downward transitional region. Six other characteristics of Atlantic Canada are:

- Atlantic Canada has a long and rich history. In 1604, Port-Royal (now Annapolis Royal) was founded in present-day Nova Scotia.

A bridge spanning Halifax's harbour, which is a principal port in Atlantic Canada. (*Corel Photos*)

- The region comprises part of the Appalachian physiographic region and includes a vast continental shelf.
- Halifax is the largest metropolitan city in Atlantic Canada.
- Offshore energy developments have sparked hopes for an economic revival in Newfoundland and Nova Scotia.
- The nearly 300,000 Acadians living in New Brunswick form the largest francophone population cluster outside of Québec.
- *Key Topic*: Given the importance of the fisheries to Atlantic Canada, this topic is explored in more detail in Chapter 9.

The Territorial North

At 3.4 million km², the Territorial North is the largest geographic region in Canada. Stretching from 60° N to the North Pole, the Territorial North includes nearly 40 per cent of the territory of Canada and four physiographic regions. The four physiographic regions are the Cordillera, the Interior Plains, the Canadian Shield, and the Arctic Lands. Settlement is concentrated in the southern Cordillera and Interior Plains. Even so, with fewer than 100,000 residents, the Territorial North has a population smaller than that of Prince Edward Island (135,000 in 1996). Aboriginal peoples form nearly half of this population. Development in

An Inuvialuit drum dance. (*Courtesy Northwest Territories Information Bureau*)

this resource hinterland has been hampered by remoteness from world markets and by a cold environment. Energy and mineral resources dominate the northern economy, while renewable resource development based on agriculture and forestry is limited. Wildlife is the major renewable resource and many Aboriginal families depend on it for much of their meat and fish. The Territorial North is a resource frontier in Friedmann's classification system. Here are some other characteristics:

- The Territorial North has a cold environment with widespread permafrost.
- Over one-third of the population lives in Whitehorse and Yellowknife, while the rest lives in small towns and villages.
- While the resource economy is the driving force behind the northern economy, most inhabitants are employed in the public sector.
- By 1997, four Aboriginal peoples had signed comprehensive land-claim agreements that provide both cash and land. The cash enables them to enter the market economy, and the land enables them to maintain their land-based economy.
- A new territory, Nunavut, became a reality in 1999.
- The emergence of a new political geography that recognizes Aboriginal peoples is a bold step that can only strengthen the social fabric and regional structure of Canada.
- *Key Topic*: Large-scale resource projects, which play a critical role in the northern economy, are covered in Chapter 10.

As a prelude to understanding the many interrelated physical and human dimensions of Canada's geographic regions, the next three chapters focus on Canada's physical geography, history, and human geography. Chapter 2 ('Canada's Physical Base') explores the wide range of natural and geomorphic processes that account for wide variations in landforms across Canada. Seven physiographic regions are described with brief accounts of their geology, soils, natural vegetation, and climate. A major theme is the physical environment's influence on human occupancy of the land, and human occupancy's effects on the physical environment. Chapter 2, therefore, is more than a mere backdrop for the remainder of this text; rather, it illuminates how human activities and human decision-making both shape and are shaped by the physical environment in which they occur.

Summary

Geography is the study of the earth's surface as the space within which the human population lives. A fundamental goal is to gain an understanding of our changing world by accurately describing, analysing, and interpreting its geographic regions.

Geographers conceptualize space into regions. A regional geographer selects critical physical and human characteristics that logically divide a large spatial unit into a series of regions. Towards the margins of a region, its main characteristics become less distinct and merge with those of a neighbouring region. For that reason, boundaries are best considered as transition zones. Canada is a country of regions. Shaped by its history and physical geography, Canada is distinguished by six geographic regions: Ontario, Québec, British Columbia, Western Canada, Atlantic Canada, and the Territorial North.

The essential foundation for studying regional geography is an ability to conceptualize places and regions as components of a constantly changing global system. The core/periphery model provides an abstract spatial framework for understanding the general workings of the modern capitalist system. It consists of an interlocking set of industrial cores and resource peripheries. This model can function at different geographic scales and serves as an economic framework for interpreting Canada's regional nature. In addition to the core region, three types of regions devised by Friedmann extend our appreciation of the diversity of the Canadian periphery. They are: (1) an upward transitional region, (2) a downward transitional region, and (3) a resource frontier.

Notes

1. Geographers have interpreted Canada's regions differently. Thirty years ago, Kenneth Hare (1968:3) attempted to capture the richness of the country by describing Canada as 'a great geographical system, whereby a vast area of land, much of it then empty, has been organized and made productive'. From his perspective, the organization of Canada (into regions) was a direct reflection of its physical geography. J. Lewis Robinson had the same approach. In 1981, however, Larry McCann de-emphasized the role of physical geography in regional geography and increased the role of history within a theoretical framework known as the core/periphery model. In doing so, McCann made a bold departure from the more traditional approach to regional geography.

2. Back in 1972, Canadian geographers used this same regional structure to prepare a series of six regional books on Canada for the 22nd International Geographical Congress. The editors of this regional series were the well-known geographers of the 1970s: L. Gentilcore, F. Grenier, A.G. Macpherson, P.J. Smith, J.L. Robinson, and W.C. Wonders.

3. As a result of the federal government's recognition of Aboriginal rights, land-claim settlements have begun modifying the cultural, economic, and political landscape of the Territorial North and the northern areas of many provinces. Already the James Bay Cree, Inuit, Naskapi, Inuvialuit, Gwich'in, Sahtu, Inuit of Nunavut, and the Indians of Yukon have obtained land-claim settlements and received land and cash in exchange for surrendering their Aboriginal rights to vast areas of land. The James Bay Cree also achieved a regional form of self-government in northern Québec. While the North remains a resource hinterland, land-claim settlements ensure that those Aboriginal peoples will have more control over resource development and will therefore be better able to protect the wildlife and their hunting lifestyle.

4. Prior to Confederation, British colonies were part of the imperial colonial system. During the 1840s, Britain abandoned the mercantile system of preferential trade with its colonies in favour of free trade. Britain believed that free trade would reduce the cost of imported raw materials and foodstuffs,

but it also meant lower prices for primary products produced in British North America. The British North American colonies sought other trade arrangements. In the late 1840s, the Province of Canada used a tariff to protect its manufacturing firms. In 1854, the Reciprocity Treaty with the United States was put into place. Under this treaty, staple goods could move freely between both countries. Trade flourished, particularly between the Maritimes and New England. In 1866, the United States cancelled this treaty, forcing the British North American colonies to find an alternative, which was to trade among themselves. From this perspective, Confederation was intended to promote trade between the four founding provinces.

5. Environmental determinism is based on the principle that the physical environment determines human affairs. Geographers reject that position, although they recognize that the environment does exert a strong influence on the nature of human activities in various regions of the world. Students can find a more complete discussion of environmental determinism and other philosophical options in geography, including possibilism, positivism, humanism, and Marxism in William Norton, *Human Geography*, 3rd edn (Toronto: Oxford University Press, 1998:31–9). The writings of most geographers reflect either one of these philosophical positions or some combination. The core/periphery model, for example, sprang from Marxist scholarship. Because of the model's powerful spatial implications, other scholars holding different philosophical positions have modified this theory by removing its economically deterministic character. Instead, they accept that external forces (such as global agreements like GATT) and internal forces (such as federal/provincial agreements like equalization payments) can modify the impact of the physical environment on regional development. Hinterlands, therefore, are not locked into a single outcome because of their physical geography.

6. In 1993, Trevor Barnes served as the guest editor for a series of short appraisals on the influence of Harold Innis's books and papers on Canadian geographers. The general conclusion was that most geographers have ignored Innis's contribution to geography. Barnes argues that Innis is too 'little read now' but 'should be' (1993:353). Barnes describes Innis as a Canadian nationalist who 'pioneered a distinctly Canadian political economy— one that sought to understand why this particular place, Canada, moved "from colony to nation to colony".' In this geographical appreciation of Harold Innis, Barnes sees Innis as an intellectual radical who broke ranks with his colleagues and challenged the colonial view of the world and the place of colonies such as Canada in that world. In short, the staple thesis offered a fresh look at regional development within Canada, one that emphasized its diverse regional geography within a core/peripheral theoretical model.

Key Terms

core

An abstract area or real place where economic power, population, and wealth are concentrated; sometimes described as an industrial core, heartland, or metropolitan centre.

core/periphery model

A theoretical concept based on a dual spatial structure of the capitalist world and a mutually beneficial relationship between its two parts, which are known as the core and the periphery. While both parts are dependent on each other, the core (industrial heartland) dominates the economic relationship with its periphery (resource hinterland) and thereby benefits the most from this relationship. The core/periphery model can be applied at several geographical levels, including international, national, and regional.

cultural regions

Geographic areas defined by one or more cultural features such as religion.

dependency theory

A Neo-Marxist interpretation of the dual spatial structure of the capitalist world that is based on an exploitive relationship between its two parts known as the core and periphery. As a consequence of this exploitive economic relationship, the periphery is eventually stripped of its natural resources and left in a state of underdevelopment. Neo-Marxist scholars see underdevelopment as resulting from capitalist exploitation.

economies of scale

A reduction in unit costs of production that results from an increase in output.

gross domestic product

An estimate of the total value of all materials, foodstuffs, goods, and services produced by a country or province in a particular year.

heartland

A geographic area in which a nation's industry, population, and political power are concentrated; also known as a core.

hinterland

A geographic area based on resource development that supplies the heartland with many of its primary products; also known as a periphery.

megaregions

Extremely large geographic units such as Canada.

mesoregions

Medium-size geographic units such as the Cordillera.

microregions

Small geographic units such as Cape Breton Island.

modernization theory

Economic theory that explains how economic growth occurs and how wealth generated by this growth is distributed throughout the world. The core/periphery model, for example, illustrates how the periphery is 'developed' by the core.

natural regions

Geographic areas defined by one or more physical features such as a soil type.

periphery

The weakly developed area surrounding an industrial core; also known as a hinterland.

physiography

A study of landforms, their underlying geology, and the processes that shape these landforms; geomorphology.

primary products

Goods derived from agriculture, fishing, logging, mining, and trapping; non-processed products.

region

An area of the earth's surface that is defined by its distinctive human or natural characteristics. Boundaries between regions are often transition zones where the main characteristics of one region merge into those of a neighbouring region. Geographers create regions to study one part of the world.

regional geography

The study of the geography of regions and the interplay between physical and human geography, which results in an understanding of human society, its physical geographical underpinnings, and a sense of place.

sense of place
> The special and often emotional feelings that people have for the region in which they live. These feelings are derived from a variety of experiences; some are due to natural factors such as climate, while others are from cultural factors such as language. Whatever its origin, a sense of place is a powerful psychological bond between people and their region.

staple thesis
> The idea that the history of Canada, especially its regional economic and institutional development, was linked to the discovery, utilization, and export of key resources in Canada's vast frontier. Harold Innis devised this thesis in the early 1930s, and his ideas continue to influence Canadian scholars.

References

Bibliography

Barnes, Trevor, ed. 1993. 'Focus: A Geographical Appreciation of Harold A. Innis'. *The Canadian Geographer* 37, no. 4:352–64.

Clawson, David L., and James S. Fisher. 1998. *World Regional Geography: A Development Approach*. Toronto: Prentice-Hall.

De Blij, J.H., and Peter O. Muller. 1994. *Geography: Realms, Regions and Concepts*. Toronto: John Wiley.

Frank, A.G. 1969. *Capitalism and Underdevelopment in Latin America*. New York: Monthly Review Press.

Friedmann, John. 1966. *Regional Development Policy: A Case Study of Venezuela*. Cambridge: MIT Press.

Hamelin, Louis-Edmond. 1979. *Canadian Nordicity: It's Your North Too*, translated by W. Barr. Montréal: Harvest House Ltd.

Hare, Kenneth F. 1968. 'Canada'. In *Canada: A Geographical Interpretation*, edited by John Warkentin, 3–12. Toronto: Methuen.

Hutchison, Bruce. 1942. *The Unknown Country: Canada and Her People*. Toronto: Longmans, Green and Company.

Innis, Harold. 1930. *The Fur Trade in Canada: An Introduction to Canadian Economic History*. New Haven: Yale University Press.

McCann, L.D., ed. 1982. *A Geography of Canada: Heartland and Hinterland*. Scarborough: Prentice-Hall.

———. 1987. *A Geography of Canada: Heartland and Hinterland*, 2nd edn. Scarborough: Prentice-Hall.

———, and Angus Gunn, eds. 1998. *Heartland and Hinterland: A Regional Geography of Canada*, 3rd edn. Scarborough: Prentice-Hall.

Malecki, Edward J. 1991. *Technology and Economic Development*. New York: Wiley.

Marsh, James H., ed. 1988. *The Canadian Encylopedia*, 2nd edn. Edmonton: Hurtig Publishers.

Matthews, Ralph. 1983. *The Creation of Regional Dependency*. Toronto: University of Toronto Press.

Norton, William. 1998. *Human Geography*, 3rd edn. Toronto: Oxford University Press.

Savoie, Donald J. 1986. *The Canadian Economy: A Regional Perspective*. Toronto: Methuen.

Stanford, Quentin H., ed. 1998. *Canadian Oxford World Atlas*, 4th edn. Toronto: Oxford University Press.

Statistics Canada. 1993. *Population Estimates by First Official Language Spoken 1991*. 1991 Census of Canada. Catalogue no. 94-320. Ottawa: Industry, Science and Technology.

———. 1994. *Canada's Aboriginal Population by Census Subdivision and Census Metropolitan Area*. 1991 Census of Canada. Catalogue no. 94-326. Ottawa: Industry, Science and Technology.

———. 1997a. 1996 Census: Nation Tables—Population by Mother Tongue, Showing Age Groups, for Canada, Provinces and Territories, 1996 Census—20% Sample Data, 2 December 1997 [online database], Ottawa. Searched 15 July 1998; <URL:http://www.statcan.ca/english/census/>:3 pp.

———. 1997b. The Daily—1996 Census: Mother Tongue, Home Language and Knowledge of Languages, 2 December 1997 [online database], Ottawa. Searched 14 July 1998; <URL:http://www.statcan.ca/Daily/English/>:15 pp.

———. 1997c. *A National Overview: Population and Dwelling Counts*. 1996 Census of Canada. Catalogue no. 93-357-XPB. Ottawa: Industry Canada.

———. 1998a. The Daily—1996 Census: Aboriginal Data, 13 January 1998 [online database], Ottawa. Searched 14 July 1998; <URL:http://www.statcan.ca/Daily/English/>:10 pp.

———. 1998b. The Daily—1996 Census: Ethnic Origin, Visible Minorities, 17 February 1998 [online database], Ottawa. Searched 16 July 1998; <URL:http://www.statcan.ca/Daily/English/>:21 pp.

———. 1998c. *Canadian Economic Observer*. Catalogue no. 11-010-XPB. Ottawa: Statistics Canada.

Wallerstein, Immanuel. 1979. *The Capitalist World-Economy*. Cambridge: Cambridge University Press.

Further Reading

French, H.M., and O. Slaymaker, eds. 1993. *Canada's Cold Environments*. Montréal–Kingston: McGill-Queen's University Press.

McCann, Larry D., and Angus Gunn, eds. 1997. *Heartland and Hinterland: A Regional Geography of Canada*, 3rd edn. Scarborough: Prentice-Hall.

Robinson, J. Lewis. 1989. *Concepts and Themes in the Regional Geography of Canada*, 2nd edn. Vancouver: Talonbooks.

Warkentin, John. 1997. *Canada: A Regional Geography*. Scarborough: Prentice-Hall.

Chapter 2

Overview

Canada has diverse physical regions, and the natural features of those regions influence human occupation of the land. Some regions offer more favoured natural environments for settlement and industrial development. This fact provides an obvious link to the core/periphery model. That is, areas with certain advantages in terms of geographical location, climate, and natural resources become wealthier, more developed core regions compared to less advantageous areas, which remain peripheral to and dependent on the core.

While many natural regions exist, two are emphasized in this chapter: physiographic regions and climatic zones, which in turn have distinct soil, natural vegetation, and wildlife zones. The interrelationship between physical geography's influence on human occupation and humans' impact on the land is also emphasized.

Objectives

- Demonstrate that physical geography forms a basic building-block for understanding the regional nature of Canada.
- Examine the geological structure, origins, and characteristics of Canada's physical base and seven physiographic regions.
- Discuss the main global factors that influence climate.
- Explore the interrelationships among environmental factors such as climate, soil, natural vegetation, and permafrost in Canada's physical and regional geography.
- Discuss why and how favourable physiographic features and environmental conditions can encourage human occupancy of the land, while unfavourable features and conditions can inhibit settlement, and relate this to the core/periphery model.
- Examine how human activities and the physical environment are interrelated and study some of the negative impacts humans have on the natural environment.

Canada's Physical Base

Introduction

The earth provides a natural home for human beings. For that reason, physical geography helps us understand the regional nature of our world. The basic question posed in this chapter is: Why is Canada's physical geography so essential to an understanding of its regional geography? The answer lies in the regional character of Canada's physical geography, and in the interrelationships between physical geography and human settlement and activity. Physical geography is an underlying factor in shaping Canada's national and regional character.[1] In this text, physical geography also provides the *raison d'être* for the physical basis of the core/periphery model. The argument is a simple one: regions with a more favourable physical base are more likely to develop into core regions, while regions with less favourable physical geographic conditions remain peripheral and dependent on the core.

Physical Variations Within Canada

Geographers recognize that the physical nature of Canada varies in a number of ways. For example, climate varies from place to place. The Maritimes have a mild, wet climate, while the Arctic has a cold, dry climate. Climate also affects landforms (mountains, plateaux, and lowlands), shaping their forms through a variety of weathering processes. Major landforms illustrate the regional nature of Canada's physical geography. For instance, the Canadian Prairies have totally different features compared to those of the Canadian Shield. The Prairies have a flat to gently rolling landscape, while the adjacent Canadian Shield consists of rugged, rocky, hilly terrain.

Geographers perceive an interaction between people and the physical world. Natural features and processes (such as landforms and climate) affect human life in many ways. In turn, human activities often have an impact on the natural environment. This interactive two-way relationship is a fundamental component of regional geography. Favourable physical conditions can make a region more attractive for human settlement. The combination of a mild climate and fertile soils in the Great Lakes–St Lawrence Lowlands encourages agricultural activities, while the St Lawrence River and the Great Lakes provide low-cost water transportation to local and world markets. The favourable physical features of this region have allowed it to become Canada's industrial heartland.

As scientists who study the spatial aspects of nature and the processes that shape nature, physical geographers are concerned with all aspects of the physical world: physiography (landforms), bodies of water, climate, soils, and natural vegetation. Regional geographers, however, are more interested in how physical geography varies and subsequently influences human settlement of the land. The Rocky Mountains, for instance, offer few opportunities for agricultural settlement, but its spectacular scenery has led to the emergence of an economy based on tourism. Regional geographers are also concerned about the effect of human activities on the natural environment. In most cases, humans have a negative impact on the environment. For example, within the Bow Valley of the Rocky Mountains, extensive land developments have reduced the size of the natural habitat of wild animals such as bears and elk. Ironically, if more land is converted into golf courses, resort facilities, and housing developments, the animals that make this wilderness region so unique and attractive to tourists may no longer be able to survive. Another example of the negative impact of human activity is the use of farmland for urban and industrial growth. In our contemporary world, therefore, humans are the most active and, some would say, the most dangerous agent of environmental change.

The discussion of physical geography in this chapter and in the six regional chapters is designed to provide basic information about the natural environment and its essential role in the regional geography of Canada. To that end, the following points are emphasized:

- While physical geography varies across Canada, it has distinct and unique regional patterns.

- Landforms are one aspect of this physical diversity.
- Climate, soils, and natural vegetation are other aspects.
- The impact of human activity is changing the natural environment and, in the case of industrial pollution, there are long-term negative implications for all life forms.
- Physical geography has a powerful impact on Canadians by making certain areas more attractive for settlement and urban/industrial development. This relationship between the natural environment and the human world forms the basis of the core/periphery model.

We begin our discussion of physical geography by examining the nature and origin of landforms.

The Nature of Landforms

The earth's surface features a variety of landforms: mountains, plateaux, and lowlands. These landforms are subject to change by various physical processes. Some processes create new landforms while others reduce them. The earth, then, is a dynamic planet whose surface is actively shaped and reshaped over time. For instance, the process known as weathering breaks down the earth's surface. Weathering consists of both chemical and physical processes that break down rocks into smaller particles. Geomorphic processes (such as moving water, ice, and waves) transport these smaller particles to other locations where they are deposited. This deposition eventually results in the formation of new landforms. Geomorphic processes also erode the land and in doing so reshape the earth's surface.

Weathered basalt columns on Axel Heiberg Island, Northwest Territories. Basalt is an igneous rock. (*Courtesy Walter Kupsch*).

The earth's crust, which forms less than 0.01 per cent of the earth and is the thin solidified shell of the earth, consists of three types of rocks: igneous, sedimentary, and metamorphic. When the earth's crust cooled about 3.5 billion years ago, **igneous rocks** were formed from molten rock known as magma. Nearly 3 billion years later, **sedimentary rocks** were formed from particles derived from previously existing rock. Through weathering, rocks are broken down and transported by water, wind, or ice and then deposited in a lake or sea. At the bottom of a water body, these sediments form a soft substance or mud. In geological time, they harden into rocks. Hardening occurs because of the pressure exerted by the weight of additional layers of sediments and because of chemical action that cements the particles together. Since only sedimentary rocks are formed in layers (called **strata**), this feature is unique to this type of rock. **Metamorphic rocks** are distinguished from the other two types of rock by their origin: they are igneous or sedimentary rocks that have been transformed into metamorphic rocks by the tremendous pressures and high temperatures beneath the earth's surface. Metamorphic rocks are often produced when the earth's crust is subjected to folding and faulting. **Faulting** is a process that fractures the earth's crust, while **folding** bends and deforms the earth's crust.

The earth's crust is broken into at least fourteen huge slabs or plates, each moving in response to the currents of molten material just below the crust. This motion is known as **continental drift**, which can result in the folding and faulting of the earth's crust.[2] For example, as these huge plates drift, they may collide with one another, thus causing earthquakes. Over sufficient geological time, these plates have compressed parts of the earth's crust into mountain chains.

Physiographic Regions

The earth's surface can be classified into a series of physiographic regions. A **physiographic region** is a large area of the earth's crust that has distinct characteristics. These are three key characteristics of a physiographic region:

- It extends over a large, contiguous area with similar relief features.
- Its landform has been shaped by a common set of geomorphic processes.
- It possesses a common geological structure and history.

Canada has seven physiographic regions (Figure 2.1). The Canadian Shield is by far the largest region, while the Great Lakes–St Lawrence Lowlands is the smallest. Perhaps the most spectacular and varied topography (i.e., the landforms of the earth's surface) occurs in the Cordillera, while the Hudson Bay Lowlands has the most uniform relief. The remaining three regions are the Interior Plains, Arctic Lands, and Appalachian Uplands.

Each physiographic region has a different geological structure. These structural differences have produced a particular set of mineral resources in each physiographic region. For example, formed from the solidification of the earth's crust about 3.5 billion years ago, the Precambrian crystalline rock that makes up the Canadian Shield contains deposits of copper, gold, nickel, iron, and uranium. The famous Sudbury nickel mines are in the Canadian Shield. Other physiographic regions were formed much later, as shown in the geological time chart (Table 2.1). The formation of the Interior Plains began about 500 million years ago when rivers deposited sediment in a shallow sea that existed in this area. Over a period

Table 2.1
Geological Time Chart

Geological Era	Geological Time (millions of years ago)	Physiographic Region(s) Formed
Precambrian	600 to 3,500	Canadian Shield
Palaeozoic	250 to 600	Appalachian Uplands, Interior Plains, and Arctic Lands
Mesozoic	100 to 250	Interior Plains
Cenozoic	0 to 100 0 to 2	Cordillera Great Lakes–St Lawrence Lowlands, Hudson Bay Lowlands

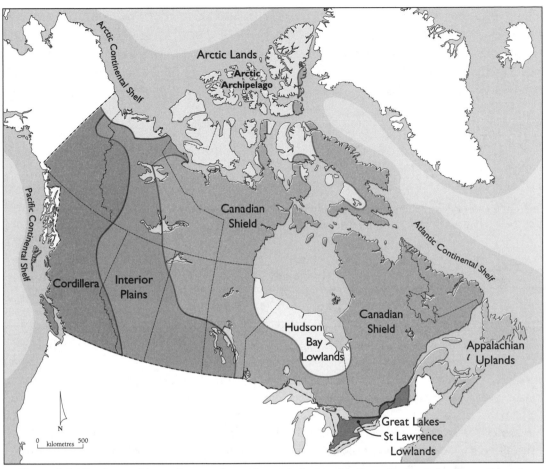

Figure 2.1 Physiographic regions and continental shelves in Canada. Arctic Lands consists of a dozen large islands and numerous small islands that together are known as the Arctic Archipelago. The Canadian Shield is the largest physiographic region and extends beneath the Interior Plains, the Hudson Bay Lowlands, and the Great Lakes–St Lawrence Lowlands.

of about 300 million years, more and more material was deposited into this inland sea, including massive amounts of vegetation and the remains of dinosaurs and other creatures. Eventually, these deposits were solidified into layers of sedimentary rocks 1 to 3 km thick. As a result, the Interior Plains have a sedimentary structure that contains oil and gas deposits. Such variation in the geological structure of each physiographic region has produced unique mineral resources for the

human occupants of these regions to extract. Furthermore, as these regions developed their various resources, differences in regional economies began to take shape.

The Canadian Shield

As noted previously, the Canadian Shield is the largest physiographic region in Canada. It extends over nearly half of the country's land mass. The Canadian Shield forms the ancient geological core of North America. More than 3 billion years ago, molten rock solidified into the Canadian Shield (Table 2.1). Today, these ancient Precambrian rocks are not only exposed at the surface of the Shield but also underlie many of Canada's other physiographic regions.

During the last ice age, the surfaces of the Canadian Shield and those of other physiographic regions were subjected to glacial erosion and deposition (Vignette 2.1). **Glacial erosion** and **deposition** are caused by giant ice sheets slowly grinding over the earth's surface.

As the ice sheet moved over the surface of the Canadian Shield, the ice scraped, scoured, and scratched its massive rock surface. During the movement of ice sheets, a variety of loose materials such as sand, gravel, and boulders are trapped within the ice sheet. As the ice sheet reached its maximum extent, the edge of the ice sheet melted, depositing rocks, soil, and other debris. This debris is called **till**. Towards the end of the ice age, these ice sheets melted *in situ*, depositing whatever debris they contained. Sometimes the water from the

Vignette 2.1

The Last Ice Age

During the earth's geological history, climatic cooling has produced a number of ice ages. Climatic cooling is not fully understood, but scientists have two theories. One is that climate cooling occurs when the distance between the earth and the sun is at its maximum every 21,000 years, due to variations in the earth's orbit. The other theory postulates that an increase in the amount of dust in the atmosphere from erupting volcanoes causes climatic cooling. Both theories offer explanations for a reduction in the amount of sunshine (solar energy) reaching the surface of the earth, which would cool the planet enough to trigger an ice age. Scientists also believe that we are living in an interglacial period and that the climate will again cool, resulting in another ice age.

The last ice age took place over the last 2 million years in the Quaternary period of the Cenozoic era (see Table 2.1). At least four times during the Quaternary period, huge

sheets of ice, perhaps over 5 km thick, spread over Canada and the northern edge of the United States. The most recent glacial advance is called the Wisconsin, which lasted approximately 100,000 years, reaching as far south as the state of Wisconsin (hence the name of this particular glacial advance). During this glacial advance, ice sheets formed, advanced, and retreated as climatic fluctuations occurred. Eventually, the climate cooled sufficiently to allow the ice to cover most of North America. The late Wisconsin ice sheet reached its maximum extent 18,000 years ago.

The late Wisconsin ice advance consisted of two major ice sheets, the Laurentide and the Cordillera ice sheets. The Laurentide ice sheet was centred in the Hudson Bay area. As its mass increased, the sheer weight of the ice sheet caused it to move, eventually covering much of Canada east of the Rocky Mountains. In the Cordillera, a series of alpine glaciers coalesced into the Cordillera ice sheet, which

Vignette 2.1 (continued)

spread westward into the continental shelf off the Pacific Coast and eastward, eventually merging with the Laurentide ice sheet.

Roughly 15,000 years ago, the climate began to warm, causing these ice sheets to retreat. Seven thousand years ago, the last main remnants of these ice sheets were in the Rocky Mountains and in the uplands of the Canadian Shield in northern Québec–Labrador and on Baffin Island. Today, the largest glaciers in Canada are in the mountains of Ellesmere Island. These ancient ice sheets and alpine glaciers have had a lasting impact on Canada's landforms.

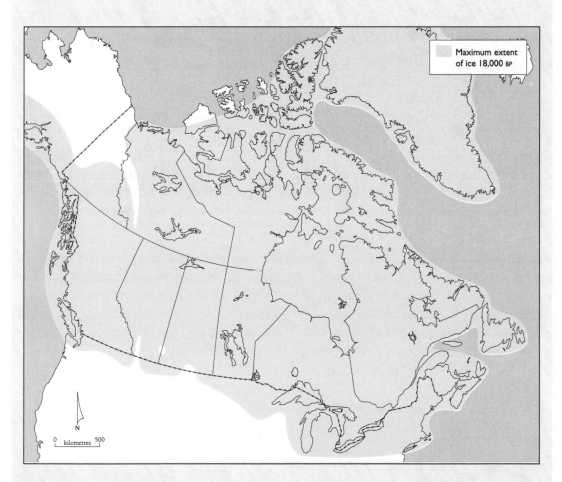

Figure 2.2 Maximum extent of ice, 18,000 BP. The last advance of the Wisconsin ice sheet covered almost all of Canada and extended into the northern part of the United States. Geologists believe that the present 'warm' climate is an interlude before the next ice advance.

A large esker across a lake in the Northwest Territories. (*Courtesy Walter Kupsch*)

melting ice was blocked from reaching the sea by the retreating ice sheet. These waters then formed temporary lakes. Once this ice was removed, these waters surged towards the sea.

Evidence of the impact of these processes on the surface of the Canadian Shield is widespread. Drumlins and eskers, both depositional landforms, are common to this region. **Drumlins** are long, low hills composed of till (material deposited and shaped by the movement of an ice sheet), while **eskers** are long, narrow mounds of sand and gravel deposited by melt water streams found under a glacier. There are also **glacial striations**, which are scratches in the rock surface caused by large rocks embedded in the slowly moving ice sheet.

The Canadian Shield consists mainly of a rugged, rolling upland. Shaped like an inverted saucer, the region's lowest elevations are along the shoreline of Hudson Bay, while its highest elevations occur in Labrador and Baffin Island, where the most rugged and scenic landforms in the Canadian Shield are found. The Torngat Mountains, for instance, provide spectacular mountain and fjord scenery. These mountains reach heights of 1600 m, making them the highest mountain range east of the Rocky Mountains. They also form the boundary between northern Québec and Labrador.

Another area of the Canadian Shield, known as the Laurentides, is located just north of Montréal. It has many lakes and hills that are the summer and winter playground for local residents as well as tourists from Ontario and New England.

Other areas of the Canadian Shield are dotted by large communities that operate as single-industry mining towns, such as the iron-mining town of Labrador City in Labrador/ Newfoundland, the nickel centre of Sudbury in Ontario, and the copper mining and smelter town of Flin Flon in Manitoba. Often located in remote places, resource towns such as these are vulnerable to closure if the mining operation ceases.

The Cordillera

The Cordillera is a complex region of mountains, plateaux, and valleys. Stretching from the American border to the Yukon, this region extends over 16 per cent of Canada's territory. The Cordillera, classified as a young geological structure, was formed about 40 to 80 million years ago (Table 2.1) when the North American tectonic plate slowly moved westward, eventually colliding with the Pacific plate. The collision slowly but steadily compressed sedimentary rocks into a series of mountains and plateaux now known as the Cordillera. These Cenozoic sedimentary rock strata can be seen along exposed mountain sides in the Rockies. Along the Pacific Coast, this process of continental drift continues, making the coast of British Columbia vulnerable to both earthquakes and volcanic activity. With the vast majority of the population of this region clustered along the coast in the cities of Vancouver, Victoria, New Westminster, and Nanaimo, the potential damage from a major earthquake (measuring 7.0 or greater on the Richter scale) might be the worst natural disaster to strike

The St Elias Mountains in the Yukon are part of the northern Cordillera. (*Courtesy R.M. Bone*)

Vignette 2.2

Alpine Glaciation

While glaciers still exist in the Rocky Mountains, they are slowly melting and retreating. During the Wisconsin ice advance about 18,000 years ago, these glaciers grew in size and eventually covered the entire Cordillera. At that time, alpine glaciers advanced down slopes, carving out hollows called **cirques**. As the glaciers increased in size, they spread downward into the main valley, creating **arêtes**, steep-sided ridges formed between two cirques. As it moved through the valley, the glacier eroded the sides of the river valleys, creating distinctive U-shaped glacial valleys known as **glacial troughs**. The Bow Valley is one of Canada's most famous glacial troughs. Cutting through the Rocky Mountains, the Bow Valley now serves as a major transportation corridor. It has also developed into an international tourist area. The centre of this tourist trade is the world-famous ski resort of Banff.

Angel Glacier, Mount Edith, in Jasper National Park, Alberta. (*Corel Photos*)

Canada. The strongest earthquake ever recorded in Canada shook the Queen Charlotte Islands in August 1949. This earthquake measured 8.1 on the Richter scale.

In more recent geological times, the Cordillera ice sheet altered the landforms of the region. Over the last 20,000 years, alpine glaciation has sharpened the features of the mountain ranges in the Cordillera and broadened its many river valleys (Vignette 2.2). The Rocky Mountains are the most well known of these mountain ranges. Most have elevations between 3000 and 4000 m. Their sharp, jagged peaks create some of the most striking landscape in North America. The highest mountain in Canada—at nearly 6000 m—is Mount Logan, part of the St Elias Mountain Range in the southwest Yukon.

The Interior Plains

The Interior Plains region is a northward extension of the Great Plains of the United States. This vast sedimentary plain covers nearly 20 per cent of Canada's land mass. The Interior Plains region is wedged between the Canadian Shield and the Cordillera. It extends from the Canada–US border to the Arctic Ocean. Within the Interior Plains, most of the population lives in the southern area where a longer growing season permits grain farming.

Millions of years ago, a huge inland sea occupied the Interior Plains. Over the course of time, sediments were deposited into this sea. Eventually, the sheer weight of these deposits produced sufficient heat and pressure

Alberta Badlands. Badlands form where sedimentary rocks are exposed to forceful erosional processes, such as rapid runoff from torrential rain. (*Courtesy Dirk de Boer*)

to transform these sediments into sedimentary rocks. The oldest sedimentary rocks were formed during the Palaeozoic era, about 500 million years ago (Table 2.1). Since then, other sedimentary deposits have settled on top of them, including those associated with the Mesozoic era when dinosaurs roamed the earth.

Tectonic forces have had little effect on the geology of this region. For that reason, the Interior Plains are described as a stable geological region. For example, sedimentary rocks formed millions of years ago remain as a series of flat rock layers within the earth's crust. Geologists used such sedimentary structures as geological time charts. In Alberta and Saskatchewan, rivers have cut deeply into these soft rocks, exposing Mesozoic strata. Archaeologists have discovered many dinosaur fossils within these Mesozoic rocks in southern Alberta and Saskatchewan.

Within the Interior Plains, valuable deposits of oil and gas are in sedimentary structures called **basins**. Known as fossil fuels, oil and gas are the result of the capture of the sun's energy by plants and animals in earlier geologic time and the storage of this energy in the form of hydrocarbon compounds in sedimentary rocks. The Western Sedimentary Basin is the largest such basin. Most oil and gas production in Alberta comes from this basin. Fossil fuels are non-renewable resources, meaning that they cannot regenerate themselves, unlike renewable resources, such as trees, which can reproduce.

As the Laurentide ice sheet melted and began to retreat from the Interior Plains about 12,000 years ago, the surface of the region was covered with as much as 300 m of debris deposited by the ice sheet. Huge glacial lakes were formed in a few places. Later, the melt water from these lakes drained to the sea, leaving behind an exposed lakebed. Lake Agassiz, for example, was once the largest glacial lake in North America and covered much of Manitoba, northwestern Ontario, and eastern Saskatchewan—its lakebed is now flat and fertile land that provides some of the best farmland in Manitoba. When glacial waters escaped into the existing drainage system, they cut deeply into the glacial till and sedimentary rocks, creating huge river valleys known as **glacial spillway**. The geological history of the Interior Plains accounts for the great variety of landforms found within the area.

Just north of Edmonton, the Interior Plains slope towards the Arctic Ocean. The Athabasca River marks this northward course as it eventually enters the Mackenzie River. Edmonton lies on the banks of the North Saskatchewan River, which flows eastward into Lake Winnipeg. These waters then enter the Nelson River, which empties into Hudson Bay. Across this section of the Interior Plains, the land slopes towards Hudson Bay. Elevations decline from 1200 m just west of the Rocky Mountains to about 200 m near Lake Winnipeg. These changes in elevation create three subregions within the Prairies: the Manitoba Lowland, the Saskatchewan Plain, and the Alberta Plateau. Typical elevations are 250 m in the Manitoba Lowland, 550 m in the Saskatchewan Plain, and 900 m in the Alberta Plateau. The Cypress Hills, however, provide a sharp contrast to the flat or rolling terrain of the Canadian Prairies (Vignette 2.3), as do the deeply incised river valleys of the Peace River country.

Vignette 2.3

Cypress Hills

The Cypress Hills, a subregion of the Interior Plains, consists of a rolling plateau-like upland that is deeply incised by fast-flowing streams. Situated in southern Alberta and Saskatchewan, this subregion is the highest point in Canada between the Rocky Mountains and Labrador. These hills are an erosion-produced remnant of an ancient higher-level plain formed in the Cenozoic era (see Table 2.1). With an elevation of over 1400 m, these hills rise 600 m above the surrounding plain formed about 50 million years ago from materials borne eastward by rivers originating in the Rocky Mountains. During the maximum extent of the Laurentide ice sheet about 18,000 years ago, the higher parts of the Cypress Hills remained above the Laurentide ice sheet. Known as **nunataks**, these areas served as refuge for animals and plants. As the alpine glacier melted, streams flowing from the Rocky Mountains deposited a layer of gravel up to 100 m thick on these hills.

Today, the Cypress Hills area is a humid 'island' surrounded by a semi-arid environment and has an entirely different natural vegetation compared to the area surrounding it. Unlike the grasslands, the Cypress Hills has a mixed forest of lodgepole pine, white spruce, balsam poplar, and aspen. The Cypress Hills also contains many varieties of plants and animals found in the Rocky Mountains.

The Hudson Bay Lowlands

The Hudson Bay Lowlands region was formed when the Laurentide ice sheet no longer blocked the Atlantic Ocean from entering what is now Hudson Bay. This inland extension of the Atlantic Ocean has been termed the Tyrrell Sea. This inland saltwater sea reached its maximum extent about 7,000 years ago, extending over much of the lowlands surrounding Hudson and James bays. With the huge weight of the ice sheet removed, the earth's crust began to rise, forcing the Tyrrell Sea to retreat. This process is called **isostatic rebound** (Vignette 2.4). Slowly the isostatic rebound caused the seabed of the Tyrrell Sea to rise above sea level, thus exposing a low, poorly drained coastal plain (most of which is called the Hudson Bay Lowlands). This process of isostatic rebound continues today, making the Hudson Bay Lowlands the youngest of the physiographic regions in Canada (Table 2.1).

The Hudson Bay Lowlands comprises about 3.5 per cent of the area of Canada. It lies mainly in northern Ontario, though a small portion stretches into Manitoba. This region extends from James Bay along the west coast of Hudson Bay to just north of the Churchill River. Much of the ground's surface consists of wet peatland known as **muskeg**. Low ridges of sand and gravel are interspersed between these extensive areas of muskeg. These ridges are the remnants of former beaches of the Tyrrell Sea. Because of its almost level surface,

Vignette 2.4

Isostatic Rebound

At its maximum extent about 18,000 years ago, the weight of the huge Laurentide ice sheet caused a depression in the earth's crust. When the ice sheet covering northern Canada melted, this enormous weight was removed, and the elastic nature of the earth's crust allowed it to return to its original shape. This process, known as isostatic rebound or uplift, follows a specific cycle. As the ice mass slowly diminishes, the isostatic recovery begins. This phase is called a **restrained rebound**. Once the ice mass is gone, the rate of uplift reaches a maximum. This phase is called a **postglacial uplift**. It is followed by a period of final adjustment called the **residual uplift**. Eventually the earth's crust reaches an equilibrium point and this isostatic process ceases. In the Canadian North, this process began about 11,000 years ago and has not yet completed its cycle.

much of the land is poorly drained. Underneath the peatland are recently deposited marine sediments mixed with glacial till. With few resources to support human activities, the region has only a handful of tiny settlements. From this perspective, the Hudson Bay Lowlands is one of the least favourably endowed physiographic regions of Canada. Moosonee (at the mouth of the Moose River in northern Ontario) and Churchill (at the mouth of the Churchill River in northern Manitoba) are the largest settlements in the region. Each has a population of just over 1,000 people. These two settlements, formerly fur-trading posts, are now the termini of two northern railways (the Northern Ontario Railway and the Hudson Bay Railway respectively).

Arctic Lands

The Arctic Lands region stretches over 20 per cent of the area of Canada. Centred in the Canadian Arctic Archipelago, this region lies north of the Arctic Circle. It is a complex composite of coastal plains, plateaux, and mountains. The Arctic Platform, the Arctic Coastal Plain, and the Innuitian Mountain Complex are the three principal physiographic subregions. The Arctic Platform consists of a series of plateaux composed of sedimentary rocks. This subregion is in the Arctic Archipelago around Victoria Island. The Arctic Coastal Plain extends from the Yukon coast and the adjacent area of the Northwest Territories into the islands located in the western part of the Beaufort Sea. The third subregion, the Innuitian Mountain Complex, is composed of ancient sedimentary rocks. Like the Rocky Mountains, its sedimentary rocks were folded and faulted. However, unlike the Rocky Mountains, the plateaux and mountains in the Innuitian subregion were formed in the early Palaeozoic era (Table 2.1). At 2616 m, Mount Barbeau on Ellesmere Island is the highest point in the Arctic Lands region.

Across these lands, the ground is permanently frozen to great depths, never thawing even in the short summer. This cold thermal condition is called **permafrost**. Frost action, a form of physical weathering, shatters bedrock and produces various forms of patterned ground. **Patterned ground** consists of rocks

Tundra polygons, a form of patterned ground, east of Horton River, Northwest Territories. (*Courtesy Walter Kupsch*)

An iceberg, formed from a glacier, along the coast of Labrador. (*Courtesy R.M. Bone*)

arranged in polygonal forms by minute movements of the ground caused by repeated freezing and thawing. Patterned ground and **pingos** (ice-cored mounds or hills) appear frequently on the landscape.

The climate in this region is cold and dry. In the mountainous zone of Ellesmere Island, glaciers are still active. That is, as these alpine glaciers advance from the land into the sea, the ice is 'calved' or broken from the glacier, forming an iceberg. On the plains and plateaux, it is a polar desert environment. The term 'polar desert' describes barren areas of bare rock, shattered bedrock, and sterile gravel. Except for primitive plants known as lichens, no vegetation grows. Aside from frost action, there are no other geomorphic processes, such as water erosion, to disturb the patterned ground.

Most people live in the coastal plain in the western part of this physiographic region. The three largest settlements are situated at the mouth of the Mackenzie River. Inuvik has a population of almost 3,000, while Aklavik and Tuktoyaktuk are smaller communities.

The Appalachian Uplands

The Appalachian Uplands region represents only about 2 per cent of Canada's land mass. Sometimes known as Appalachia, this physiographic region consists of the northern section of the Appalachian Mountains (stretching south to the eastern United States), though few mountains are found in the Canadian section. With the exception of Prince Edward Island (Vignette 2.5), its terrain is a mosaic of rounded uplands and narrow river valleys. These weathered uplands are the remnants of ancient mountains that underwent a variety of erosional processes almost 500 million years ago. A combination of weathering and geomorphic processes (including water, wind, and ice) have worn down these mountains, creating a much lower mountain landscape. The highest elevations are in the Gaspé Peninsula. Here, Mount Jacques Cartier, at an elevation of 1268 m, is found. The coastal area has been slightly submerged; consequently, ocean waters have invaded the lower valleys, creating bays or estuaries. The result is a number of excellent small harbours and a few large ones,

Vignette 2.5

Prince Edward Island

Unlike other areas of Appalachia, Prince Edward Island has a flat to rolling landscape. While the island is underlain by sedimentary strata, these rocks are a relatively soft, red-coloured sandstone that is quickly broken down by weathering and erosional processes. Occasionally, outcrops of this sandstone are exposed but, for the most part, the surface is covered by reddish soil that contains a large amount of sand and clay. The heavy concentrations of iron oxides in the rock and soil give the island its distinctive reddish-brown hue. Prince Edward Island, unlike the other provinces in this region, has an abundance of arable land.

such as Halifax Harbour. The island of New-foundland consists of a rocky upland with only pockets of soil found in valleys. Like the Maritimes, it has an indented coastline where small harbours abound. The nature of this physiographic region has favoured settlement along the heavily indented coastline where there is easy access to the fishing banks.

The Great Lakes–St Lawrence Lowlands

The Great Lakes–St Lawrence Lowlands physio-graphic region is small but important. Extend-ing from the St Lawrence River near Québec City to Windsor, this narrow strip of land is wedged between the Appalachians, the Cana-dian Shield, and the Great Lakes. Near the eastern end of Lake Ontario, the Canadian Shield extends across this region into the United States where it forms the Adirondack Mountains in New York state. Known as the Frontenac Axis, this part of the Canadian Shield divides the Great Lakes–St Lawrence Lowlands into two distinct subregions.

As the smallest physiographic region in Canada, the Great Lakes–St Lawrence Low-lands comprises less than 2 per cent of the area of Canada. As its name suggests, the land-scape is flat to rolling. This topography reflects the underlying sedimentary strata and glacial deposits. In the Great Lakes Lowlands sub-region, flat sedimentary rocks are found just below the surface. This slightly tilted sedi-mentary rock, which consists of limestone, is exposed at the surface in southern Ontario, forming the Niagara Escarpment. A thin layer of glacial and lacustrine (i.e., lake) material, deposited after the melting of the Laurentide ice sheet in this area about 12,000 years ago, forms the surface, covering the sedimentary rocks.

In the St Lawrence Lowlands subregion, the landscape was shaped by the Champlain Sea, which occupied this area for about 2,000 years. It retreated about 10,000 years ago and left surface materials on these exposed low-lands. These include former beaches, which are now broad terraces that slope gently towards the river (Vignette 2.6). The sandy to

---

Vignette 2.6

Champlain Sea

About 12,000 years ago, vast quantities of glacial water from the melting ice sheets around the world drained into the world's oceans. Sea levels rose because of this addi-tional water, causing the Atlantic Ocean to surge into the St Lawrence and Ottawa val-leys, perhaps as far west as the edge of Lake Ontario. Known as the Champlain Sea, this body of water occupied the depressed land between Québec City and Cornwall and extended up the Ottawa River Valley to Pem-broke. These lands had been depressed earlier by the weight of the Laurentide ice sheet. About 10,000 years ago, the earth's crust rebounded sufficiently to cause the Champlain Sea to retreat. However, the sea left behind marine deposits, which today form the basis of the fertile soils in the St Lawrence Lowlands.

clay surface materials are a mixture of recently deposited sea, river, or glacial materials. For the most part, this subregion's soils are fertile, which, when combined with a long growing season, allow agricultural activities to flourish.

The physiographic region lies well south of 48° N. The Great Lakes subregion extends from 42° N to 45° N, while the St Lawrence subregion lies somewhat further north, reaching towards 47° N. As a result of its southerly location, its proximity to the industrial heartland of the United States, and its favourable physical setting, the Great Lakes–St Lawrence region is home to Canada's population and manufacturing core.

The Impact of Physiography on Human Activity

Not only do physiographic regions provide a basic understanding of the physical shape and geological structure of Canada, they have also exerted a powerful influence over the geographic pattern of early settlers' land selection. In some instances, settlers were attracted to certain types of land while they avoided other types. Two examples illustrate this point. In the seventeenth century, the St Lawrence Lowlands was an attractive area for the establishment of a French colony because of its agricultural lands and its accessibility by water to France. Few settlers ventured beyond this favoured area. To the north was the rocky Canadian Shield, while to the south lay the Appalachian Uplands. Neither of these surrounding physiographic regions offered attractive land for farming. Instead, Indians occupied these lands where they hunted game and trapped furs in order to barter with French traders for European goods.

Physical features can also create barriers to settlement. The Rocky Mountains were such a barrier in the nineteenth century. In 1867, when the Dominion of Canada was formed, the Rocky Mountains isolated the small British colony on the southern tip of Vancouver Island from the settled area of Canada. At that time, communications and trade with adjacent American settlements along the Pacific Coast proved much easier and quicker than the overland route used by fur traders to reach Montréal. Until the completion of the CPR in 1885, the Rocky Mountains were such an imposing physical barrier that many residents of Vancouver Island favoured joining the United States rather than the Dominion of Canada.

In our contemporary world, such physical barriers are no longer the obstacles they once were. Technological advances in transportation and communications have greatly reduced the **friction of distance**, the term used to describe how interactions between two points decrease as the distance between them increases. The obstacles presented by physical barriers and distance have been greatly diminished.

Geographic Location

Canada's location in the northern half of North America is a critical factor in its physical geography. For instance, Canada has a much cooler environment than the country to its south, the United States of America. A measure of geographic location on the earth's surface is provided by latitude and longitude. Because the earth is a spherical body, this measure is given in degrees. By **latitude**, we mean the measure of distance north and south of the equator. For example, Ottawa is 45 degrees 24 minutes north of the equator. Degrees and minutes are expressed as ° and ' respectively. Since the distance between each degree of latitude is about 110 km, Ottawa is about 5000 km north of the equator. By **longitude**, we mean the distance east or west of the prime meridian. As the equator represents zero latitude, the prime meridian represents zero longitude. It is an imaginary line that runs from the North Pole to the South Pole and passes through the Royal Observatory at Greenwich, England. Canada lies entirely in the area of west longitude. Ottawa, for example, is 75°28' west of the prime meridian. Within Canada, latitude and longitude vary enormously. The variation in latitude has considerable implications for climate, which in turn affects the types of soils, natural vegetation, and wildlife found in each climatic zone. The possible range of latitudes and longitudes found in Canada is shown in Table 2.2.

Climate

Our physical world encompasses more than just landforms, physiographic regions, and geographic location. Climate, for instance, plays a key role in our physical world. **Climate** represents the average weather conditions of a region over a long period. As such, climate is a particularly important element of the natural environment and also has a direct effect on soils, natural vegetation, and the creation of zonal patterns of soils and natural vegetation. Extreme weather events—such as blizzards, droughts, and ice storms—are also part of climate and often have very powerful impacts on humans.

Climatic conditions vary around the world and within Canada. In the Köppen climatic classification scheme for the world, for example, there are twenty-five climate types that reflect different temperature patterns and seasonal precipitation patterns. Seven of these climatic types are found in Canada (Table 2.3).

Table 2.2
Selected Centres with Their Latitude and Longitude

Centre	Latitude	Longitude
Windsor, Ontario	42°18' N	83° W
Alert, Northwest Territories	83°63' N	60°05' W
St John's, Newfoundland	47°34' N	52°43' W
Victoria, British Columbia	48°26' N	123°20' W
Whitehorse, Yukon	60°41' N	135°08' W

Climatic Controls

These are the three dominant climatic controls that affect Canada's weather and climate and that are related to the global atmospheric and oceanic circulation system:

- Variations in the amount of solar energy reaching different parts of the earth's surface correspond with latitude and temperature. That is, lower latitudes receive more solar energy and therefore have higher temperatures than higher latitudes.
- The global circulation of air masses causes a westerly flow of air across Canada, although invasions of air masses

Table 2.3
Climatic Types

Köppen Classification	Canadian Climatic Zone	General Characteristics
Marine West Coast	Pacific	1. Warm to cool summers; mild winters 2. Precipitation throughout the year with a winter maximum
Highland	Cordillera	1. Cooler temperature at similar latitudes because of higher elevations
Steppe	Prairies	1. Cool summers and cold winters 2. Low annual precipitation
Humid continental	Great Lakes–St Lawrence Lowlands	1. Warm, humid summers and short, cold winters 2. Moderate annual precipitation with little seasonal variation
Humid continental, cool	Atlantic Canada	1. Cool, humid summers and short, cool winters
Subarctic	Subarctic	1. Short, cool summers; long, cold winters 2. Low annual precipitation
Tundra	Arctic	1. Extremely cool and very short summers; long, cold winters 2 Very low annual precipitation

Sources: Adapted from Robert W. Christopherson, *Geosystems: An Introduction to Physical Geography*, 2nd edn (Englewood Cliffs: Macmillan, 1994); F. Kenneth Hare and Morley K. Thomas, *Climate Canada* (Toronto: Wiley 1974).

from the Arctic and the Gulf of Mexico can temporarily disrupt this general pattern of air circulation.

• Distance from oceans plays an important role in affecting temperature and precipitation. That is, as distance from oceans increases, the annual range of temperatures increases and the annual amount of precipitation decreases.

Global Circulation System

Regional climates are controlled by the amount of solar energy that is absorbed by the earth and its atmosphere and then converted into heat. The amount of energy received at the earth's surface varies by latitude. In low latitudes around the equator, there is a net surplus of energy (and therefore high temperatures), but in high latitudes around the North and South poles, more energy is lost through re-radiation than is received, and therefore temperatures are extremely low. Canada, which extends from 42° N (Windsor) to 83° N (Alert), experiences great variation in the amount of solar energy received (and therefore great variation in temperatures).

The **global circulation system** redistributes this energy (i.e., energy transfers) from low latitudes to high latitudes through the circulation system in the atmosphere (system of winds and air masses) and the oceans (system of ocean currents). For example, the Japan Current warms the Pacific Ocean, bringing milder weather to British Columbia. On Canada's East Coast, the opposite process occurs as the Labrador Current brings Arctic waters to Atlantic Canada. While Halifax, at 44°40' N, lies about 500 km closer to the equator than Vancouver, at 49°13' N, Halifax's winter temperatures, on average, are much lower than those experienced by Vancouverites.

The atmospheric circulation system travels in a west-to-east direction in the higher latitudes of the Northern Hemisphere, causing air masses that develop over large water bodies to bring mild and moist weather to adjacent land mass. Such air masses are known as **marine air masses**. In this way, energy transfers ultimately determine the patterns of global weather and climate (see Tables 2.3 and 2.4). Air masses originating over large land masses are known as **continental air masses**. These air masses are normally very dry and vary in temperature depending on the season. In the winter continental air masses are cold, while in the summer they are associated with hot weather.

Canada experiences warmer and moister weather in its lower latitudes and colder and drier conditions in its higher latitudes. However, Canada's coastal areas (particularly its Pacific Coast) experience smaller ranges of seasonal temperatures and more annual precipitation than do inland or continental areas at the same latitude (Figures 2.3, 2.4, and 2.5). Winnipeg, for example, experiences a much greater daily and annual range in temperature than does Vancouver, even though both lie near 49° N. The principal reason is Vancouver's greater proximity to the ameliorating effects of the Pacific Ocean.

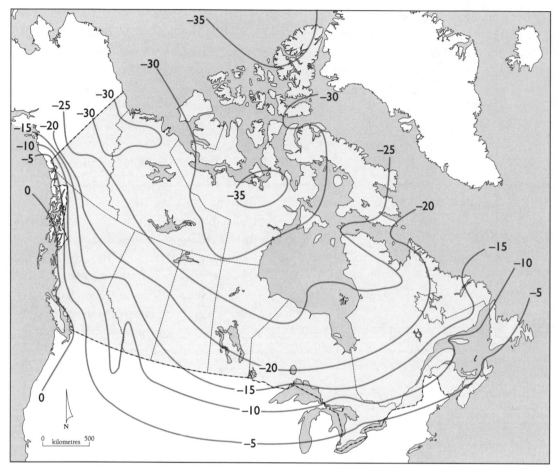

Figure 2.3 Seasonal temperatures in Celsius, January. The moderating influence of the Pacific Ocean and its warm air masses is readily apparent in the 0 to -5° C January isotherm.

Air Masses

Air masses are large bodies of air with similar temperature and humidity characteristics. They form over large areas that have uniform surface features and relatively consistent temperatures. Such areas are known as source regions. The Pacific Ocean is a marine source region, while the interior of North America is a continental source region. During a period of about a week or so, an air mass may form over a source region, taking on the temperature and humidity characteristics of that source region. Canada's weather is affected by five air masses associated with the Northern Hemisphere. For example, Pacific air masses bring mild, wet weather to British Columbia's coast for most of the year. These air masses are much stronger in the winter, so British Columbia experiences a winter maximum for precipitation. In some years, British Columbia can have a relatively

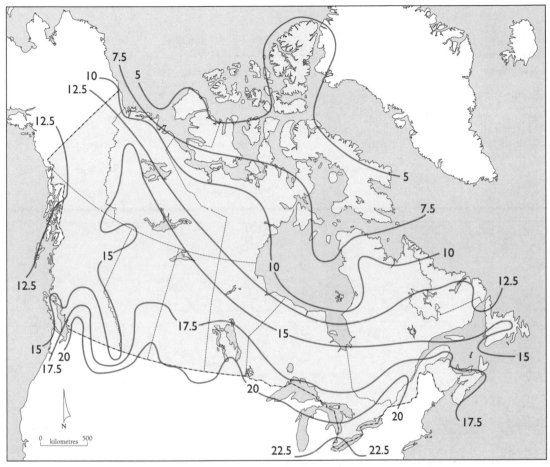

Figure 2.4 Seasonal temperatures in Celsius, July. The continental effect results in very warm summer temperatures that extend into high latitudes, as illustrated by the 15° C July isotherm.

dry summer. The general characteristics of the five major air masses affecting Canada's weather are shown in Table 2.4.

Climate, Soils, and Natural Vegetation

As noted earlier, climate affects the development of soils and the growth of natural vegetation. In fact, the interdependency between climate, soils, and natural vegetation is so strong that physical geographers have identified an orderly and interrelated global pattern of climatic, soil, and natural vegetation zones. Climate determines to a large extent the **soil order** and native vegetation that exist in a given region and hence influences the utilization of the land, whether it be crop cultivation, forestry, or grazing. Together with topography, climate determines the land's ability to support a population.

Table 2.4
Air Masses Affecting Canada

Air Mass	Type	Characteristics	Season
Pacific	Marine	Mild and wet	All
Atlantic	Marine	Cool and wet	All
Gulf of Mexico	Marine	Hot and wet	Summer
Southwest US	Continental	Hot and dry	Summer
Arctic	Continental	Cold and dry	Winter

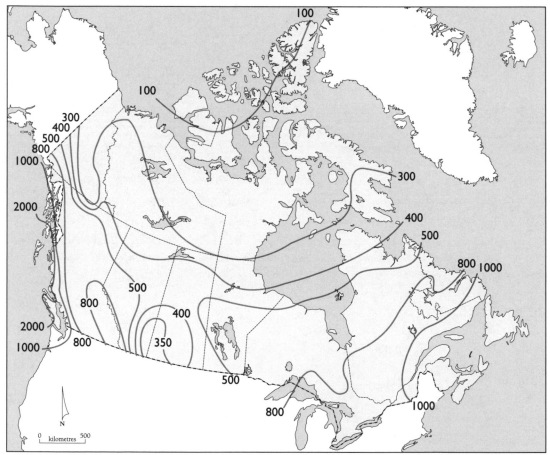

Figure 2.5 Annual precipitation in millimetres. The lowest average annual precipitation occurs in the Territorial North, indicating the dry nature of the Arctic air masses that originate over the ice-covered Arctic Ocean.

Climate has a direct impact on many economic activities. Long, cold winters cause people to use more energy; Canadians are among the highest consumers of energy in the world. Precipitation too has an impact. Below-normal precipitation often has a negative effect on agriculture, forestry, and hydroelectric production. For example, the 1988 drought in Canada cost the national economy (approximately $1.8 billion) in decreased agricultural and hydroelectric output, increased costs of fighting forest fires, and a loss of commercial timber and wildlife habitat (Shabbar, Bonsal, and Khandekar 1997:3016).

Climatic Zones

The earth's atmosphere is a perpetually moving global system of air circulation that attempts to adjust the differences in pressure and temperature over different parts of the globe. This global circulation system, together with ocean bodies and major topographic features, affect Canada's weather. There is a climatic order within our complex and dynamic atmosphere. This order is expressed in several ways, including climatic zones. A **climatic zone** is an area of the earth's surface where similar weather conditions occur. Long-term data describing annual, seasonal, and daily temperatures and precipitation are used to define the extent of a climatic zone. Similar weather conditions occur in a particular area for complex reasons. For example, land near large water bodies usually receives more precipitation than land far from large water bodies. As well, coastal settlements have a small range of seasonal temperatures due to

the sea's cooling effect in the summer and its warming effect in the winter. Land-locked places, however, do not benefit from the influence of the sea and have a much wider range of annual, seasonal, and daily temperatures.

Canada lies in the northern half of North America. This geographical location has several consequences for Canadians:

- Canada, located in the middle and high latitudes, receives much less solar energy than the continental United States and Mexico. It therefore has shorter summers and longer winters.
- Canada is noted for its long, cold winters, which affect Canadians in many ways. 'Coldness', wrote French and Slaymaker, 'is a pervasive Canadian characteristic, part of the nation's culture and history' (1993:i). They go on to state that winter's effects include not only low absolute temperatures but also exposure to wind chill, snow, ice, and permafrost.
- Continental climates are widespread in Canada. These climates are formed over large areas such as the Canadian Prairies and are characterized by cold, dry winters and warm, dry summers.
- Marine climates are limited in their geographic extent to the Pacific Coast of British Columbia and to Atlantic Canada.

Canada has seven climatic zones (Figure 2.6): the Pacific, Cordillera, Prairies, Great Lakes–St Lawrence, Atlantic, Subarctic, and Arctic. Each of these climatic zones has a particular natural vegetation type and soil (Table 2.5). Between each climatic zone, transition zones

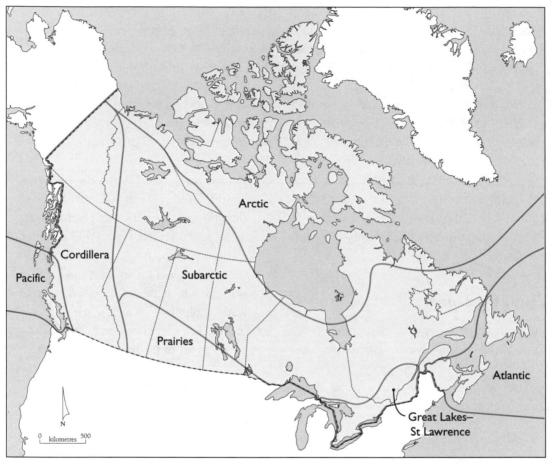

Figure 2.6 Climatic zones of Canada. Canada's most extensive climatic zone, the Subarctic, is associated with the boreal forest and podzolic soils.

exist. The characteristics of each of these climatic zones will now be examined.

Pacific Zone

The Pacific Coast of British Columbia has the most temperate climate in Canada, dominated by the constant flow of moist Pacific air masses across its terrain. The result is mild and often wet, cloudy weather. Because eastward-moving Pacific air masses must rise above the Vancouver Island Mountains and the Coast

Mountains, orographic precipitation frequently occurs (Vignette 2.7). This, combined with frontal precipitation, gives the West Coast a high level of annual precipitation. Known locally as 'liquid sunshine', the annual precipitation on the West Coast of Vancouver Island frequently exceeds 300 cm. Further east, the annual precipitation declines. At Vancouver, located near the Coast Mountains, the annual precipitation declines to about 100 cm. Victoria, however, lies in the partial rainshadow

Table 2.5

Canadian Climatic Zones

Canadian Climatic Zone	Natural Vegetation Type	Soil Order
Pacific	Coastal rainforest	Podzolic
Cordillera	Montane and boreal forests	Mountain complex
Prairies	Grassland and parkland	Chernozemic
Great Lakes–St Lawrence	Broadleaf and mixed forests	Podzolic
Atlantic	Mixed and boreal forests	Podzolic
Subarctic	Boreal forest	Podzolic
Arctic	Tundra and polar desert	Cryosolic

Note: See Figures 2.6, 2.7, and 2.8. Also see Key Terms at end of chapter for definitions of soil orders.

Vignette 2.7

Types of Precipitation

As air masses rise, their temperature drops. This cooling process triggers condensation of water vapour contained in the air masses. With sufficient cooling, water droplets are formed. When these droplets reach a sufficient size, precipitation begins. Precipitation refers to rainfall, snow, and hail. There are three types of precipitation. **Convectional precipitation** results when moist air is forced to rise because the ground has become particularly warm. Often this form of precipitation is associated with thunderstorms. **Frontal precipitation** occurs when warm air masses are forced to rise over colder (and denser) air masses. **Orographic precipitation** results when air masses are forced to rise over high mountains. However, as those same air masses descend along the leeward slopes of those mountains (that is, the slopes that are sheltered from the wind), their temperature rises and precipitation is less likely to occur. This phenomenon is known as the **rain shadow effect**.

of the Vancouver Island Mountains and thus avoids the heavy orographic rainfall that affects Vancouver. Consequently, the capital city receives nearly 40 per cent less rainfall than Vancouver.

Moderate temperatures, high rainfall, and mild but cloudy winters make the Pacific climatic zone unique in Canada. Unlike those in other climatic zones in Canada, winters here are extremely mild and freezing temperatures are uncommon. Summer temperatures, while warm, are rarely as high as temperatures com-

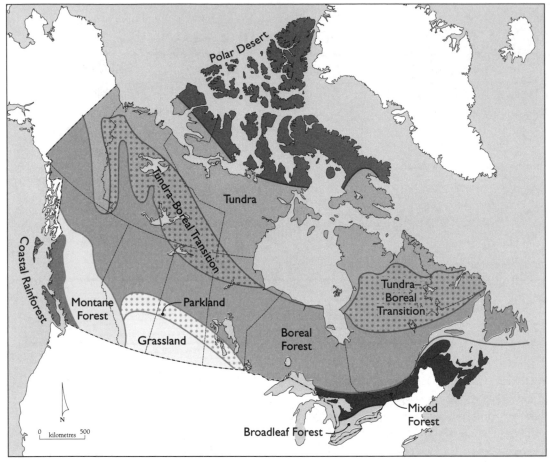

Figure 2.7 Natural vegetation zones. Transitions exist between natural vegetation zones. Two major transition zones shown here are the Tundra–Boreal Transition and the Parkland.

mon in the more continental and dry climate of the Interior Plateau of British Columbia.

Along the West Coast, the mild, wet climate encourages a rainforest. Lush evergreen and deciduous trees grow on the windward slopes. Characteristic tree species are western hemlock, Douglas fir, western red cedar, and Sitka spruce. Under this vegetation cover, **podzolic** soils are common. These soils are strongly acidic and low in plant nutrients, making it necessary for farmers to use fertilizers. Fertil-

izers are used because nutrients are washed away by heavy rainfall, and the needles of the coniferous trees, when dropped, add to the ground cover and produce an acidic soil.

Cordillera Zone

As a mountain climate, the Cordillera climatic zone consists of a number of microclimates. Sudden changes in elevation (high above sea level), latitude (from southern to northern British Columbia), and distance from

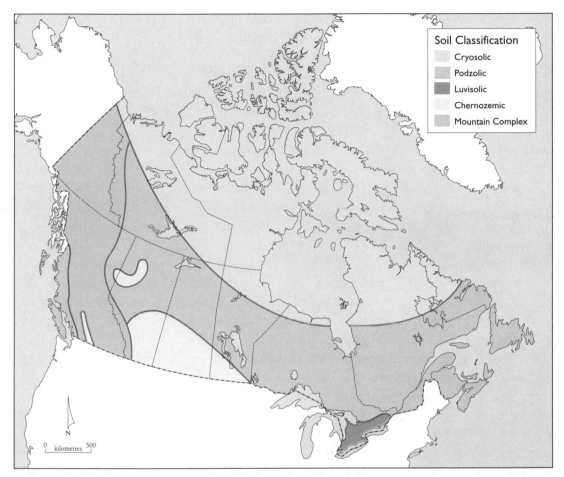

Figure 2.8 Soil zones. Most agricultural land is in luvisolic and chernozemic soil zones that together comprise about 5 per cent of Canada's land base.

the Pacific Ocean control each microclimate, which in turn affects microvegetation and microsoil conditions. In general, these microclimates become drier as distance from the Pacific Ocean increases and cooler as either elevation or latitude increases. The range of natural vegetation and soil types is enormous. In Thompson Valley near Kamloops, an arid microclimate results in a desert-like grassland vegetation with **chernozemic** soils. In sharp contrast, huge trees in the coastal rainforest grow in podzolic soils. Different still are the higher elevations of the Rocky Mountains extending beyond the treeline, where **cryosolic** soils support grasses, mosses, and lichens. For much of northern British Columbia, the northern coniferous forest (evergreen, needle-leaf trees) is the most common natural vegetation.

The Cordillera exerts a powerful influence on the distribution of precipitation both within and beyond its zone. Along the western slopes of the Coast Mountains, rainfall exceeds 100 cm per year, while in southern Alberta and Saskatchewan annual precipitation is less than 30 cm. Such a large difference is due to two factors: orographic rainfall along the Coast Mountains and the rain shadow effect of the mountain ranges in the Cordillera. The low amounts of precipitation in Alberta and Saskatchewan are partly attributed to the rain shadow effect of the Rocky Mountains (Vignette 2.7).

Because the Cordillera stretches from southern British Columbia to the Yukon, the climate turns into a Subarctic mountain climate north of 58° N. Higher elevations in these latitudes have Arctic climatic conditions. South of 58° N, however, summers are longer and warmer, partly because the eastward-moving Pacific air masses are more dominant and partly because of greater annual solar energy. Places that have a rain shadow effect have extremely low annual precipitation. Just east of the Coast Mountains, for example, the annual precipitation drops from 100 cm per year to less than 25 cm. In the Thompson Valley and Okanagan Valley of the Interior Plateau, hot, dry conditions result in an arid climate with sagebrush in the valleys and ponderosa pines on the valley slopes. Chernozemic soils are common but given the extremely dry conditions, these soils are almost desert-like. Further north, the higher latitude causes a general decline in average temperatures. As a result, the natural vegetation of the forest cover includes more lodgepole pine, aspen, and white spruce. Chernozemic soils are common in the valleys of central British

Columbia, but they give way to a podzolic soil in higher latitudes.

Prairies Zone

A dry, continental climate prevails over the interior of Western Canada. Known as the Prairies climatic zone, this area has low annual precipitation, warm to hot summers, and great seasonal changes in temperature. Due to a combination of low annual precipitation and high summer temperatures, this zone has a 'water deficit'. This zone's low precipitation is primarily the result of the eastward-moving Pacific air masses losing most of their moisture as they rise over the Cordillera's mountain chains. By the time they descend the eastern slopes of the Rocky Mountains, these Pacific air masses have turned into dry air masses. As a result, the Prairies receives little rain or snow from the Pacific Ocean. This water deficit is often measured in terms of potential evapotranspiration, which is the amount of water vapour that can potentially be released from an area of the earth's surface through evaporation and transpiration (the loss of moisture through the leaves of plants). There is a water deficit if the evapotranspiration rate is greater than the average annual precipitation. For example, most of the Prairies zone usually receives less than 40 cm annually, while its evapotranspiration rate exceeds 50 cm. The difference indicates a water deficit of over 10 cm.

During the winter, the cold, dry Arctic air masses dominate weather conditions in the Prairies, placing the region in an Arctic 'deep-freeze'. On a cold, clear January night, the temperature can drop to -40° C. During the day, the temperature may recover to a high of -25° C. Yet the worst and more unsettled weather occurs when strong winds and sub-zero temperatures produce bitterly cold temperatures. Under those circumstances, wind chill (the transfer of heat from the warmer human body to the colder atmosphere) can freeze exposed flesh in a matter of minutes. Major snowstorms accompanied by strong winds and low visibility are the most feared winter storms on the Prairies. These storms are called blizzards, which often strike suddenly, blocking highways and rail lines, causing electrical power disruptions, stranding motorists, and killing livestock that graze on the open range.

The Prairies climatic zone has grassland and parkland natural vegetation (Figure 2.7). Soil moisture determines the type of natural vegetation. The short grassland vegetation occurs in the southern area where the evapotranspiration rate is greater due to higher temperatures. Here, a brown type of chernozemic soil is common, while further north in the parkland, a black soil has formed because of a lower evapotranspiration rate and a greater accumulation of vegetative material in the soil.

Although rainfall is not plentiful, the soils are thick and fertile because the natural vegetation returns nutrients to the soil. Also, the meagre rainfall has little effect on soluble soil nutrients. That is, rainfall does not wash away the nutrients from the upper layers of the soil in the Prairies zone as it does in wetter climatic zones such as the Pacific climatic zone.

Great Lakes–St Lawrence Lowlands Zone

Southern areas of Ontario and Québec enjoy a moderate continental climate. Located south of 46° N, this zone is the most southerly area of Canada. Its seasonal weather patterns are strongly influenced by the major air masses. In the summer the warm, moist air masses that originate in the Gulf of Mexico extend over this area, which result in long, hot, and humid summers. Arctic air masses that sweep southwards produce a short but sometimes cold winter. Such weather is associated with clear, sunny days. However, much of the winter consists of unsettled, cloudy weather, with temperatures hovering around the freezing point. Depending on the temperature, precipitation takes the form of rain, freezing rain, hail, or snow. Such weather is associated with frontal disturbances caused by the mixing of cold and warm air masses. The ice storm of January 1998 that devastated eastern Ontario and southern Québec was due to a large flow of warm, moist air from the Gulf of Mexico that met a colder and denser air mass from the Arctic (Vignette 2.8).

The Great Lakes modify temperatures and funnel winter storms into this zone. During the winter, the Great Lakes–St Lawrence Lowlands zone experiences a great variety of weather conditions. Occasional invasions of Arctic air masses bring cold, stormy weather. High winds that are funnelled along the Great Lakes storm track often accompany these cold spells, driving the wind chill to a dangerous level. Fortunately, most storms are short-lived and followed by relatively mild weather.

Vignette 2.8

The Great Ice Storm

On 5 January 1998, the worst ice storm in Canada's history struck eastern Ontario, southern Québec, and the Maritime provinces. It also affected the neighbouring New England states. While freezing rain occurs regularly in Canadian winters, this ice storm was unusually long. On the first day of the storm, about 20 mm of freezing rain fell on the Montréal area. Similar amounts fell in Kingston, Ottawa, Sherbrooke, and Trois-Rivières. These five cities and their surrounding rural communities felt the full fury of nature. But, unlike the one-day ice storm of 1961 that affected the Montréal area, the storm of 1998 continued throughout the next five days. The weight of the ice snapped hydroelectric poles and caused high-voltage transmission towers to collapse. The electrical system in eastern Ontario and southern Québec failed. More than 4 million people were without electrical power for at least thirty-six hours. About 500,000 people in the Estrie (the Eastern Townships) were without power for several weeks. The physical damage was enormous, perhaps reaching $500 million. Economic losses to farmers and businesses also totalled millions of dollars. Hydro-Québec faced considerable reconstruction of its damaged transmission system. The storm impacted negatively both human activity and the natural landscape (many trees were lost or cut down as a result).

Annual precipitation in this zone is about 100 cm. Warm, moist air masses from the Gulf of Mexico provide most of the moisture for the region, which falls as convectional and frontal precipitation. Convectional and frontal precipitation account for most rain and snow. The greatest amounts occur in the eastern areas of the Great Lakes where winter snowfall is particularly heavy. The Great Lakes also affect regional precipitation patterns because air masses absorb moisture from their surface, thus increasing local precipitation.

A mixed forest vegetation flourishes under these mild and wet climatic conditions. This forest is a transition zone between the boreal forest and the broad-leaved forest. Characteristic species are eastern white pine, red pine, eastern hemlock, yellow birch, sugar and red maples, red oak, basswood, and white elm. Grey-brown **luvisolic** soils predominate. These soils are rich in plant nutrients because of the ground cover created by the fallen leaves of deciduous trees.

Atlantic Zone

The climate in Atlantic Canada is far from homogeneous. The great north-south extent of this region is one reason why it has two climatic zones—the Atlantic and Subarctic zones. For example, the distance from the southern tip of Nova Scotia (44° N) to the northern extremity of Labrador (60° N) is over 2000 km. Because of this great latitudinal difference, daily maximum temperatures usually vary by 20° C or more between the Atlantic and Subarctic zones.

The general atmospheric circulation system brings several air masses to the Maritimes and Newfoundland, where they interact and result in unsettled weather conditions. The main air masses originate in the interior of North America, the Gulf of Mexico, and the North Atlantic Ocean. Consequently, summers are usually cool and wet, while winters are short but often bring heavy snow and rainfall. Most precipitation falls in the winter. Temperature differences between inland and coastal locations are striking. Coastal locations are usually several degrees warmer in the winter than inland locations. During the summer, the reverse is true; that is, coastal areas are usually several degrees cooler than inland areas.

Along the narrow coastal zone of Atlantic Canada, the climate is strongly influenced by the Atlantic Ocean. The summer temperatures of coastal settlements along the shores of Newfoundland are markedly cooled by the very cold water of the Labrador Current (a cold ocean current in the North Atlantic Ocean), which brings cold water and icebergs. Another effect of the sea occurs in the spring and summer—fog and mist result when the warm Gulf Stream waters (which originate in the Gulf of Mexico) mix with the cold Labrador Current. In the winter, the clash of warm and cold air masses sometimes results in severe winter storms characterized by heavy snowfall. However, such storms are usually short-lived and more normal weather conditions return quickly. Annual precipitation is abundant throughout Atlantic Canada. It averages around 100 cm in New Brunswick and 140 cm in Newfoundland.

Topography influences local conditions and the development of natural vegetation and soils. The Acadian mixed forest is in the Appalachian Uplands. The Acadian forest region, which is closely related to the Great Lakes–St Lawrence forest, contains red spruce, balsam fir, yellow birch, and sugar maple. The Atlantic zone's boreal forest covers the island of Newfoundland. White and black spruce are the principal species in this coniferous forest. Because soils are found only in the lowlands and at the bottom of valleys, there is little agricultural land in Newfoundland and Labrador. The rocky Labrador coast has an Arctic climate where tundra vegetation (shrubs, mosses, and lichens) struggle to survive. There is little soil under this vegetation.

Subarctic Zone

The Subarctic zone has a continental climate that extends in a wide band from Labrador to Yukon. The Subarctic climatic zone extends into much higher latitudes in northwest Canada than in northeast Canada because of warmer temperatures in the northwest. In northwest Canada, the average July temperature often reaches or exceeds 10° C, thus permitting the growth of trees. In similar latitudes of northeast Canada, summer temperatures are much lower. In northern Québec, for example, the Arctic climate prevails, thus preventing tree growth. The Subarctic climatic zone therefore has a southeast to northwest alignment. This alignment is caused by three factors:

• A continental effect means that land heats up more rapidly than water in the summer, resulting in higher summer temperatures in continental areas, such as the Yukon and the Mackenzie Valley, compared to coastal areas, such as the coast of Hudson Bay and Baffin Island, at the same latitude. Also, oceans warm up more slowly than land because oceans reflect more solar energy and because solar energy is distributed throughout the water body.

- The snow cover in the western section of the Subarctic is much thinner than in the eastern half. Pacific air masses dominate the weather pattern in this area in the spring and bring warmer weather. A thinner snow cover and warmer spring temperatures cause snow to disappear more quickly in the western Subarctic. Once the snow is gone, temperatures rise sharply.
- The Atlantic Ocean (including Hudson Bay) has cooling effects on northern Québec and Labrador. Part of that cooling effect is due to the Labrador Current, which brings Arctic waters to the middle latitudes of Atlantic Canada, and to the marine air masses that originate over the Atlantic Ocean. Combined with the deeper snow pack, this keeps spring temperatures low in the eastern subregion.

The Subarctic climatic zone experiences the greatest seasonal variation in temperatures of all the climatic zones in Canada; that is, long, cold winters and short, warm summers. As is typical of continental climates, extremely cold winter temperatures occur. January minimum daily temperatures often drop to -40° C and sometimes even to -50° C. Winters are influenced by Arctic air masses and are therefore extremely dry. In the short summer period, daily temperatures often exceed 20° C and occasionally reach 30° C. During the summer, Pacific air masses usually dominate this weather pattern, providing most of the precipitation in the Subarctic zone. Under these air mass conditions, the annual temperature range is quite broad, perhaps reaching 80° C.

There are also important variations in annual precipitation within the Subarctic zone. In the western subzone of the Subarctic, annual precipitation is low—about 40 cm—due to the rain shadow effect of the Cordillera. In the eastern subzone, annual precipitation is much higher, sometimes exceeding 80 cm, most of which is provided by the Atlantic and Gulf of Mexico air masses.

The warm but short summers provide adequate growing conditions for coniferous trees. For example, average monthly summer temperatures exceed 10° C, thereby promoting tree growth. Black and white spruce are the most common species in the Canadian boreal forest. Birch and poplar also occur, especially along the southern edge of the boreal forest. Stands of jack pine trees indicate an area that is recovering from a forest fire. Beneath this coniferous forest, there are podzolic soils. Wetlands are widespread: much of the land is poorly drained due to the disrupted drainage pattern caused by glaciation and permafrost. Wetlands contain numerous lakes, peat bogs, and marshes. Canadians often refer to this type of poorly drained land as muskeg.

Arctic Zone

The Arctic zone has an extremely cold and dry climate. Distinguished by extremely long winters and a brief summer, the Arctic climate is normally associated with high latitudes and lower levels of solar energy. The ice-covered Arctic Ocean keeps summer temperatures cool even though the sun remains above the horizon for most of the summer. Arctic air masses dominate the weather patterns throughout the year. Cool summer temperatures, which Köppen defined as an average mean of less than 10° C in the warmest month, prevent normal tree growth. For that reason, the Arctic climate region has tundra vegetation, which includes lichens, mosses, grasses, and low shrubs.

Beyond 70° N, growing conditions become very limited. With lower temperatures and less precipitation than in the lower latitudes of this climatic zone, the tundra vegetation cannot sustain itself. In these high latitudes, much of the land is barren. Such land is often called a polar desert. The Arctic climate, however, does extend into lower latitudes in two areas: along the coasts of Hudson Bay and of the Labrador Sea. In both cases, these cold bodies of water chill the summer air along the adjacent land mass. In this way, the Arctic climate extends along the coasts of Ontario, Québec, and Labrador well below 60° N, sometimes extending as far south as 55° N.

Arctic vegetation with fireweed and mountain avens. (*Courtesy Alec Aitken*)

Permafrost

A particularly distinctive feature of Canada's physical geography is permafrost. As noted earlier in this chapter, permafrost is permanently frozen ground with temperatures at or below zero for at least two years. The vast extent of permafrost in Canada provides a measure of the size of the country's cold environment (Figure 2.9). Permafrost exists in the Arctic and Subarctic climatic zones and occurs at higher elevations in the Cordillera zone.

Overall, permafrost is found in just over two-thirds of Canada's land mass.

Permafrost extends deeply into the ground. North of the Arctic Circle, permafrost may extend more than several hundred metres into the ground. Further south, permafrost is less frequent and where it occurs, it rarely penetrates more than 10 m into the ground. Permafrost is found in all six of Canada's geographic regions and reaches its most southerly position along 50° N in central Québec. Along the

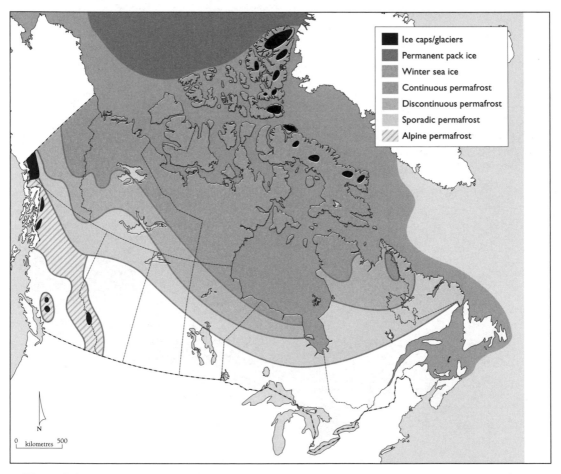

Figure 2.9 Permafrost zones. Canada's cold environment is best revealed by the permanently frozen ground that extends over three-quarters of the country.

southern edge of permafrost, there is a transition zone where small pockets of frozen ground have a depth of less than 1 m. Further south, these pockets of permafrost disappear.

Permafrost is divided into four types. **Alpine permafrost** is found in mountainous areas and takes on a vertical pattern as elevations of a mountain increase. Over most of Canada, however, permafrost follows a zonal pattern, which does not correspond to latitude but rather to the annual mean temperatures that fall below zero.[3] The zonal pattern has a northwest to southeast alignment; that is, from the Yukon to central Québec (see Figure 2.9).

As the mean annual temperature varies, the type of permafrost also changes. **Continuous permafrost** occurs in the higher latitudes of the Arctic climatic zone, where at least 80 per cent of the ground is permanently frozen, although it also extends into northern Québec. Continuous permafrost is associated with very low mean annual air temperatures of -15° C or less. **Discontinuous permafrost** occurs when 30 to 80 per cent of the ground is permanently frozen. It is found in the Subarctic climatic zone where mean annual air temperature ranges from -5° C in the south to -15° C in the north. **Sporadic permafrost** is found mainly in the northern parts of the provinces, where less than 30 per cent of the area is permanently frozen. Sporadic permafrost is associated with mean annual temperatures of zero to -5° C.

Human Impacts on the Natural Environment

Humans exist within a natural environment that has been modified by both individual and collective actions. Since the Industrial Revolution, humans have placed greater and greater demands on their natural environment, including harnessing the hydroelectric potential of large rivers by constructing powerplants; burning fossil fuels; cutting forests; and discharging toxic wastes into lakes, rivers, and oceans. Many serious environmental problems are the consequence of such human actions (Dearden and Mitchell 1997: Chapter 1). In this section some human impacts on the physical world are examined, focusing on those affecting Canada's natural environment.

Major Drainage Basins and Hydroelectric Projects

A **drainage basin** is land that slopes towards the sea and is separated from other lands by topographic ridges. These ridges form drainage divides. The continental divide of the Rocky Mountains, for example, separates those streams flowing to the Pacific Ocean from those flowing to the Arctic and Atlantic oceans. While each stream has a drainage basin, those streams flowing to the same ocean combine their basins to form a major drainage basin.

Canada has four major drainage basins (Figure 2.10): the Atlantic Basin, the Hudson Bay Basin, the Arctic Basin, and the Pacific Basin. In addition, a small portion of southern Alberta and Saskatchewan forms part of the Mississippi River system, which drains into the Gulf of Mexico. Since this area is extremely arid, there is little water from Canada that flows south.

Historically, the rivers in each basin have played major roles in the development of the country. For example, Aboriginal peoples and

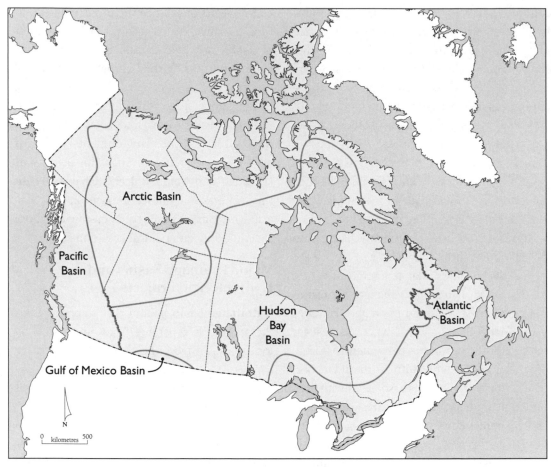

Figure 2.10 Drainage basins of Canada. The Hudson Bay drainage basin is by far the largest of the five basins in Canada. It also serves as a boundary between southern Alberta and British Columbia, and between northern Québec and Labrador.

Europeans both used the St Lawrence and Mackenzie rivers as transportation routes during the fur trade. These rivers remain important waterways today.

Atlantic Basin

The Atlantic Basin is centred on the St Lawrence River and its tributaries, but the basin also includes Labrador. The Atlantic Basin has the third largest drainage area and also the third greatest stream-flow. The largest hydroelectric development in this drainage basin is located at Churchill Falls in Labrador. Development of the lower Churchill River may take place in the next century. Other hydroelectric developments are located along the

St Lawrence River in southern Québec and along its tributary rivers that flow out of the Laurentide Upland of the Canadian Shield. Rivers such as the Manicouagan River originate in the higher elevation of the Laurentide Upland. Here, the combination of abundant precipitation, natural lakes, and a sharp increase in elevation provides ideal natural conditions for the generation of hydroelectric power. Because there is a large market for electrical power in the St Lawrence Lowlands, virtually all potential sites in the Laurentides have been developed.

The Hudson Bay Basin

The Hudson Bay Basin is the largest drainage basin in Canada (Table 2.6). It occupies about 3.8 million km^2. In the west, its rivers originate in the Rocky Mountains and flow to Hudson Bay. In the east, the headwaters of its rivers in the uplands of northern Québec flow westward into James Bay. In northern Ontario and Manitoba, rivers drain into James and Hudson bays.

The combination of large rivers and sudden drops in elevation that occur in the Canadian Shield makes this part of the basin ideal for developing hydroelectric powerplants (Vignette 2.9). In fact, most of Canada's hydroelectric power is generated in the Canadian Shield area of the Hudson Bay Basin—the largest installations are on the La Grande Rivière in northern Québec and on the Nelson River in northern Manitoba. The La Grande Rivière's hydroelectric developments are the

Table 2.6
Canada's Drainage Basins

Drainage Basin	Area (million km²)	Stream-flow (m³/second)
Hudson Bay	3.8	30 594
Arctic	3.6	20 491
Atlantic	1.6	21 890
Pacific	1.0	24 951
Gulf of Mexico	<0.1	—
Total	10.0	105 135

Sources: A.H. Laycock, 'The Amount of Canadian Water and Its Distribution', in *Canadian Aquatic Resources*, edited by M.C. Healey and R.R. Wallace (Ottawa: Department of Fisheries and Oceans, 1987):32; Philip Dearden and Bruce Mitchell, *Environmental Change and Challenge: A Canadian Perspective* (Toronto: Oxford University Press, 1997):142.

Vignette 2.9

Waterpower Resources

Hydroelectric developments depend on three factors: climate, topography, and access to market. The development of hydroelectric power in Canada varies considerably from region to region. In the 1960s, most power sites were built close to Canadian markets. Since then, high-voltage transmission lines have facilitated the transmission of electricity over long distances, permitting more distant sites to be developed. Several large-scale hydroelectric projects have been developed in more remote areas, namely in northern parts of Québec and Manitoba. Canada's principal hydroelectric generating stations and their installed capacity in megawatts are:

- LG–2 on the La Grande Rivière, Québec (5328 MW)
- Churchill Falls on the Churchill River, Labrador (5225 MW)

- LG–4 on the La Grande Rivière, Québec (2650 MW)
- Gordon M. Shrum on the Peace River, BC (2416 MW)
- Kettle Rapids on the Nelson River, Manitoba (1225 MW).

The main advantages of hydroelectric developments are: the long life of the facilities, low operating costs, and zero air pollution. However, there are drawbacks: the initial high capital investment, the long construction period, and the socio-environmental costs. When river valleys used for hunting and fishing by Aboriginal peoples are flooded, the loss of prime wildlife habitat and fishing grounds has serious impacts on their economy and culture. Such losses are a very high hidden social cost.

first stage in the James Bay Project. The Great Whale River Project was to follow the completion of the hydroelectric projects on the La Grande Rivière, but a variety of circumstances (low energy demand, low prices in New England, and strong opposition from environmental groups and the Cree Indians of northern Québec) stalled its development.

The Arctic Basin

The Arctic Basin is Canada's second largest drainage basin. The Mackenzie River dominates the drainage system in this basin. Along with its major tributaries (the Athabasca, Liard, and Peace rivers), the Mackenzie River is the second longest river in North America. However, because of low precipitation in the Arctic, this basin has only the fourth largest stream-flow. There are few hydroelectric projects in the Arctic Basin because of the long distance to markets, with the exception of the hydroelectric development on the Peace River in British Columbia. Here, power from the Gordon M. Shrum generating facility is transmitted to the population centres in southern British Columbia and to the United States, primarily to the states of Washington, Oregon, and California.

The Pacific Basin

The Pacific Basin is the smallest basin. However, it has the second highest volume of water draining into the sea. Heavy precipitation along the coastal mountains of British Columbia accounts for this unusually high streamflow. As a result, the Pacific Basin is the site of one of Canada's largest hydroelectric projects. Located at Kemano, this facility is owned and operated by Alcan, which uses the electrical generating station to supply power to its aluminum smelter at Kitimat. The ice-free, deepwater harbour at Kitimat and low-cost electric power generated at Kemano make Kitimat an ideal location for an aluminum smelter.

Resource conflicts do occur during hydroelectric developments. Often the conflict pits non-renewable resource development against renewable resource development. The Pacific Basin provides such an example. This basin forms an integral part of the natural biological cycle for salmon (a renewable resource), which live most of their lives in the Pacific Ocean. When they reach maturity, however, they return to their spawning grounds in the headwaters of the various tributaries that flow into the major rivers of the Pacific Basin, including the Nechako River, which is an important tributary of the Fraser River. Alcan Canada's construction of a dam on this river (for hydroelectricity) flooded prime salmon spawning grounds and thus, created a resource conflict. This particular conflict is discussed more fully in the chapter on British Columbia.

Acid Rain

Acid rain is precipitation that has an unusually acidic chemical composition. Acidity levels are described in terms of the pH factor, which measures the acidity or alkalinity of a substance on a scale of 0 to 14, with 0 being extremely acidic and 14 being extremely alkaline. The average pH of normal rainfall is 5.6, making it slightly acidic. Acid rain, however, has a pH of 2.4 or less. At this pH level, acid rain is a serious environmental problem in many parts of the world, including eastern Canada.

Most acid rain results from the chemicals derived from the burning of fossil fuels in industrial plants and from automobile exhaust. Sulphur dioxide and nitrogen oxides are emitted in the smoke from coal-burning plants, while automobile engines are particularly prolific producers of nitrogen oxides. Acid rain is created when oxides of sulphur and nitrogen change chemically as they dissolve in water vapour in the atmosphere, and then return to earth as acidic rain, snow, or fog.

The most visible impact of acid rain is in urban areas where it slowly corrodes limestone buildings and defaces marble sculptures. Less visible but just as destructive are acid rain's effects on forests, lakes, and soils. In eastern Canada, acid rain has damaged many forests and caused a sharp decline in fish stocks in many lakes. The principal areas in Canada affected by acid rain are located in Ontario, Québec, and the Maritime provinces.

Global Warming

How exactly does our atmosphere obtain its energy? Only a few gases (called greenhouse gases) in our atmosphere can absorb solar energy. These gases (including carbon dioxide, methane, and water vapour) form less than 0.1 per cent of the atmosphere. Carbon dioxide is the principal greenhouse gas. It is transparent to short-wave radiation from the sun but opaque to the long-wave radiation emitted by the earth. In this way, carbon dioxide and other greenhouse gases heat the atmosphere, causing air temperatures to rise. This process is known as the **greenhouse effect**.

Some experts predict a major change in our climate within fifty years due to global warming. More specifically, they anticipate a sharp rise in the world's mean air surface temperature by the middle of the next century. Some believe that these changes have already begun. Unlike past climatic changes (Vignette 2.10), global warming is caused by human actions, particularly the burning of fossil fuel. Such human activities are adding more and more carbon dioxide to our atmosphere, thereby increasing the atmosphere's capacity to absorb solar energy. The increase in the mean world temperature by 2050 is estimated at a minimum of 2° C and a maximum of 5° C (Cohen 1997:1).

Although no one can predict the particular impacts of global warming on Canada, climatic change would translate into longer, hotter summers and shorter, milder winters.[4] As well, climatic zones in North America would

---

Vignette 2.10

The Little Ice Age

Both major and minor climatic changes have occurred during the earth's history. A major cooling of the world's climate occurred in the Quaternary period of the Cenozoic era (Table 2.1). During that geological period, four major ice advances occurred, the last being the Wisconsin.

However, minor cooling of the world's climate has also occurred. Minor cooling refers to relatively short periods of time when the global climate is slightly cooler than normal. A minor cooling, known as the Little Ice Age, took place between 1450 and 1850. The Little Ice Age had a dramatic impact on human beings living in the Arctic. During the Little Ice Age, the global climate was much cooler than it is today. This period was characterized by lower temperatures and longer winters in higher latitudes. In northern Canada, the ice cover over the Arctic Ocean was more extensive, which prevented bowhead whales from entering these waters. The consequences for the Thule inhabitants, who had developed a hunting economy based on the bowhead whale, were devastating. They were forced to hunt smaller game—seals and caribou. The results were twofold: the new hunting system could not support as many people and it required smaller, more mobile hunting groups. Archaeologists believe that the Inuit are the descendants of the Thule people.

shift poleward. Those zones now present in the northern United States would extend into Canada, while those in southern Canada would move northward. The Arctic climatic zone would diminish in size. The long-term impact on Canada's environment might include the disappearance of the ice cover on the Arctic Ocean, the flooding of cities along Canada's coastlines (as melting ice and snow cause ocean levels to rise), and the appearance of a desert-like climate in the Canadian Prairies.

Is global warming underway? Nobody knows for certain, but as more carbon dioxide and other greenhouse gases are added to the atmosphere, more solar energy will be trapped in the atmosphere, thus causing air temperatures to rise. Certainly the mean world air temperature has increased over the last 250 years, probably due to the burning of fossil fuels. Supporting evidence for this argument is based on the smaller amount of carbon dioxide in the atmosphere before the Industrial Revolution compared with the amount in 1985. Before the Industrial Revolution, carbon dioxide was estimated to form 0.02 per cent of the atmosphere, while by 1985 it comprised over 0.03 per cent. Most meteorologists predict that by 2050 the amount of carbon dioxide in the atmosphere will increase to 0.06 per cent (Christopherson 1994:137). However, temperature increases have so far fallen within the so-called normal temperature variation associated with the earth's climate; that is, within plus or minus 1 per cent of the world mean air surface temperature. Until world temperatures exceed this limit, global warming has not 'officially' begun. While the

physics of global warming in the greenhouse model are elementary, the actual process of climate change is extremely complex and remains unclear.

The economic consequences of global warming for Canada would be far-reaching. Some consequences would be favourable while others would not. Agricultural activities could take place further north, but, on the negative side, grain agriculture in the Canadian Prairies might be subject to greater risk of drought. Water transportation would greatly benefit from longer navigation seasons. The Great Lakes and the St Lawrence River, for instance, would be navigable all year round. The ice-free shipping season in Hudson Bay and the Arctic Ocean would be greatly extended. Coastal areas, however, might be flooded as sea ice and glaciers melt, causing sea levels to rise. Global warming could thaw the permanently frozen ground known as permafrost. Melting of the ice in permafrost could cause massive ground **subsidence**, resulting in an irregular relief referred to by physical geographers as 'thermokarst topography'. Such ground subsidence could disrupt transportation and pipeline systems and play havoc with foundations for buildings, bridges, and other human-made structures (Bone, Long, and McPherson 1997:265–74).

Global warming seems inevitable, but other events might diminish the amount of solar energy reaching the earth's surface and thereby reverse the temperature trend. For example, volcanic eruption, by releasing huge amounts of dust into the air, could reflect significant amounts of incoming solar energy to outer space and thereby generate a cooling effect on the world's climate. But can we count on such an event occurring to reverse the current warming trend?

Summary

Physical geography varies across Canada. This variation is critical in understanding Canada's regional character. At the macrolevel, physiographic regions represent large areas with similar landforms and geological structures. Climate creates a similar zonal arrangement of soils and natural vegetation. Climate therefore determines to a large extent the type of soil and native vegetation in a given region and hence influences the utilization of the land. Together with topography, climate partly determines the land's ability to support a population. This link between the physical and human worlds identifies those regions having a more favourable mix of physical characteristics for economic development, and translates the abstract core/periphery model into a geographic reality. The Great Lakes–St Lawrence Lowlands region is the most favoured physical region in Canada. Physical barriers, such as the Rocky Mountains, and extreme climatic conditions, such as the very long and cold winters in northern Canada, have also affected the historical settlement of the country and continue to influence contemporary economic activities.

Canada has several physiographic regions and climatic zones. The seven physiographic regions are: the Canadian Shield, the Cordillera, the Interior Plains, the Hudson Bay Lowlands, the Arctic Lands, the Appalachian Uplands, and the Great Lakes–St Lawrence Lowlands. The seven climatic zones are: the Pacific, Cordillera, Prairie, Great Lakes–St Lawrence, Atlantic, Subarctic, and Arctic.

Because Canada lies in the northern latitudes, permafrost is common in the northern areas of provinces (Newfoundland, Québec, Ontario, Manitoba, Saskatchewan, Alberta, and British Columbia) and in the Territorial North. Within Canada, rivers flow into four major drainage basins: the Atlantic, Hudson Bay, Arctic, and Pacific. Within southern Alberta and Saskatchewan, a tiny area slopes southward into the Mississippi Basin; streams in this area eventually drain into the Gulf of Mexico.

In our contemporary world, humans are the most active and dangerous agents of environmental change. The cultivation of the land, exploitation of its renewable and non-renewable resources, and processing of its primary products have forever changed our natural environment in many ways. The construction of the Confederation Bridge, which joins Prince Edward Island to the mainland of Canada, is one example. Often, however, human activities have negative environmental consequences such as acid rain and global warming. Acid rain is a serious environmental problem today, while global warming may prove to be *the* environmental challenge of the next century.

In the next chapter, our attention is focused on historical geography, particularly on the territorial evolution of Canada. The principal objectives of Chapter 3 are:

- to emphasize the impact that physical geography had in shaping the historical and territorial evolution of Canada

- to introduce the notion of two visions of Canada and to show how they influenced our history and geography
- to examine the historic basis of the three main issues facing Canadians: the Aboriginal/non-Aboriginal issue, the centralist/decentralist issue (often manifest as a core/periphery dichotomy), and the French/English issue.

Notes

1. Canadians have various visions of themselves, their region, and their country. For the most part, these visions are rooted in the physical nature and historical experiences that affected Canada and its regions. For example, people see Canada as a northern country because of its location in the North American continent and because of its climates, which are often noted for long, cold winters. Hamelin's concept of nordicity (as discussed in Chapter 1) exemplifies this northern perspective. Artists too have found this northern theme appealing. Gilles Vigneault, one of Québec's best-known *chansonniers*, wrote the song *Mon pays*. Though referring to Québec, the opening line, *Mon pays ce n'est pas un pays, c'est l'hiver* (My country is not a country, it is winter), resonates equally well in all regions of Canada.

2. Tectonic forces press, push, and drag portions of the earth's crust in a slow but steady movement. This process is the basis for the theory of continental drift. In 1912, Alfred Wegener suggested that long ago all the earth's continents formed one huge land mass. Tectonic action caused the breakup of this land mass into huge slabs. These slabs of the earth's crust drifted slowly on the molten mass (magma) beneath the earth's crust.

3. The mean annual temperature of a location on the earth's surface is a measure of the energy balance at that point. Solar energy is the source of heat for the earth, and this energy is returned to the atmosphere in a variety of ways. Therefore, a global energy balance exists. However, there are regional energy surpluses and deficits in different parts of the world. For example, the Arctic has an energy deficit, while the Tropics has a surplus. These energy differences drive the global atmospheric and oceanic circulation systems. When the mean annual temperature is below zero, it indicates that an energy deficit exists.

4. Slaymaker and French discuss the effects of global warming and climatic change on Canada's North. They maintain that climatic change caused by the greenhouse effect would alter Canada's cold environments more dramatically than the country's other environments. The authors describe possible changes to sea ice, permafrost, snow cover, sea level, and natural vegetation. There are four key factors for such a remarkable climatic change in northern Canada:

- The percentage of carbon dioxide in the atmosphere over the Territorial North is much greater than that over southern Canada.
- The thinning of the ozone layer over the Territorial North permits more solar radiation to enter the atmosphere.
- The release of methane gases from the muskeg in northern Canada will add more greenhouse gases to the northern atmosphere.

- The reduction in the duration of snow cover will expose the northern lands to solar radiation for a longer time.

In the same book, Ledrew presents the nature of climatic change, while Smith describes the impact of climatic change on permafrost. See chapters 11, 12, and 13 in *Canada's Cold Environments*, edited by Hugh M. French and Olav Slaymaker (Montréal–Kingston: McGill-Queen's University Press, 1993).

Key Terms

alpine permafrost
Permanently frozen ground that is found at high elevations.

arêtes
Sharp mountain ridges that are formed between two cirques.

basins
Structural depressions in sedimentary rock that are caused by a bending of sedimentary strata into huge bowl-like shapes. Petroleum may accumulate in sedimentary basins.

chernozemic
A soil order identified by a well-drained soil that is often dark brown to black in colour; associated with the grassland and parkland natural vegetation types and located in the Prairies climatic zone.

cirques
Large, shallow depressions found in mountains caused by the plucking action of alpine glaciers.

climate
An average condition of weather over a very long period of time.

climatic zone
Geographic areas where similar types of weather occur.

continental air masses
Homogeneous bodies of air that have taken on moisture and temperature characteristics of the land mass of their origin. Continental air masses are normally dry and cold in the winter and dry and hot in the summer.

continental drift
The movement of the earth's crust. This concept is also known as plate tectonics.

continuous permafrost
Extensive areas of permanently frozen ground in the Arctic, where at least 80 per cent of the ground is permanently frozen.

convectional precipitation
An upward movement of moist air causes the air to cool, resulting in condensation and then precipitation.

cryosolic
A soil order associated with permafrost and poorly drained land; soil is either lacking or extremely thin; associated with the tundra and polar desert vegetation types and located in the Arctic climatic zone.

deposition
The deposit of material on the earth's surface by various processes such as ice, water, and wind.

discontinuous permafrost
Permanently frozen ground mixed with unfrozen ground in the Subarctic. At its northern boundary about 80 per cent of the ground is permanently frozen, while at its southern boundary about 30 per cent of the ground is permanently frozen.

drainage basin
> Land sloping towards the sea; an area drained by rivers and their tributaries.

drumlins
> Landforms created by the deposit of glacial till and shaped by the movement of the ice sheet.

eskers
> Long, sinuous mounds of sand and gravel that were deposited on the bottom of a stream flowing under a glacier.

faulting
> The breaking of the earth's crust.

folding
> The bending of the earth's crust.

friction of distance
> The effect of distance on spatial interaction; that is, as distance increases, the number of spatial interactions (such as telephone calls) diminishes.

frontal precipitation
> When a warm air mass is forced to rise over a colder air mass, condensation and then precipitation occur.

glacial erosion
> The scraping and plucking action of moving ice on the surface of the land.

glacial spillway
> A deep and wide valley formed by the flow of massive amounts of water originating from a melting ice sheet or from water escaping from glacial lakes.

glacial striations
> Scratches or grooves in the bedrock caused by rocks embedded in the bottom of a moving ice sheet or glacier.

glacial trough
> A U-shaped valley carved by an alpine glacier.

global circulation system
> The movement of ocean currents and wind systems that redistribute energy around the world.

greenhouse effect
> The absorption of long-wave radiation from the earth's surface by the atmosphere.

igneous rocks
> Rock formed when the earth's surface first cooled or when magma or lava reached the earth's surface.

isostatic rebound
> The uplifting process of the earth's crust following the removal of an ice sheet that, because of its weight, depressed the earth's crust. Also known as postglacial uplift.

latitude
> A line parallel to the equator that encircles the globe.

longitude
> A line that runs through both the North and South poles.

luvisolic
> A soil order identified by a well-drained soil that is often grey-brown in colour; associated with the broadleaf and mixed forest natural vegetation types in the Great Lakes–St Lawrence climatic zone.

marine air masses
> Large homogeneous bodies of air with moisture and temperature characteristics similar to the ocean where they originated. Marine air masses are normally moist and mild in both winter and summer.

metamorphic rocks
> Formed from igneous and sedimentary rocks by means of heat and pressure.

muskeg
> A wet, marshy area found in areas of poor drainage, such as the Hudson Bay Lowlands. Muskeg contains peat deposits.

nunataks
> An unglaciated area of a mountain that stood above the surrounding ice sheet.

orographic precipitation

Rain or snow created when air is forced up the side of a mountain, thereby cooling the air and causing condensation followed by precipitation.

patterned ground

The arrangement of stones and pebbles in polygonal shapes. Patterned ground occurs in the Arctic where continuous permafrost exists and where frost shattering is the principal erosion process.

permafrost

Permanently frozen ground.

physiographic region

A large geographic area where a single landform, such as the Interior Plains, is found.

pingos

Hills or mounds that have an ice core and that are found in areas of permafrost.

podzolic order

A soil order identified by a poorly drained soil that is often grey in colour; associated with the boreal forest and the coastal rainforest and with climates that have large amounts of precipitation, such as the Pacific, Atlantic, and Subarctic climatic zones.

postglacial uplift

The slow rising of the earth's crust following the removal of an ice sheet that, because of its weight, depressed the earth's crust. Also known as isostatic rebound.

rain shadow effect

Results in a dry area on the lee side of mountains where air masses descend, causing them to become warmer and drier.

residual uplift

The final stages of isostatic rebound.

restrained rebound

The first stage of isostatic rebound.

sedimentary rocks

Rocks formed from the accumulation, in a layered sequence, of sediment deposited in the bottom of an ocean.

soil order

Classes of soil based on observable soil properties and soil-forming processes. In Canada there are nine soil orders, including chernozemic, cryosolic, and podzolic.

sporadic permafrost

Pockets of permanently frozen ground mixed with large areas of unfrozen ground. Sporadic permafrost ranges from a trace of permanently frozen ground to an area having up to 30 per cent of its ground permanently frozen.

strata

Layers of sedimentary rock.

subsidence

A downward movement of the ground. In areas of permafrost, subsidence occurs when large blocks of ice within the ground melt, causing the material above to sink or collapse.

till

Unsorted glacial deposits.

References

Bibliography

Bone, R.M., Shane Long, and Peter McPherson. 1997. 'Settlements in the Mackenzie Basin: Now and in the Future 2050'. In *The Final Report of the Mackenzie Basin Impact Study*, edited by Stewart J. Cohen, 265–74. Downsview: Environment Canada.

Canada. 1997. *Canada Year Book 1998*. Ottawa: Minister of Supply and Services.

Christopherson, Robert W. 1994. *Geosystems: An Introduction to Physical Geography*, 2nd edn. Englewood Cliffs: Macmillan Publishing Company.

Cohen, Stewart J., ed. 1997. *The Final Report of the Mackenzie Basin Impact Study*. Downsview: Environment Canada.

Dearden, Philip, and Bruce Mitchell. 1997. *Environmental Change and Challenge: A Canadian Perspective*. Toronto: Oxford University Press.

French, H.M., and O. Slaymaker, eds. 1993. *Canada's Cold Environments*. Montréal–Kingston: McGill-Queen's University Press.

Hare, F. Kenneth, and Morley K. Thomas. 1974. *Climate Canada*. Toronto: Wiley.

Laycock, A.H. 1987. 'The Amount of Canadian Water and Its Distribution'. In *Canadian Arctic Resources*, edited by M.C. Healey and R.R. Wallace, 13–42. Ottawa: Department of Fisheries and Oceans.

Marsh, James H., ed. 1988. *The Canadian Encyclopedia*, 2nd edn. Edmonton: Hurtig Publishers.

Norton, William. 1998. *Human Geography*, 3rd edn. Toronto: Oxford University Press.

Shabbar, Amir, Barrie Bonsal, and Madhav Khandekar. 1997. 'Canadian Precipitation Patterns Associated with Southern Oscillation'. *Journal of Climate* 10:3016–27.

Further Reading

Archibold, O.W. 1995. *Ecology of World Vegetation*. London: Chapman & Hall.

Bone, R.M. 1992. *The Geography of the Canadian North: Issues and Challenges*. Toronto: Oxford University Press.

Briggs, David, Peter Smithson, and Timothy Ball. 1993. *Fundamentals of Physical Geography*. Toronto: Copp Clark Pitman.

Cohen, Stewart J. 1990. 'Bringing the Global Warming Issue Closer to Home: The Challenge of Regional Impact Studies'. *American Meteorological Society* 71, no. 4:520–6.

Dyke, Arthur S., and Victor K. Prest. 1987. 'Late Wisconsinan and Holocene History of the Laurentide Ice Sheet'. *Géographie physique et quaternaire* XLI, no. 2:181–6.

Fulton, Robert J., and Victor K. Prest. 1987. 'The Laurentide Ice Sheet and Its Significance'. *Géographie physique et quaternaire* XLI, no. 2:181–6.

Trenhaile, Alan S. 1998. *Geomorphology: A Canadian Perspective*. Toronto: Oxford University Press.

Young, Steven B. 1989. *To the Arctic: An Introduction to the Far Northern World*. New York: Wiley.

Chapter 3

Overview

Canada's historical geography began about 30,000 years ago when the first humans arrived in North America. Since then, history and geography have combined to forge Canada's historical geography. Among the many important historical events, three have shaped the character of each region and given the country a special identity. These three are: (1) the establishment of Indian and Inuit societies prior to European contact; (2) the arrival of French and British settlers in the seventeenth, eighteenth, and early-nineteenth centuries and their impact on the land; and (3) the territorial evolution of Canada since 1867 to the much larger Canada of today. In combination, history and geography have moulded a complex society—a country with two visions of itself and three principal tensions that affect national unity. These tensions are described as Aboriginal/non-Aboriginal, centralist/decentralist, and French/English faultlines.

Objectives

- Describe the arrival of Canada's first people.
- Examine the settlement of Canada by the French and the British.
- Present the territorial evolution of Canada.
- Identify three tensions within Canada.
- Discuss the notion of 'One Country, Two Visions'.
- Suggest 'solutions' to complex problems.

Canada's Historical Geography

Introduction

In one sense, Canada is a young country. Its formal history began in 1867 with the passing of the British North America Act by the British Parliament. In another sense, Canada is an old country whose history goes back much further than 1867. As an old country, its history has followed many twists and turns, but two events stand out because they continue to have a profound impact on the nature of Canadian society. These events are the arrival of the first people in North America, and the colonization of North America by France and England. After 1867, a third historical event continued to shape the country: the territorial expansion of Canada. This expansion continued and only ceased in 1949, when Newfoundland joined Canada. As the country was spread over such a huge territory, Canada developed a distinct regionalized character shaped by history and physical geography.

The political reality of Canada includes a division of powers between the central and provincial governments,[1] which adds to the country's regional character. Therefore, much of this great political experiment called Canada involved the difficult task of forging a unity among its regional societies. While Can-

ada is admired by most peoples of the world, it has seeds of regional discontent. Regional discontent, often related to economic disparities and squabbling over natural resources, has generated ongoing tensions between the federal government and the provinces, as well as tensions among the provinces over issues such as language laws and constitutional amendments. To use *The Globe and Mail* columnist Jeffrey Simpson's metaphor, these tensions can be likened to faultlines—large cracks in the earth's crust caused by tectonic forces. Like geological faultlines, Canada's economic, social, and political faultlines are powerful opposing regional forces that threaten to destabilize Canada's integrity as a nation. Both now and in the past, these faultlines have, from time to time, split the country into wrangling factions.

Three key tensions are examined in this chapter: (1) Aboriginal versus non-Aboriginal Canadians, (2) centralists versus decentralists, and (3) English-speaking versus French-speaking Canadians. In each case, these disagreements occur between a core (for example, Central Canada) and a periphery (for example, the rest of Canada). Central Canada is often the symbol of the political power held by Ottawa, and at other times the economic power held by Ontario and Québec. Although the periphery refers in general to the rest of Canada, it can also represent a particular geographic region or even a group of people with a common complaint against the core.

Of the three faultlines, the French/English rift is by far the most threatening to Canada. For example, if the next Québec referendum points conclusively to sovereignty, it could lead to a reorganization of Canada's territorial boundaries. Yet at the same time, the historic and contemporary interaction between French- and English-speaking Canadians is the very process that makes Canada so unique. This give and take between the two founding peoples often spills over to affect the Aboriginal and regional issues that threaten Canada's unity. Oddly enough, because of the interaction between Aboriginal and non-Aboriginal Canadians, centralists and decentralists, and English-speaking and French-speaking Canadians, each faultline is dependent on the other two. Events affecting one faultline often have a positive or negative effect on the other two.

Chapter 3 begins by tracing the deep historical roots and the geographic realities that, in combination, have split the nation along the three major faultlines that will be examined later in the chapter.

The First People

The first people to set foot on North American soil were Old World hunters who crossed a land bridge (known as Beringia) into Alaska and Yukon. Archaeologists believe some Old World hunting societies that existed in northeast Asia about 40,000 years ago wandered across a landbridge, following herds of woolly mammoths. There is some evidence (although not conclusive) that these Old World hunters reached the northeastern tip of North American about 25,000 to 30,000 years ago, around the time the last ice advance began. Because such huge amounts of water were frozen in the Wisconsin ice sheet, the sea level dropped by about as much as 100 m, exposing the Beringia land bridge, which reached its maximum extent about 18,000 years ago. Archaeologists believe that between the time of the last ice advance and the maximum extent of Beringia, a number of Old World hunting groups used the land bridge to cross from Siberia to Alaska. However, because most of Canada was covered by the Wisconsin ice sheet at the time of this migration, these groups were blocked by ice that was perhaps 4 km thick. They had to wait until an ice-free corridor developed before they were able to migrate southward into the heart of North America (Figure 3.1). Geologists have estimated that this ice-free corridor, which was just east of the Rocky Mountains, appeared about 12,000 to 13,000 years ago, although the precise date of this migration is unknown. By this time, the Beringia land bridge was again under water, which made further migration from Asia to North America virtually impossible until the marine technology of the Palaeo-Eskimos emerged about 5,000 years ago.

Some archaeologists even speculate that Old World hunters may have arrived in North America much earlier. According to one theory, these early hunters made a coastal migration along the edge of the Cordillera ice sheet

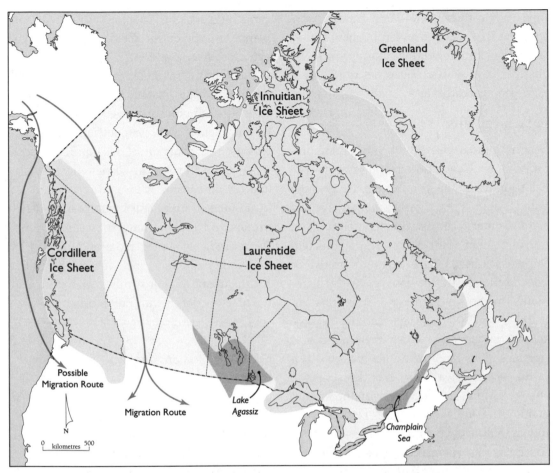

Figure 3.1 Migration routes into North America. The Wisconsin ice sheet had retreated by 12,000 BP, leaving an ice-free corridor between the two remaining parts of the Wisconsin ice sheet (the Cordillera and the Laurentide ice sheets). In places where the melt water could not flow to the sea, water collected in low-lying areas and created glacial lakes such as Lake Agassiz. As explained in Vignette 2.6, however, the Champlain Sea was not a glacial lake but an extension of the Atlantic Ocean.

that covered all the mountains and valleys of British Columbia. By slowly moving southward along the unglaciated islands adjacent to the ice-covered British Columbia coast, these early people could have circumvented the Cordillera ice sheet by island-hopping, thereby reaching the unglaciated area of North America much earlier than 12,000 years ago. However, archaeological evidence of such a

route is lacking because these ancient campsites, if they existed, are now well below the current sea level.

Although the question of when humans first reached North America is still unsettled, archaeologists are certain that people occupied the unglaciated areas of North America by at least 11,500 years ago. The Clovis (or Palaeo-Indian) culture was well established by this time. While these New World hunters were spread across temperate North America, they shared a common hunting culture, which was characterized by its uniquely designed fluted-point stone spearhead.

Palaeo-Indians

How did these Old World hunting cultures evolve into a New World hunting system? The Palaeo-Indians, the people who devised the fluted points, were descendants of the Old World hunters. The oldest fluted points found in North America are about 11,500 years old. These spearheads have been discovered in Canada, along with the bones of woolly mammoths, in the southern part of the Canadian Prairies. By 9,000 years ago, many of the large species, such as the woolly mammoths and the mastodons, became extinct, possibly as a result of excessive hunting and/or climatic change. After the extinction of the woolly mammoths and mastodons, later Palaeo-Indian cultures, which archaeologists refer to collectively as Plano, developed a variety of unfluted stone points with stems for attachment to spear shafts, which made weapons more suitable for hunting buffalo and caribou. About 8,000 years ago, these hunters in the grasslands of the interior of Canada hunted buffalo, while those in the tundra and forest lands of northern and eastern Canada hunted caribou. However, these smaller prey species could not support large numbers of people, so the Palaeo-Indians had to develop new survival strategies. These strategies involved remaining in one area (and presumably keeping other peoples out of that area), developing effective hunting techniques for the local game, and making extensive use of fish and plants to supplement their principal diet of game.

This link between geographic territory and hunting societies marked the development of Palaeo-Indian **culture areas**, which were geographic regions with the following two characteristics: (1) a common set of natural conditions that resulted in similar plants and animals, and (2) inhabitants who used a common set of hunting, fishing, and food-gathering techniques and tools. Under these conditions, Palaeo-Indians formed more enduring social units which became the forerunners of the North American Indian tribes who existed at the time of contact with Europeans.

Indians

Climatic conditions vary across North America, ranging from a tropical climate in Mexico to a colder climate north of the Great Lakes–St Lawrence Lowlands. These varied conditions provided different opportunities for Indians, some of whom were descendants of the Palaeo-Indians. Most archaeologists support the idea that Algonquians are direct descendants of Palaeo-Indians, but they are less certain about Athapaskans, whose ancestors may have arrived in North America some time after the Palaeo-Indian culture emerged about 11,500 years ago.

About 5,000 years ago, Indians living in the tropical climate of Mexico began to domesticate plants and animals. This agricultural system and its people gradually spread northward into areas with more restrictive growing conditions. These climatic differences required Indians to adapt their agricultural system accordingly. About 3,000 years ago, Indians in the eastern United States planted corn, beans, and squash, which supplemented their diet of game and fish.

Agriculture was not possible north of the Great Lakes–St Lawrence Lowlands because of its colder climate. Algonquian-speaking Indians, who lived north of the Great Lakes had to hunt big game animals, particularly caribou, for sustenance. They also traded with those Indians, such as the Iroquois, who practised agriculture in the Great Lakes–St Lawrence Lowlands and in the Ohio Valley. In northwestern Canada, Athapaskan-speaking Indians, whose ancestors probably came from Asia much later (perhaps between 7,000 and 10,000 years ago), occupied the forest lands of northwestern Canada and made summer hunting trips to the tundra where the caribou had their calving grounds.

Arctic Migration

Arctic Canada was settled much later than the forested lands of the Subarctic. Before people could occupy the Arctic lands, two developments were necessary. The first was the melting of the ice sheets that covered Arctic Canada. About 8,000 years ago, only small remnants of the great Laurentide ice sheet remained in northeastern Canada, including what is now northern Québec. The second development was the emergence of a hunting technique that

A Thule tent ring—the remains of a Thule house—near Igloolik. (*Courtesy Alec Aitken*)

would enable people to live in an Arctic environment. About 5,000 years ago, both these developments took place. By this time, the Palaeo-Eskimos, who were living along the northeast coast of Asia and in Alaska, had developed an Arctic sea-based hunting technique. This Arctic hunting culture of Palaeo-Eskimos was known as the Denbigh. Unlike previous hunting societies, these people invented a harpoon and other tools that enabled them to hunt seals and other marine mammals along the Arctic Coast, though they also relied heavily on terrestrial game such as the caribou.

As the Palaeo-Eskimo population increased, they slowly spread eastwards along the Alaskan Arctic Coast and then into Arctic Canada. About 3,000 years ago, this culture was replaced by another one known as the Dorset culture. Whether the Dorset people were among another wave of immigrants from Asia or evolved from the Denbigh culture is unclear. The final Arctic migration took place roughly 1,000 years ago, when the Thule people spread eastward from Alaska and gradually succeeded their predecessors. The Thule, the ancestors of the Inuit, hunted the bowhead whale and the walrus. However, with the cooling of the climate (known as the Little Ice Age), whales no longer entered the Arctic Ocean in large numbers because this ocean was covered by ice for most of the year, so the Thule turned to smaller game such as the seal and the caribou.

Initial Contact

By the time of European contact, the descendants of Old World hunters occupied all of North and South America. They were the North American Indian and Inuit tribes who met the European explorers searching for a trade route to the Orient. While the population of these tribes can only be estimated, many

A skirmish between Frobisher and some Inuit, drawn by John White. (*Courtesy © British Museum 202756*)

scholars now believe there may have been as many as 500,000 Indians and Inuit living in Canada at the time of contact.

John Cabot, the first European explorer, reached the East Coast in 1497. He was followed by other explorers, including Jacques Cartier and Martin Frobisher. In 1534, Cartier made contact with two Indian tribes along the Gaspé Coast, and in 1576, Frobisher encountered an Inuit encampment along the Arctic Coast of southern Baffin Island. Both explorers were searching for a route to Asia, but instead

discovered a new continent and peoples. In both cases, contact with North Americans ended badly. Lives were lost on both sides, and the ore that both explorers took back to Paris and London respectively proved worthless (Vignette 3.1). Instead of gold, they had found fool's gold (iron pyrites).

Culture Regions

At the time of contact with Europeans, Aboriginal peoples in Canada obtained their food from hunting, fishing, and gathering within a

Vignette 3.1

Contact between Cartier and Donnacona

In 1534, Jacques Cartier made contact at the Gaspé Coast with Donnacona, the leader of the St Lawrence Iroquois. The Iroquois, whose village (Stadacona) was located at the site of Québec City, came to the Gaspé to fish for mackerel. The next year Cartier returned. After reaching the Iroquois village of Hochelaga, which later became the site of Montréal, Cartier realized that the St Lawrence did not provide a sea passage to China. Like the Spanish in Mexico and Peru who discovered gold and silver, Cartier hoped to find wealth in the New World. In 1541, on his third voyage to Stadacona, Cartier hoped to find diamonds and gold. He left France with several hundred colonists. By then, Donnacona and all but one of the Iroquois captives that Cartier had taken with him when he returned to France in 1536 were dead. The French authorities, who wished to conceal this fact, decided not to

allow the surviving Indian woman to return to Stadacona with Cartier. The French arrived at Cap Rouge where they established a settlement just west of Stadacona. When the St Lawrence Iroquoians realized that Cartier was not going to return their chief and the other Iroquois captives, relationships quickly soured. Over the winter, the Iroquois killed at least thirty-five of the French settlers. Cartier and his surviving party left for France as soon as the river ice melted. He took along rocks that he believed contained gold and diamonds, but, like the ore mined on Baffin Island by Frobisher's men, Cartier brought back fool's gold and quartz. As there seemed to be no prospect for mineral wealth, the king of France lost interest in the New World. Sixty years went by before the French made another attempt at establishing a settlement at Stadacona. In 1608, Samuel de Champlain founded Québec City.

Source: Adapted from Ramsay Cook, *The Voyages of Jacques Cartier* (Toronto: University of Toronto Press, 1993).

Table 3.1
Timeline: Old World Hunters to Contact with Europeans

Date (BP = Before present)	Event
30,000–25,000 BP	Old World hunters of the woolly mammoth cross the Beringia land bridge into the unglaciated areas of Alaska and southern Yukon.
18,000 BP	The Wisconsin ice sheet reaches its maximum geographic extent, covering virtually all of Canada.
15,000 BP	The Wisconsin ice sheet retreats rapidly in western Canada, exposing a narrow ice-free area along the foothills of the Rocky Mountains. Ice persists in northern Québec until about 6,000 years ago.
13,000–12,000 BP	Descendants of the Old World hunters migrate southwards through the ice-free corridor east of the Rocky Mountains.
9,000 BP	Mammoths and mastodons become extinct, forcing early inhabitants of North America to adjust their hunting practices and thereby become more mobile and less numerous.
5,000 BP	Palaeo-Eskimos (known as the Denbigh) cross the Bering Strait to the Arctic Coast of Alaska. Later they move eastwards along the Canadian section of the Arctic Coast.
3,000 BP	The Dorset people, who were descendants of the Denbigh, develop a more advanced technology suited for an Arctic environment.
1,000 BP	A second wave of marine hunters, known as the Thule, migrate across the Arctic, eventually reaching the Arctic Coast of Labrador. At the same time Vikings reach Greenland and North America, where they establish a settlement on the north coast of Newfoundland (L'Anse aux Meadows).
1497	John Cabot lands on the East Coast (Newfoundland or Nova Scotia).
1534	Jacques Cartier plants the flag of France near Baie de Chaleur.

specific geographic area. Within each area, these people developed survival techniques suitable for the local environment. Such areas are called culture regions. Aboriginal peoples adapted to seven culture regions: eastern Woodlands, eastern Subarctic, western Subarctic, Arctic, Plains, Plateau, and Northwest Coast (Figure 3.2). The Inuit occupied the Arctic cultural region. In the eastern Subarctic cultural region, the Cree were the principal Algonquian tribe. The Cree had developed a technology—snowshoes—to hunt moose in deep snow in the eastern Subarctic region. In the western Subarctic region, the Athapaskans,

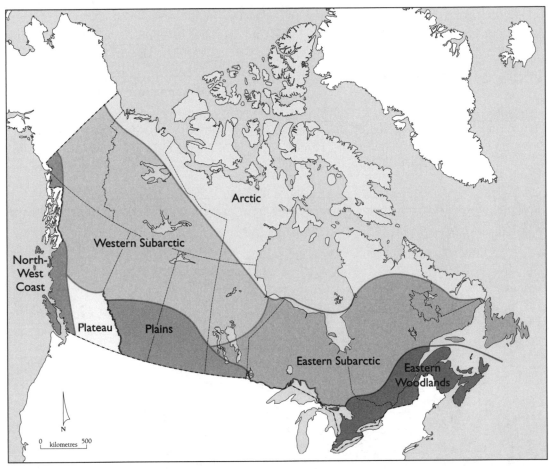

Figure 3.2 Culture regions of Aboriginal peoples. Common resources and natural conditions in certain geographic areas of North America were associated with similar ways of hunting wildlife and organizing economic and social activities.

including a number of Dene tribes, hunted caribou and other big game animals. Northwest Coast Indians harvested the rich marine life found along the Pacific Coast. Tribes such as the Haida, Nootka, and Salish comprised the Northwest Coast cultural region. In the southern interior of British Columbia, the Plateau Indians—including the Carrier, Lillooet, Okanagan, and Shuswap—occupied the valleys of the Cordillera, forming the Plateau cultural

region. Across the grasslands of the Canadian West, Plains Indians hunted the buffalo. The Assiniboine, Blackfoot, Sarcee, and Plains Cree hunted bison. The Iroquois and Huron lived in the eastern Woodlands of southern Ontario and Québec. Both tribes combined agriculture with hunting. In the Maritimes, the Micmac and Malecite also occupied the eastern Woodlands where they hunted and fished. The complexity and diversity of Aboriginal peoples can be gleaned from the spatial arrangement of their languages (Figure 3.3).

The Second People

The colonization of North America by the French and the British was the second major development in Canada's early history. France and England established colonies in North America in the seventeenth century. Québec City, founded in 1608 by Samuel de Champlain, was the first permanent settlement in Canada. By 1663, the French population was about 3,000 compared to a population of about 10,000 Indians (mainly Huron and Iroquois), who lived in the St Lawrence Valley, the Great Lakes, and the Ohio Basin. By 1750, the French Canadians constituted most of the

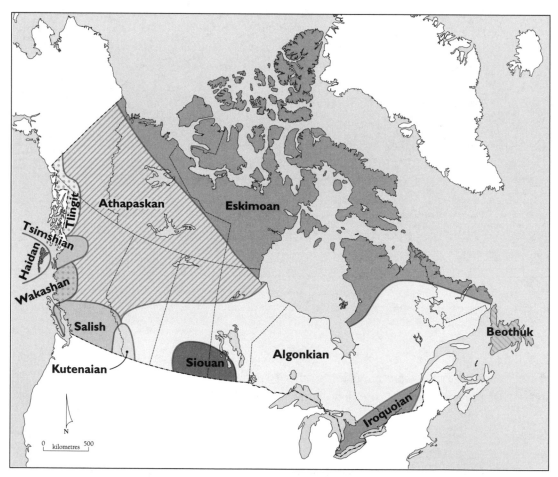

Figure 3.3 Aboriginal language families. Aboriginal peoples in Canada form a very diverse population. At the time of contact, there were over fifty distinct Aboriginal languages spoken. These languages formed twelve language families. Six of these language families are found in one natural cultural region, the Northwest Coast.

population in New France, where they numbered about 60,000, while the Indian population had dropped sharply because of disease and warfare. Following the British conquest of New France in 1759, British immigrants began to move to Canada. In doing so, they increased the number of British settlers. Meanwhile, the French-Canadian population now had to depend entirely on natural population increase to maintain their numbers.

The first large wave of immigrants consisted of British refugees from the United States. These Loyalists had supported Britain during the American War of Independence (1775–83). After the defeat of the British army, they sought refuge in other parts of the British Empire, including its North American possessions. In North America, most Loyalists settled in Nova Scotia, while a smaller number moved to the Eastern Townships and Montréal. A few settled in what is now southern Ontario.

The statue of Champlain overlooking the Ottawa River. (Courtesy R.M. Bone)

The second wave of immigrants came from Britain. From 1790 to 1860, almost 1 million people migrated from the British Isles to British North America, mostly because of deteriorating economic conditions in Great Britain. Land was in short supply in Britain, forcing many people to move to the cities. However, even in the cities, there were few

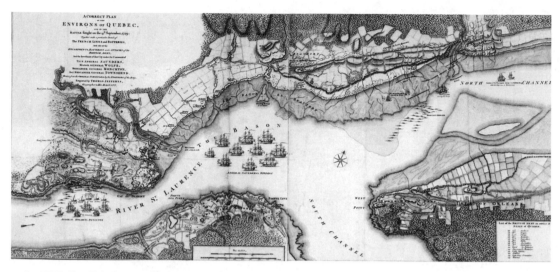

In 1759, English forces, under the command of General James Wolfe, defeated the French, led by Louis-Joseph, Marquis de Montcalm, on the Plains of Abraham, Québec City. The plains are depicted in the left of this battle map, drawn a year later. (*Courtesy National Archives of Canada 128079*)

jobs available. One solution was emigration to the New World. In the 1840s, the potato famines in Ireland added to the problems in the Old World, causing terrible hardships for the Irish people. Thousands fled the countryside and many left for the New World to settle in the towns and cities of British North America.

These two waves of migrations greatly changed Canada. At the time of Confederation in 1867, the population of British North America had reached 3 million. Approximately 80 per cent of these British subjects lived in the Great Lakes–St Lawrence Lowlands, while about 20 per cent inhabited Atlantic Canada. Across the rest of British North America, Aboriginal peoples made up most of the population. In the Red River Colony, a new Aboriginal peoples, the Métis, who were of Native and European descent,

had emerged. By 1867, the Métis, who totalled about 10,000, greatly outnumbered White settlers and fur traders in the Red River Colony.

But more important, the old balance of demographic power between the French and English had been reversed. British migration to Canada had, over the course of 100 years, changed the demographic balance between French-speaking and English-speaking Canadians. Approximately 60 per cent of the European population was English-speaking. The new demographics also resulted in large English-speaking populations in the principal cities of Québec. Migration had also changed the ethnic composition of the English-speaking population from almost entirely English to a mixture of English, Irish, Scottish, and Welsh. Moreover, by the 1860s, Canada's ethnic character varied by region. In Atlantic Canada, the Scottish and the Irish outnumbered the English. In Québec, the English and Irish formed a sizeable minority in the towns and cities, though rural Québec remained solidly French-speaking, except for the Eastern Townships. Ontario, like Atlantic Canada, was decidedly British.

The separation of the British and French in different geographic parts of British North America ensured the continued existence of two visions of Canada. In 1867, 92 per cent of the population was either British or French. Each linguistic group developed its own vision of Canada: French Canadians saw Canada as two founding peoples, while English-speaking Canadians favoured the notion of equal provinces. In English Canada the notion of equal provinces grew out of the following factors:

- the nature of Confederation was such that provincial powers were shared equally
- the British formed the majority of the population in three of the four provinces, thereby dominating the political affairs of those provinces
- the British, while a minority within Québec, were the dominant business group

The Territorial Evolution of Canada

A brief presentation of the territorial evolution of Canada marks the start of our account of Canada's formal history as a nation. The British North America Act of 1867 united the colonies of New Brunswick, Nova Scotia, and the Province of Canada (formerly Upper Can-

ada and Lower Canada) into the Dominion of Canada (Vignette 3.2). Thus, Canada began as a small country, consisting of what is now known as southern Ontario, southern Québec, New Brunswick, and Nova Scotia (Figure 3.4).

The British government was a strong advocate of uniting its colonies in North America. For Britain, the union of its North American colonies had three advantages: (1) a better chance for their political survival against the growing economic and military strength of the United States; (2) an improved environment for British investment, especially in railways; and (3) a reduction in British expenditures for the defence of its North American colonies. The British colonies perceived unification differently. The Province of Canada, led by John A. Macdonald, pushed hard for a united British North America because it would have a larger domestic market for its growing manufacturing industries and a stronger defensive position against a feared American invasion. The Atlantic colonies showed little interest in such a union. As they were part of the British Empire, the attraction of joining the Province of Canada had little appeal. Furthermore, unlike the Canadians, Maritimers continued to base their prosperity on a flourishing transatlantic trading economy with Caribbean countries and Great Britain. From 1840 to 1870, the backbone of the Maritime economy was the construction of wooden sailing-ships. Shipbuilding was so important that this period was known as the 'Golden Age of Sail' in the Maritimes. Even diplomatic pressure from Britain to join Confederation had little effect on Maritime politicians. But the Fenian raid into New Brunswick in 1866

Vignette 3.2

The British North America Act, 1867

Under the terms of the British North America Act, the Dominion of Canada was composed of four provinces (Ontario, Québec, New Brunswick, and Nova Scotia). Modelled after the British parliamentary and monarchical system of government, the newly formed country had a Parliament made up of three elements: the head of government (a governor-general who represented the monarch), an upper house (the Senate), and a lower house (the House of Commons). This act was modified several times to accommodate Canada's evolving political needs and its gradual movement to independent nationhood. The patriation of Canada's Constitution in 1982 removed the last vestige of Canada's political dependence on the United Kingdom, although Canada still recognizes the British monarch as its symbolic head.

The British North America Act was based on the highly centralized government of the United Kingdom in the 1860s. However, this act assigned specific powers to the provinces in order to satisfy Québec's demand for control over its culture. The Canadian political system that emerged, therefore, allowed for regionalized politics. For example, political parties in the House of Commons sometimes serve regional interests. In the 1920s, the Progressive Party represented the concerns of farmers in Western Canada, while the pro-independence Bloc Québécois, which was formed in 1990, not only serves the interests of Québec but is also active in the separatist movement. Furthermore, while the House of Commons is based on the principle of representation by population, Senate membership is based on the principle of equal regional representation. However, because senators are appointed by the prime minister and not elected by the people in the different regions of the country, the Senate fails to provide a regional counterweight to the House of Commons.

and the termination of the Reciprocity Treaty with the United States quickly changed public opinion in the Maritimes (Vignette 3.3).[2] Shortly after the Fenian raids, the legislatures of both New Brunswick and Nova Scotia voted to join Confederation.

Within a decade, the territorial extent of Canada expanded from four British colonies to the northern half of North America. The new Dominion grew in size with the addition of other British colonies and territories. Manitoba and the North-West Territory (1870), British Columbia (1871), and Prince Edward Island (1873) joined the new Dominion (Figure 3.5), while the British government transferred its claim to the Arctic Archipelago to Canada in 1880. By then, Canada stretched from the Atlantic to the Pacific and north to the Arctic. While still part of the British Empire, Canada had begun the slow journey to independence and nationhood. In 1905, the provinces of Alberta and Saskatchewan were created, then in 1949, Newfoundland joined Canada, completing the union of British North

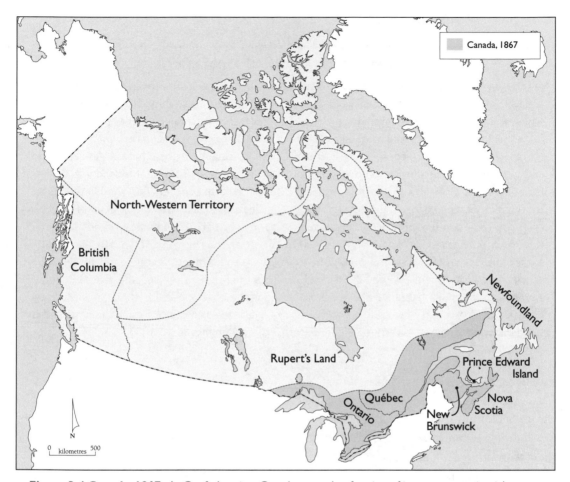

Figure 3.4 Canada, 1867. At Confederation, Canada was only a fraction of its current territorial extent.

America into a single political entity (The territorial evolution of Canada is illustrated in Figures 3.4 to 3.8 and Table 3.2).

Within a decade after Confederation, Canada's territorial extent had burgeoned, making it the second largest country in the world, which posed a serious problem for the nation's leaders, who had to somehow transform this vast territory into a nation. Sir John A. Macdonald, Canada's first prime minister, resolved this in part by authorizing the construction of a transcontinental railway, the Canadian Pacific Railway.

National Boundaries

Well before Confederation in 1867, wars and treaties between Britain and the United States shaped many of Canada's boundaries. The southern boundary of New Brunswick, Québec, and Ontario was settled in 1783 when Britain and the United States signed the Treaty of Paris. Under this treaty, the United States

---

Vignette 3.3

America's Manifest Destiny

In the nineteenth century, many Americans still believed in the doctrine of Manifest Destiny, which was based on the belief that the United States would eventually expand to all parts of North America, thus incorporating Canada and Mexico into the American republic. From its beginnings along the Atlantic seaboard, the United States had greatly increased its territory by a combination of force, negotiation, and purchase. To Americans, such an expansion was an expression of their right to North America. As well, it would rid North America of the much-hated European colonial powers.

Not surprisingly, the Fathers of Confed-eration were concerned about American designs on British North America. First, in 1866, the Fenians raided Upper Canada, Lower Canada, and New Brunswick with the intention of seizing British North America and holding it for ransom until Ireland was free of British rule. Second, in 1867, the United States purchased Alaska from the tsar of Russia, leaving British Columbia wedged between American territory to its north and south. When the United States purchased Alaska from the Russians in 1867, the exact boundary along the coastline south of 60° N was uncertain. Canada and the United States settled this final border dispute in 1903.

---

gained control of the Indian lands of the Ohio Basin. Earlier, Britain had formally recognized the rights of Indians to these lands in the Royal Proclamation of 1763, which provided the constitutional framework for negotiating treaties with Aboriginal peoples. This recognition was the basis of Aboriginal rights in Canada (see 'Aboriginal Rights' in this chapter). Based on the fur trade route to the western interior, the boundary of 1783 passed through the Great Lakes to the Lake of the Woods and then west to the Mississippi River.

In 1818, Canada's southern boundary was adjusted; it was set at 49° N from the Lake of the Woods to the Rocky Mountains. As for the northwestern boundary, there was some question as to where British territory ended and Russian territory began. British and Russians had come into contact through the fur trade.

In the eighteenth century, Russian fur traders established trading posts along the Alaskan coast. Indians travelled along the Yukon River from the interior of Alaska to trade at these Russian forts. By the early-nineteenth century, the Hudson's Bay Company had reached the tributaries of the Yukon River. In 1825, Britain and Russia set the northern boundary at 141° W (the Treaty of St Petersburg). Russia also agreed to relinquish to Britain her claims to the coastal regions south of 54°40' N to 42° N. In Atlantic Canada, the boundary between Maine and New Brunswick had not been precisely defined in 1783. Minor adjustments took place in 1842 when Britain and the United States concluded the Webster-Ashburton Treaty.

The last major territorial dispute between Britain and the United States took place over

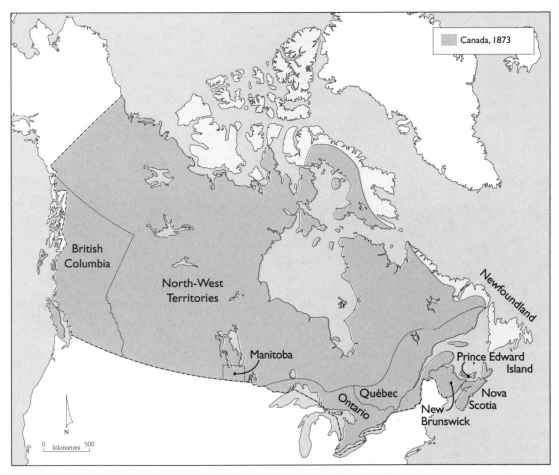

Figure 3.5 Canada, 1873. Within seven years of Confederation, Canada had obtained the Hudson's Bay Company's lands (including the Red River Colony), and two British colonies (British Columbia and Prince Edward Island) had joined the new Dominion.

the Oregon Territory. The Oregon Territory stretched along the Pacific Coast from Alaska to Mexico—Mexico at that time extended northward to 42° N. In the early part of the nineteenth century, the Hudson's Bay Company established a fur-trading post at the mouth of the Columbia River, where it conducted trade with the local Indian tribes. In the 1830s, American settlers crossed the Rocky Mountains into the Columbia Valley where they proceeded to cultivate the fertile soils of the Willamette Valley. Because both Britain and the United States had a foothold in the Oregon Territory, sovereignty was uncertain. Great Britain's claim was based on two factors: that (1) the Hudson's Bay Company had established the first settlement (Fort Vancouver) in the region, and (2) the Hudson's Bay Company exerted control over the land where the Indians trapped fur-bearing animals. The

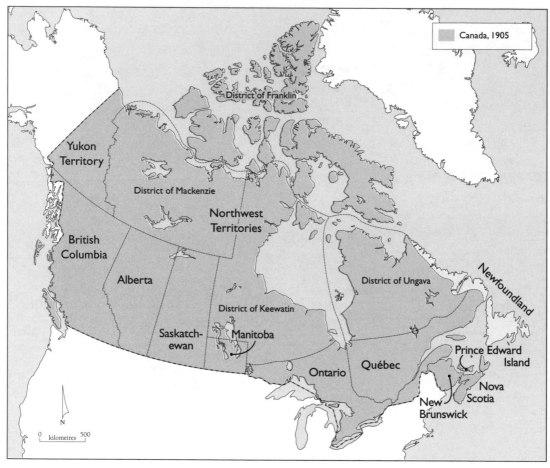

Figure 3.6 Canada, 1905. By 1905, two new provinces (Alberta and Saskatchewan) and one territory (Yukon) had joined. Ontario, Québec, and Manitoba expanded their boundaries.

American claim was centred on the fact that the vast majority of settlers were Americans. In the final outcome, there was no doubt that occupancy was a more powerful claim to disputed lands than claims based on exploration and the presence of a fur-trading economy. The boundary between British and American territory was set at 49° N with the exception of Vancouver Island, which extended south of this parallel.

Internal Boundaries

Since Confederation, the internal boundaries of Canada have changed (Figures 3.4 to 3.8). These changes have created new provinces (Alberta and Saskatchewan) and expanded the area of existing provinces (Manitoba, Ontario, and Québec). In all cases, these political changes took land away from the Northwest Territories.

In 1870, the boundaries of Manitoba formed a tiny rectangle comprising little more than the Red River Colony. The province's boundaries, while extended in 1881 and 1884, did not reach their present limits until 1912. By this time, Manitoba spread northward as far as Hudson Bay and 60° N.

Québec too received northern territories. In 1898, its boundary was extended northward to the Eastmain River and then eastward to Labrador. At that time, Canada believed that the province of Québec extended to the narrow coastal strip along the Labrador coast, while the colony of Newfoundland contended that Newfoundland owned all the land draining into the Atlantic Ocean. In 1927, this dispute between two British dominions (Canada and Newfoundland) went to London. The British government decided that the boundary between Québec and Labrador was not the

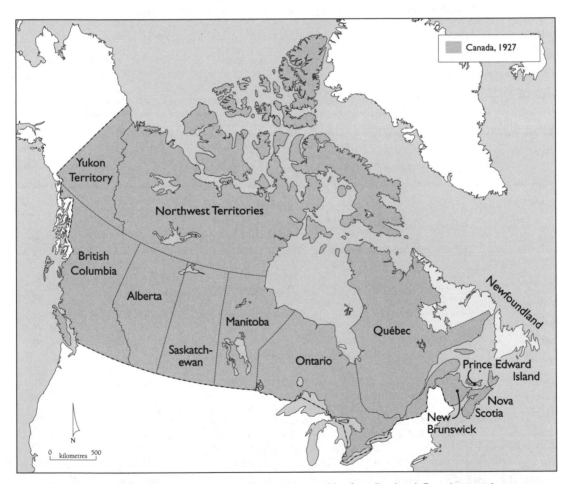

Figure 3.7 Canada, 1927. The boundary dispute between Newfoundland and Canada over the eastern boundary of Québec was resolved in Newfoundland's favour. Ontario, Québec, and Manitoba expanded again and attained their present boundaries.

Table 3.2
Timeline: Territorial Evolution of Canada

Date	Event
1867	Ontario, Québec, New Brunswick, and Nova Scotia unite to form the Dominion of Canada.
1870	Canada purchases the Hudson's Bay Company's lands. The Red River Colony enters Confederation as the province of Manitoba.
1871	British Columbia joins Canada.
1873	Prince Edward Island becomes the seventh province of Canada.
1880	Great Britain transfers its claim to the Arctic Archipelago to Canada.
1905	Alberta and Saskatchewan are carved out of the Northwest Territories to become provinces.
1949	Newfoundland and Labrador join Canada to become the tenth province.

Table 3.3
Timeline: Evolution of Canada's Internal Boundaries

Date	Event
1881	Ottawa enlarges the boundaries of Manitoba.
1898	Ottawa approves of extending Québec's northern limit to the Eastmain River.
1899	Ottawa decides to set Ontario's western boundary at the Lake of the Woods and extend its northern boundary to the Albany River and James Bay.
1905	Ottawa announces the creation of two new provinces, Alberta and Saskatchewan.
1912	Ottawa redefines the boundaries of Manitoba, Ontario, and Québec, which extend their provincial boundaries to their present position.
1927	Great Britain sets the boundary between Québec and Labrador. Québec has never accepted this decision.
1999	The government of the new territory, Nunavut, began to function.

narrow coastal strip proposed by Canada but rather the watershed of those rivers flowing into the Atlantic Ocean. While the Québec government has never formally accepted this ruling, Québec has conducted its affairs with Newfoundland as if Labrador were part of

Figure 3.8 Canada, 1999. On 1 April 1999, Nunavut became a territory.

Newfoundland. For example, Hydro-Québec has made arrangements to purchase hydroelectric power produced in Labrador and the two provincial governments are currently discussing plans to produce more hydroelectric power in Labrador, sell it to Hydro-Québec, and then transmit the power through Québec to markets in New England.

In the years following Confederation, Ontario gained two large areas. In 1899, Ontario had its western boundary set at the Lake of the Woods (previously this area belonged to Manitoba), while its northern boundary was extended to the Albany River and James Bay. In 1912, Ontario obtained its vast northern lands which stretch to Hudson Bay.

In 1905, Canada formed two new provinces, Alberta and Saskatchewan. The final adjustment to Canada's internal boundaries occurred in 1999 with the establishment of the government of the territory of Nunavut (Figure 3.8 and Table 3.3).

The Aboriginal/
Non-Aboriginal Faultline

In 1867, the British North America Act made Ottawa responsible for the Indian tribes. Later Ottawa's responsibility was extended to all Aboriginal peoples to include the Inuit and the Métis. For that reason, the Aboriginal/non-Aboriginal faultline is cast into an Ottawa versus Aboriginal peoples framework. In this model, Ottawa acts like a core, while Aboriginal Canadians living on the edge of Canadian society serve as the periphery. Originally, Ottawa's objective was the assimilation of Indian peoples into Canadian society. Instead, Aboriginal peoples were marginalized. Since the 1970s, Ottawa has adopted a more enlightened policy, recognizing two facts: Aboriginal rights, and the failure of the federal government's assimilation policy. Aboriginal peoples' struggle for power is rooted in two questions: Who are the Aboriginal peoples of Canada? What are Aboriginal rights?

Aboriginal Peoples

The Canadian Constitution Act of 1982 refers to Indians, Métis, and Inuit under the umbrella term **Aboriginal peoples**, that is, those now living in Canada who can trace their ancestry to the original inhabitants who were in North America before the time of contact with Europeans in the fifteenth century. Indians are further distinguished between status, non-status, and treaty Indians. People legally defined as **status (or registered) Indians** are registered or entitled to be registered as Indians, according to the Indian Act as amended in June 1985, and have certain rights acknow-

ledged by the federal government, such as tax exemption for income generated on a reserve. In 1991, 70 per cent of Canadians who identified themselves as Indians were status Indians. By 1995, the number of status Indians had grown to almost 600,000. **Non-status Indians** are people of Indian ancestry who are not registered as Indians and therefore have no rights under the Indian Act. **Treaty Indians** are status or registered Indians who are members of (or can prove descent from) a band that signed a treaty. They have a legal right to live on a reserve and participate in band affairs. About 55 per cent of treaty Indians live on reserves. The **Métis** are people of European and North American Indian ancestry. The **Inuit** are Aboriginal people located mainly in the North. In 1996, 1,101,960 people reported Aboriginal ancestry: 867,225 Indians, 220,740 Métis, and 49,845 Inuit (Statistics Canada 1998).

The Indians, Inuit, and Métis are a highly diversified population. One indication of their cultural diversity is linguistic classification. There are fifty-three distinct Aboriginal languages (of eleven language families) spoken in Canada. Six of the language families are spoken along the Pacific Coast, while only two are spoken east of Manitoba and one (Inuktitut) in the Arctic (Figure 3.3). The largest language family is Algonquian. There are fifteen distinct Algonquian-based languages, the most common of which are Cree and Ojibwa. Inuktitut, the Inuit language, has regional dialects and is spoken across the Canadian Arctic.

Another measure of Aboriginal diversity is self-identification. Many Indians prefer to identify themselves with the name of their

FORKS OF THE RED AND ASSINIBOINE
FOURCHE DE LA ROUGE ET DE L'ASSINIBOINE

CANADA

Strategically located at the junction of two major rivers which form part of a vast continental network, this spot has witnessed many of the key events of Western Canadian history. This was a traditional native stopping place and for this reason La Verendrye erected Fort Rouge near here in 1738. It has been a centre for trade and exploration, a focus for the first permanent European settlement in the Canadian West, cradle of the province of Manitoba, nucleus of the city of Winnipeg, a hub of rail and road transport, and the gateway for the settlement of the prairies.

Par sa situation stratégique à la jonction de deux rivières importantes

Once part of the Red River Colony, the confluence of the Red and Assiniboine rivers in downtown Winnipeg, and a plaque describing the historical importance of these waterways. (*Courtesy R.M. Bone*)

tribal group, such as Cree or Iroquois, while others use the name of their First Nation (band) for a more precise identification. For example, the Cree occupy a vast territory that stretches from Labrador to Alberta. There are many Cree tribes within that territory. A Cree living in northern Saskatchewan might identify himself or herself as a member of a Cree band, such as the Lac la Ronge First Nation.

According to the Department of Indian Affairs, seven First Nations had populations exceeding 6,000 in 1995. Southern Ontario's Six Nations of the Grand River, with its population of almost 19,000 in 1995, was the largest First Nation in Canada. The next largest, the Mohawk of Akwesasne (8,400) and Kahnawake (8,300), are located in Ontario and southern Québec. The others are the Blood (8,200) and the Saddle Lake (6,800) of Alberta, the Mohawk of the Bay of Quinte in Ontario (6,300), and the Lac la Ronge Cree in northern Saskatchewan (6,100).

Aboriginal peoples are reclaiming their identity and place names. Some are relinquishing the names given to them by Europeans in favour of their original names, such as Anishinabe (for Ojibwa) or Gwich'in (for Kutchin). The landscape is also being reclaimed. For example, the Arctic community of Frobisher Bay, named after the English explorer, Martin Frobisher, is now called Iqaluit, which means 'the place where the fish are'.

Aboriginal Rights

Aboriginal rights are group or collective rights which stem from Aboriginal peoples' original occupation of the land in the period before contact. Aboriginal peoples' traditional attitudes and values towards land and wildlife are strikingly different from those held by most non-Aboriginal Canadians. For Canada's First Peoples, the land has not only economic value but also cultural, political, and spiritual value. Aboriginal attitudes and values are based on their former economic and social systems; that is, the subsistence hunting system that was in place before contact with Europeans.

Land rights are the most fundamental Aboriginal rights. Indeed, it is from land rights that most other collective rights flow, such as self-determination and self-government. Aboriginal peoples first began to secure land rights during the periods, before and after Confederation, in which numerous treaties were signed between Aboriginal groups and Crown authorities. Some Aboriginal peoples, including the Métis, are still negotiating with Ottawa.[3] Treaties set aside land for Indian bands, land that is called a **reserve** and is held collectively by and for the benefit of the band.

The reasons for signing treaties varied depending on the historical context. Authorities acting on behalf of the Crown often signed treaties to secure Aboriginal peoples as allies during times of turmoil, such as the War of 1812, or simply to procure land for the growing numbers of settlers coming to Canada. During the expansion to the West, treaties were signed to provide a place for Indian tribes so that Indian wars common south of the border would not erupt in Canada. For Aboriginal peoples, treaties often promised protection from encroaching settlers and support or benefits during times of transition in their way of life, such as during the demise of the great buffalo herds in Western Canada, which threatened their principal food supply. Treaties offered protection to Indian tribes from the anticipated flood of settlers and some guarantee that the federal government would care for them.

The terms of each treaty varied. However, they generally included cash gratuities and presents at the signing of the treaty, annual payments in perpetuity, the promise of educational and agricultural assistance, the right to hunt and fish on Crown land until such land was required for other purposes, as well as land reserves to be owned by the Crown in trust for the Indians. In Treaty 6, for example, the amount of land assigned to each tribe was determined by its population size, i.e., each family of five received one square mile. Reserves are collectively owned by First Nations bands, though legally they are owned by the Crown in trust for them.

Conflicting ideas as to the significance of treaties between the signing parties has largely shaped Aboriginal and non-Aboriginal relations in Canada during the twentieth century. When treaties were signed, Crown authorities viewed them as vehicles for extinguishing Aboriginal rights and titles to land. Aboriginal peoples, however, viewed them as agreements between 'sovereign' powers to share land and resources. This lack of understanding, coupled with the facts that some Aboriginal groups never signed treaties and that the terms of other treaties have never been fulfilled, contributed to mounting frustration among Aboriginal peoples. The latter part of the twentieth century has witnessed various movements to repair injustices related to Aboriginal land rights (for example, through land claims), and to recognize Aboriginal

peoples' inherent right to self-determination (for example, through constitutional reform). Through these attempts and through the signing of modern treaties, Aboriginal peoples are seeking a new place in Canadian society (Vignette 3.4) in the hope of participating in its economy while still retaining their culture. The first step in that process is often to obtain a land base or homeland.

From Hunting Rights to Modern Treaties

History sometimes makes strange allies. Shortly after Pontiac, chief of the Ottawa, led a successful uprising against the British in 1763, Britain decided to form an alliance with him and other Indian leaders.[4] For strategic reasons, George III issued the Royal Proclamation of 1763, which identified a part of British territory west of the Appalachian Mountains as Indian lands. At that time, Britain believed that it could claim 'uninhabited land', which the British defined as land without permanent occupation (that is, no cultivated land or permanent settlements). However, the British also believed that Indians had a limited ownership over the lands they inhabited, and that therefore such lands must be purchased from the owners. This somewhat ambiguous concept was the basis of land claims by Canadian Aboriginal peoples (Figure 3.9).

The legal meaning of Aboriginal title to land has evolved over time. Until the 1970s, Ottawa recognized two forms of ownership. Reserve lands were one type of ownership, which the Canadian government held for Indian people. The second type of ownership was the right to use Crown land for hunting

---

Vignette 3.4

Modern Treaties

Modern treaties began in 1975 with the signing of the James Bay and Northern Québec Agreement (Figure 3.10). Since then, all modern treaties are either specific or comprehensive agreements between an Aboriginal group and the federal government. **Specific land-claim agreements** attempt to rectify shortcomings in the original treaty agreement with a band. By 1990, over 500 specific claims had been filed with Ottawa. Most claims involved relatively small amounts of land. However, in 1992, a major agreement was signed between twenty-six First Nations in Saskatchewan and Ottawa. Called the Settlement of Treaty Land Entitlement in Saskatchewan, it involved a payment of $445 million over twelve years to allow these First Nations to purchase land that they should have received at the time of treaty or that they lost through Ottawa's mishandling of their affairs.

Comprehensive land-claim agreements take place when a group of Aboriginal peoples, who have not yet signed a treaty, can demonstrate a claim to land through past occupancy. In 1984, the Inuvialiut became the first Aboriginal peoples to settle a comprehensive land claim with the federal government. Their land claim stretched over the Western Arctic. In exchange for surrendering their Aboriginal claim to all this land, they received 91 000 km², $45 million (in 1977 dollars) in financial compensation, and guaranteed rights over resource management. Since 1984, four other comprehensive agreements have been signed between the government and the Gwich'in (1992), Sahtu (1993), Inuit of the Nunavut Settlement Area (1993), and the Yukon First Nations (1993). In July 1998, officials of three governments—Ottawa, Victoria, and the Nisga'a—signed an agreement-in-principle that, if ratified, would not only settle the Nisga'a land claim but also mark the first modern land-claim agreement in British Columbia.

---

and trapping. At that time, Crown lands (both provincial and federal) included most of Canada's unsettled areas. Indian, Inuit, and Métis families lived on Crown lands, continuing to hunt, trap, and fish. However, federal and provincial governments could sell such lands to individuals and corporations or grant them a lease to use the land for a specific purpose, such as mineral exploration or logging, without compensating the Aboriginal users of those lands. By the 1960s, many Aboriginal groups still did not have a treaty with the Canadian government. Atlantic Canada, Québec, the Territorial North, and British Columbia contained huge areas where treaties had not yet been concluded. As a consequence, Aboriginal peoples had no control over developments on these lands.

A combination of events radically changed this situation. One factor was the emergence of Native leaders who understood the political and legal systems. They used the courts to force the federal and provincial governments to address the issue of Aboriginal rights and

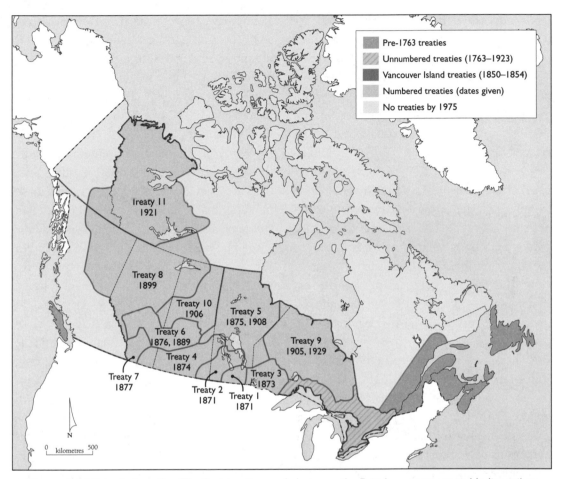

Figure 3.9 Historic treaties. The first treaties, made between the British government and Indian tribes, were 'friendship' agreements. In Upper Canada the Robinson (or unnumbered) treaties set aside reserve lands in exchange for title to the remaining lands. With the settlement of lands in the Canadian West, Indians became concerned about their future, so many numbered treaties included provisions for agricultural supplies. By 1921, when the last numbered treaty was signed, many Aboriginal peoples in Atlantic Canada, Québec, and British Columbia were still without treaties.

land claims. The first major event took place in the late 1960s. In 1969, the proposed reforms of the Indian Act (known as the White Paper) galvanized treaty Indians into action. The White Paper proposed to treat all Canadians equally. For Indians, it meant the abolition of their treaty rights and the reserve land system. About the same time, the Nisga'a Indians

in northern British Columbia took their land claim, known as the Calder case, to court. In 1973, the Supreme Court of Canada narrowly ruled against (by a vote of four to three) the Nisga'a's argument that the tribe still had a land claim to territory in northern British Columbia. However, in their ruling, six of the seven judges agreed that Aboriginal title to the land had existed in British Columbia at the time of Confederation. Furthermore, three judges said that Aboriginal title still existed in British Columbia because it had not been extinguished by the British Columbia government, while three other judges stated that Aboriginal title had been extinguished by the various laws passed by the British Columbia government since 1871. The seventh judge ruled against the Nisga'a's claim on a legal technicality. The Supreme Court of Canada's narrow verdict and the legal opinion of three judges that Aboriginal title still existed changed the course of Aboriginal land claims in Canada. Now the federal government agreed that Aboriginal peoples who have not signed a treaty may very well have a legal claim to Crown lands.

The James Bay Project and the Mackenzie Valley Pipeline Project added fuel to the political fire over Aboriginal rights. The possible impact of these industrial projects on Aboriginal peoples was made clear through the Berger Inquiry into possible environmental and socio-economic impacts and the media. It was obvious that Aboriginal organizations were prepared to take action to defend their land claims. Their position in the 1970s was 'no development without land-claims settlements'. All these events changed both the public's views of Aboriginal rights and the government's position. At first grudgingly and then more willingly, governments, corporations, and Canadian society recognized the validity of Aboriginal land claims. The James Bay and Northern Québec Agreement in 1975 was the first modern land-claim agreement in Canada (Figure 3.10). Since then, five comprehensive

claims have been finalized in northern Canada. The five land claim agreements are: (1) the Inuvialuit Final Agreement (1984), (2) the Gwich'in Final Agreement (1992), (3) the Sahtu Final Agreement (1993), (4) the Nunavut Final Agreement (1993), and (5) the Yukon Umbrella Final Agreement (1993).

Many claims remain unsettled, including those in British Columbia and Labrador. Until they are concluded, relations between those pursuing claims and the federal government will remain strained. For example, virtually the entire province of British Columbia, except for Vancouver Island, is claimed by First Nations. Until 1992, the provincial government believed that British occupancy had extinguished Aboriginal title. However, in 1992, the British Columbia government accepted the principle of Aboriginal land claims. The following year, Ottawa and Victoria agreed to a formula for paying outstanding land claims. The federal government would pay 90 per cent of the money needed to settle outstanding claims and the province would provide the land. In July 1999, the Nisga'a signed an agreement-in-principle with Ottawa and Victoria, while other First Nations are negotiating with the British Columbia Treaty Commission.

In the meantime, those Aboriginal groups who have concluded modern treaties are moving forward. They are focusing on economic and cultural developments rather than expending their energies on land-claim negotiations. As a result, a gap within the Aboriginal community between the have (those who have a modern treaty) and the have-nots (those who do not have a modern treaty) is emerging.

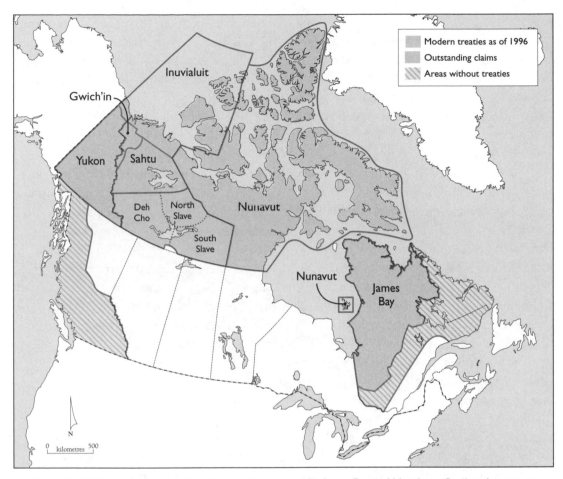

Figure 3.10 Modern treaties. The first modern treaty, the James Bay and Northern Québec Agreement, was signed in 1975. Since then, modern treaties have fallen into two categories: comprehensive and specific. By 1996, the two main areas without treaties were much of British Columbia and all of Labrador. (The original inhabitants of Newfoundland, the Beothuk, had perished from disease, encroachment, and slaughter by the early-nineteenth century.)

Bridging the Aboriginal/ Non-Aboriginal Faultline

Aboriginal peoples are taking the control of their affairs away from Ottawa. Some Indian and Inuit peoples have made substantial advances in economic development, while others have moved into the area of self-government. Unfortunately, some Aboriginal peoples, including the Métis, have not yet begun this process and remain on the margins of Canadian society. For most, fortunately, the process of change has started.

For historic reasons, while some Aboriginal peoples have made treaties, others have not. This difference is significant because a treaty, particularly a modern treaty (a comprehensive or specific land-claim agreement), provides the land, capital, and an administrative organization necessary to initiate this process of economic, social, and political change. In 1996, the report of the Royal Commission on Aboriginal Peoples identified two major goals: Aboriginal economic development and self-government. The gap between Aboriginal and non-Aboriginal societies will not be bridged until these goals are achieved. The key factor is transferring power from Ottawa (the political power core) to the various Aboriginal communities (the politically weak periphery). How well each Aboriginal community will manage its affairs is unknown, but breaking its dependency on the federal government will be an important step. Such a step will result in a new and more positive relationship between Aboriginal and non-Aboriginal Canadians. The new relationship is a working one as demonstrated by the sharing of power to assess industrial impacts on the land and people in the Inuvialuit Settlement Area rather than an adversarial one as illustrated by the Innu's opposition to proposed developments at Voisey's Bay and on the lower Churchill River.

The Centralist/ Decentralist Faultline

Geography poses powerful challenges to Canada's national unity. These challenges stem from Canada's vast space, economic competition between regions, economic attraction to the United States, and internal political differences between the provinces and Ottawa. In combination, these challenges to national unity are manifest as regional self-interest that often results in regional tensions. Sometimes these internal tensions are expressed by a group of individuals, such as fishers in Atlantic Canada, or through common interests among provinces, such as the oil-producing provinces. Whatever their form or geographic extent, these internal forces produce regional pressures that can divide the country.

Regional self-interest and resulting tensions (whether cultural, economic, or political) are a natural outcome of Canada's physical and human geography. For better or worse, regionalism is a fact of life in Canada and it may well be the most telling characteristic of Canada's national character. The regional challenge to Canada's east-west alignment, and therefore to Canada's national unity, may be summed up in four ways, namely, those centrifugal forces that:

- separate Canadian regions from each other by great distances, making trade and commerce between those regions more difficult (geographic forces)
- lead regions to compete economically with each other (forces derived from regional self-interest)
- cause political competition between the provinces and Ottawa (forces derived from the division of political powers in the Canadian Constitution)
- draw Canadian regions into the economic orbit of the United States (continental economic forces)

The almost irresistible attraction of the United States is the basis of economic continentalism; that is, the natural markets of North America have a north-south orientation.

The Maritimes are geographically tied to New England, while Southern Ontario is linked to Michigan, Ohio, Pennsylvania, and New York. The Cordillera best represents the north-south pattern of Canada's physiography, drawing British Columbia into the American Pacific Northwest.

Regional Tensions

The challenge of geography, therefore, has forced Ottawa to seek solutions to the problems bred by regional tensions. Political solutions attempted to overcome geography by binding the new territories to the core of the country, and by balancing regional economic disparities.

The first challenge facing the government of Sir John A. Macdonald was clear: a railway must span the country from the Atlantic Ocean to the Pacific Ocean. Without a more substantial presence, Macdonald's Conservative government feared that the US might annex the unoccupied parts of Western Canada (Vignette 3.3). From Ottawa's perspective, the railway would exert effective sovereignty over its western territories. Though the magnitude of this project was nearly beyond the capacity of the country, Ottawa proceeded. Its goals for this transcontinental railway were:

- to link the West with the rest of Canada
- to settle the Canadian Prairies
- to provide an export route for Prairie grain
- to create a market in the West for eastern industries

At the same time as Ottawa secured its western borders, it launched a new economic initiative, the National Policy, which reinforced the centralist/decentralist faultline. The National Policy and other federal policies achieved economic growth, but this growth took place at different rates across Canada. Rightly or wrongly, Canadians living outside Central Canada believe that the concentration of political power in Central Canada has had an unfair influence over national policies and therefore favoured economic development in Central Canada over developments in the rest of the country. In the following pages, these two themes—economic and political power struggles—are explored from the perspective of Central Canada (the core) and the rest of Canada (the periphery). In the case of economic power, Central Canada refers to the industrial core in southern Ontario and southern Québec. In the case of political power, Central Canada refers to Ottawa and indirectly to the influence of Ontario and Québec in Ottawa. Together these two power struggles are the basis of the centralist/decentralist conflict or faultline.

Centralists advocate a strong central government, national policies that exert a political dominance over provinces, and a strong national economy (which means an industrial core). Decentralists seek to strengthen the powers allocated to provinces. In particular, decentralists call for a devolution of federal powers to the provincial governments and the expansion and diversification of regional economies. Economic diversification is deemed necessary to generate more jobs and thereby encourage population growth. Eventually, it is thought that economic diversification will lead to larger populations in hinterland regions and thus more political representation, and therefore power, in Ottawa.

Economic Power Struggle

In 1879, following the creation of the National Policy, Canada's manufacturing industries were protected from foreign goods by high tariffs and bolstered by lower customs' duties on raw materials. The goal was to strengthen Canada's economy by creating a national industrial base. In fact, this economic policy fostered the development of an economic core and a resource hinterland. For Central Canada, the benefits from the policy were considerable as it ensured the region's role as Canada's industrial heartland. It also resulted in the concentration of financial power in Montréal and Toronto. Other regions were not so fortunate. The hinterland had no such advantage as a result of the federal economic policy. Instead, it was compelled to sell its raw materials and foodstuffs on the world market and buy its manufactured goods in the domestic market. For the hinterland, this arrangement translated into 'selling their products cheap' and 'buying expensive goods from Central Canada'. Feelings ran high because people in the hinterland regarded the National Policy as an arrangement that benefited Central Canada at the expense of the rest of the country. They became suspicious of other national programs announced by Ottawa because they appeared to be designed to favour Central Canada and thereby exploit the other regions.

The National Policy accentuated the economic differences in various parts of the country. In doing so, it increased regional tensions, which resulted in the centralist/decentralist

(or heartland/hinterland) divide. While each hinterland region faced a different set of economic relations with the industrial heartland of Canada, the basic economic relationship was the same—Central Canada produced the manufactured goods, while hinterlands produced raw materials and foodstuffs. But what prevented hinterlands from developing a manufacturing base under the National Policy?

Geography prevented the manufacturing companies in the Maritimes from reaching markets in Central Canada and the growing market in the West. While the steel mill in Cape Breton did supply some of the steel rails for the construction of the Canadian Pacific Railway, most Maritime firms could not overcome the distance-cost barrier to reach the West and Central Canada. Maritime industrialists were also no longer able to sell their products in the nearby market of New England because of high American tariffs. The combination of a small local market, great distance to the continental markets in Canada, and high American tariffs stalled economic development in the Maritimes. Before 1867, most Maritimers were uncertain about the benefit of Confederation and afterwards, many became convinced that union with Canada was a 'bad deal' for their region.

At the same time, unrest in Western Canada was widespread in rural areas. Under the national economic policy, western farmers had to sell their grain on the world market where prices were low, but they had to purchase their farm machinery from manufacturers in Central Canada where prices were high. By the 1920s, this unrest took the form of a political protest when the Progressive Party argued for free trade to allow farmers access to lower-priced American farm machinery. Geography compounded the problems facing grain farmers, who were located in the heart of North America. The cost of transporting their grain to foreign markets was very high, thereby making their returns even lower. How could farmers overcome the great distance to their grain markets, particularly those in Europe? They thought that the solution to this transportation problem lay in the construction of a railway to the nearest tidewater point—Hudson Bay! The Hudson Bay Railway, completed in 1929, never fulfilled the high expectations of western farmers because the savings derived from the shorter rail distance were offset by higher marine insurance costs due to the danger of icebergs. Farmers, however, believed the higher insurance charges were simply another trick by Central Canada to hold onto the grain trade at the expense of westerners.

British Columbia, though linked to the rest of Canada by the Canadian Pacific Railway, remained beyond the economic pull of Central Canada. British Columbia's economic order was driven by its geographic position on the West Coast. The region's natural markets were overseas. At first, its raw materials such as fish, forests, and mineral products were shipped to markets in the western United States, the British Isles, and the Far East. As well, because of British Columbia's geographic proximity to Western Canada, many of the Prairies' products were transported to the port of Vancouver for transshipment overseas. In 1914 the opening of the Panama Canal had a great impact on the forest industry in British Columbia because it provided a 'shorter' shipping route to markets in the British Isles, Western Europe, and the East Coast of the United States. Like Canadians in Western Canada, British Columbians resented the higher cost of manufactured goods produced in Central Canada when lower-cost American goods were readily available just across the border.

The Territorial North suffered a different fate. Until the Second World War, Ottawa simply ignored it. Regarded by the federal government as a remote wilderness inhabited by Indians and Eskimos (later known as Inuit), the Territorial North remained a fur economy and did not participate in the events transforming Canada's emerging industrial society until well into the twentieth century. Ottawa's *laissez-faire* approach limited federal expenditures to Aboriginal peoples in the Territorial North. The few northern Royal Northwest Mounted Police posts (later the Royal Canadian Mounted Police) served as Ottawa's main expression of administration. Only when resource developments occurred, such as the Klondike gold rush, did Ottawa hurry to ensure a broader federal presence. After the Second World War, the federal government's *laissez-faire* policy was replaced by state involvement in northern development. This change was sparked by Prime Minister Diefenbaker's 'northern vision', which translated into investments in northern infrastructure, especially highways leading to resources.

Political Power Struggle

While the core/periphery model emphasizes the core's economic dominance over the periphery, the core also exerts other forms of dominance, including political dominance. Within the context of Canada, Ottawa not only represents the core but, by favouring Central Canada (Ontario and Québec), extends the advantages of this political dominance to a particular area of Canada. This leads to a sense of alienation and frustration among the remaining regions. For example, in the late 1970s, provincial-federal relations soured during debates over constitutional reform. The western provinces (Alberta, British Columbia, Manitoba, and Saskatchewan), supported by Nova Scotia and Newfoundland, had developed their own agenda for constitutional reform to obtain more powers for the provinces, including a reform of the Senate and the Supreme Court, to reflect regional interests. Québec Premier René Lévesque did not participate in these discussions, leaving it to the anglophone provinces to 'dismantle' Confederation. Ottawa's most

consistent ally was Ontario, which confirmed hinterland charges that Ontario and Québec were the chief beneficiaries of old federalism.

But does Ottawa really favour Central Canada or is it just a problem of balancing national against regional interests? The reality is in the division of powers between the two levels of government (see Vignette 3.5). Can such a power imbalance exist in a federal state? How much power does a province need? How much power can the central government give up? These questions go to the heart of a federal state and the fragile relations between the central and regional governments. This tug of war is also played by the separatist movement in Québec but for an end (independence) that is different from that sought by provincial premiers (more powers). In 1968, Jacques Parizeau likened the Canadian federation to a chicken being plucked by provincial governments led, of course, by Québec. When

Vignette 3.5

Federal/Provincial Powers

Under Canada's federal system, the powers of government are shared between the federal government and the ten provincial governments. All provinces have the same powers. At the time of Confederation, the powers of the provinces were carefully enumerated (and thus limited), while those of the federal government were not limited. Provincial governments are responsible for education, health and welfare, highways, civil law (property and civil rights), local government, and natural resources, while Ottawa has a much wider mandate, including authority over defence and external affairs, criminal law, money and banking, trade, transportation, citizenship, and Indian affairs. The two levels of government were assigned joint jurisdiction over agriculture, immigration, and taxation. Territorial governments are assigned their powers from Ottawa, while municipal governments obtain theirs from the provincial governments. First Nations are seeking self-government. While First Nations' status is not yet defined by the federal government, Indian leaders often refer to a 'third level of government', meaning a political level somewhere between municipal and provincial.

Figure 3.11 The regions of Canada. This political cartoonist has aptly captured the regionalized tensions that are manifested in Canada.
Source: The Globe and Mail, 11 December 1995:A14. Reprinted with permission from The Globe and Mail.

the last feather is plucked, the chicken will perish. *Voila*, Québec has achieved its goal of independence (Nemni 1994:175).

At the federal level, political representation in Ottawa is determined by its electoral system. From a geographic perspective, this system means that Central Canada, with the greatest number of voters and ridings, provides the House of Commons with most of its members. This concentration of electoral power is further intensified by a concentration of political power within the governing party, which is naturally concerned with national issues and priorities, not regional ones. From the standpoint of those outside Central Canada, national policies appear to favour Ontario and

Québec. The Senate, which theoretically provides a regional balance in Canada's political system by attempting to give more voice to Canada's hinterland, represents seven 'regions': Ontario, Québec, the West, the Maritimes, Newfoundland, Yukon, and the Northwest Territories. However, since its members are appointed by the prime minister, largely on the basis of long-time service to the political party currently in power, senators are more concerned about party loyalty than regional interests. In short, the Senate fails to provide a regional counterweight to the House of Commons. Instead, this role falls to the provincial governments.

The struggle over political power takes

place within the six geographic regions of Canada (Ontario, Québec, British Columbia, Western Canada, Atlantic Canada, and the Territorial North), which are far from equal. Ontario and Québec have most of Canada's population, its manufacturing industries, and national corporate headquarters. Of the six geographic regions, Ontario and Québec (Central Canada) are at the top of the economic and political hierarchy. In sharp contrast, the Territorial North is at the bottom. It has no industrial base; it has the smallest population of any geographic region; and it is far from the decision-making centres in Canada. Using population size and economic power (as measured by gross domestic product) as two indicators of the strength of regional power, the remaining regions fall between Ontario and the Territorial North (Table 1.2).

The distribution of political power takes other forms—indicators of regional power include the richness and sustainability of a resource base, and access to the national, American, and international markets. Perhaps one of the more notable indicators of power is tax rates. The British North America Act gave Ottawa unlimited taxing powers, while provincial governments' ability to raise revenue through taxation is limited. Given that the provinces' levels of economic development vary, the ability to raise revenue through taxation is also uneven. For that reason, the rate of personal taxation varies from province to province. The have provinces—Alberta, British Columbia, and Ontario—have the lowest personal tax rates, while the have-not provinces have much higher rates. Newfoundland, for example, has the highest personal income

tax rate in Canada, while Alberta has the lowest. The same geographic pattern holds true for provincial sales tax rates: Newfoundland has the highest rate, while Alberta has no sales tax.

These spatial variations in political and economic power make Canada a 'troublesome country to govern'. Each prime minister from Sir John A. Macdonald to Jean Chrétien has had to balance national economic interests against regional ones. Federal intervention in the marketplace are attempts to protect national interests, but they sometimes result in regional alienation. One such example is the National Energy Policy.

National Energy Program

Resource development falls under provincial powers, yet the federal government made two initiatives into the petroleum industry. The first took place in the 1960s when the Diefenbaker government attempted to stimulate oil and gas developments in oil-producing provinces by allocating the Ontario market to the West. The second initiative, the National Energy Program, developed under the Trudeau government, was Ottawa's attempt to achieve three goals: energy security, greater Canadian ownership of the oil industry, and a greater amount of wealth produced by the oil industry for Ottawa. The National Energy Program lasted only four years (1980–4), but its chief legacy was the western oil-producing provinces' distrust of the federal government. Why was this so?

Without a manufacturing base, the Canadian hinterland has had to depend on its natural resources as an extremely important source of regional development and provincial tax revenue. Thanks to its geography, Alberta has abundant petroleum deposits. This fact, plus the sudden rise in oil prices in the 1970s, turned Alberta from a have-not province to a have province. Since natural resources fall under the jurisdiction of the provincial governments, the federal government has no authority to 'tax' them, yet with its National Energy policy, Ottawa was able to force oil-producing provinces to share their rapidly increasing oil revenue with the federal government. Why did Ottawa act in this way, knowing full well that it would lead to western alienation? It did so because Ottawa believed the National Energy Program was in the national interest.

Before the National Energy Program, the report of the Royal Commission on Energy (the Borden Commission) in 1959 provided the basis for a new federal policy on energy. Under the Diefenbaker government, Ottawa created a dual market for oil to promote oil development in Western Canada. The western market, extending from eastern Ontario (including the Toronto area) to the Pacific Coast, was reserved for oil produced from Canadian wells (primarily from Alberta). The eastern market consisted of eastern Ontario (including the Ottawa area), Québec, and Atlantic Canada. Oil producers from the Middle East and Venezuela supplied the eastern market. The purpose of the dual market was to create a larger market for western oil even though its price in Toronto was 10 to 15 per cent higher than that of offshore oil refined in Montréal. Perspectives on this policy varied: Ottawa saw a decreased dependency on foreign oil; Alberta saw an expanded oil industry; Ontario saw slightly higher oil costs; and Québec saw

its natural market area reduced (southern Ontario). Naturally, Québec was not pleased with this federal policy because its Montréal-based petroleum refineries could not sell their products in the Toronto market. As a result, oil refining expanded in Alberta (Edmonton) and Ontario (Sarnia) instead of Québec (Montréal).

In the 1970s, the price of world oil rose rapidly because of the Organization of Petroleum Exporting Countries (OPEC) strategy: by curtailing their supply of oil, the world price would rise. From 1972 to 1980, the world price of oil increased from US $2 a barrel to over US $20 a barrel. From 1973 to 1978, agreements were reached between the oil-producing provinces and the federal government to match the domestic price with the world price. Rising oil prices generated a huge profit for the oil companies and greatly increased oil revenues for Alberta and, to a much lesser degree, Saskatchewan and British Columbia. In 1979, however, Ottawa refused to match domestic oil prices with world prices, thereby creating a lower domestic price. For Central Canada, such a national policy gave its industrial firms a decided advantage compared to those across the US border. With world prices continuing to rise, subsidization of oil refineries in Atlantic Canada and Québec became more expensive. The federal government needed a new energy strategy that addressed two key concerns. How could Ottawa pay for this price differential? Should not part of this 'windfall' that companies and provincial governments receive be redirected to the federal government?

In 1980, the federal government announced the National Energy Program, which enabled

Ottawa to obtain a larger share of oil revenues through new taxes on the oil industry and other measures. Alberta objected vigorously along with the other two oil-producing provinces, Saskatchewan and British Columbia. Albertans believed that Ottawa forced oil-producing provinces to accept this program in order to ensure supplies of low-cost oil to Ontario and Québec. In 1982, Peter Smith, a professor of Geography at the University of Alberta, presented the Albertan position by denouncing the policy as 'one more manifestation of the customary heartland outlook; what is good for Ontario and Québec has to be best for the rest.' Smith went on to state that:

> to Canada's misfortune, the issue brings the negative side of regionalism to the fore by causing regional sentiment to be focused on a siege mentality rather than on the sense of organized community responsibility a functional region should possess if it is to assume a political identity. The individualism of political regions is not just an Albertan reality; it is a Canadian one. And no misreading of the heartland-hinterland relationship should be allowed to obscure that historical and geographical fact (Smith 1982:301–2).

Current Struggles

The nature of Canadian federalism demands a high level of cooperation between the federal and provincial governments because of shared or overlapping responsibilities, authority, and funding in many areas of public policy. National purposes can often only be achieved with provincial cooperation, yet tensions often arise between Ottawa and provincial governments because of shared powers, interdependency, and conflicting goals. Added to this situation, provincial governments are demanding more and more powers that now fall under the federal government's jurisdiction. All of this complicates federal-provincial relations, sometimes leading to bitter political confrontations. Negotiations between the two levels of government have come to an impasse more than once. At these times, politicians have accused those on the other side of bad faith.

A major bone of contention between Ottawa and the provinces has been funding for a series of 'national' social programs. Social programs are the responsibility of provincial governments, but because these programs are so costly, only the have provinces can afford them. In some cases, Ottawa is directly involved and delivers the program, for example, the Employment Insurance Program and the Canada Pension Program. In other cases, Ottawa provides financial support to the less affluent provinces and territories through transfer payments, notably equalization payments and block cash transfers for Medicare and postsecondary education under the Established Programs Financing Act. In 1996–7, transfer payments, including equalization payments, totalled $25 billion. Such financial support has enabled all provinces and territories to provide similar public services to its citizens, but the level of fiscal dependence is very high for certain regions. The Territorial North, Atlantic Canada, and Québec (Vignette 3.6) are heavily dependent on transfer payments from the federal government.

Vignette 3.6

Equalization Payments

Regional equity is an important theme in Canada. In 1957, the federal and provincial governments recognized the importance of providing similar public services in each of the provinces. Equalization payments would provide the funds necessary to achieve this goal. This program has ensured that per capita revenues of all provinces from shared taxes (personal income taxes, corporate income taxes, and succession duties) matched those of the wealthiest provinces. Until 1995, $128 billion had been transferred from Ottawa to the have-not provinces. About half of these payments had gone to Québec, 35 per cent to Atlantic Canada, and 15 per cent to Western Canada. The territorial governments also received funds from the fed-

eral government, but these transfers fell under a separate program.

Equalization payments and a similar program for the territories form a significant portion of the gross revenues for the five provinces and two territories. While Québec receives the largest amount of equalization payments, these federal transfers amount to about 30 per cent of Québec's gross general revenues. While the four Atlantic provinces obtain far less payments than Québec, these funds form about half of their gross general revenues. Fiscal dependency is even greater in the territories, where transfer payments amount to about 70 per cent of territorial revenues.

Federal funding for programs operating within provinces began in the 1950s with an agreement for universities (1951), national hospital insurance (1958), the Canada Assistance Plan (1966), and universal health-care insurance (1968). Canada has created an elaborate social safety net through the implementation of these social programs. Canada did not come by this social safety net easily because the division of powers between the federal and provincial governments often stood in the way. The provinces have the sole power to establish social programs, but only the federal government has the ability to pay for them. This seemingly insurmountable political conundrum was not solved until after the Second World War. By then Ottawa wielded unprecedented political and financial power

over the provinces. By offering to pay for part of these programs through grants, Ottawa was able to 'induce' the provinces to deliver programs with 'national standards'. Such financial dependence forms the basis of the current tension between Ottawa and the provinces.

By early 1990, Ottawa could no longer avoid dealing with its massive debt. While previous federal governments had begun the process of reducing this debt by raising taxes and reducing services, the Liberal government began the job in earnest. Much of the federal debt was reduced through a reduction in transfer payments that forced provinces to pay a larger share of their social expenditures, including health-care costs. Ottawa's 1997–8 budget for transfer payments, including equalization payments, was reduced by $5 billion (from $25 billion to $20 billion). Excluding equalization payments, federal transfers now fall under the Canada Health and Social Transfer Act. These transfers fell from $18.5 billion in 1996–7 to $12.5 billion in 1998–9. By 2001, these transfers were to stabilize at $11.1 billion. Before the 1997–8 budget year had ended, the severity of these cuts were felt by Canadians supported by social welfare programs, Canadians seeking medical services, and Canadians enrolled in postsecondary education programs.

Provincial governments were furious with these 'unilateral' changes in transfer payments because they now had to make up the shortfall or reduce their services. Provinces chose to reduce services rather than increase taxes, so their health-care programs, social services, and universities suffered. At the end of its 1997–8 budget, the federal government had eliminated its operating debt and was now producing a modest surplus. Ottawa was now promising to increase transfer payments to provincial governments, but the damage to federal-provincial relations was done. As Canada heads into the next century, provinces are demanding that Ottawa hand over more of its taxing authority to the provinces, thereby reducing the provinces' fiscal dependency on Ottawa. While these political struggles between Ottawa and the provinces continue, Canadians are suffering by paying higher taxes but receiving lower-quality health-care services and facing higher tuition fees for postsecondary education.

The French/English Faultline

Although the ancestors of Aboriginal peoples were the first to occupy North America,

two European powers—the French and the British—colonized Canada. This historic fact underscores the vision of Canada as two founding nations. Relations between the French and English in North America began nearly 400 years ago. These two cultures have come to represent a major faultline in Canadian society. Since 1841, however, these two communities have had to work together, each dependent on the other. This interaction has done much to shape the cultural and political nature of Canada. Over the years, they have accomplished much together. Nevertheless, significant differences between the two communities exist and, from time to time, these differences flare into serious misunderstandings. Without a doubt, Canadian unity is dependent on the continuation of this relationship and the need for compromise. A brief examination of that relationship, as outlined in the following pages, will lead to a fuller understanding of the contemporary version of French/English differences, conflicts, and compromises.

The serious nature of this rift has profound geopolitical consequences for Canada. It is therefore crucial that we understand the origins and nature of this faultline. A well-known Canadian political columnist, Jeffrey Simpson, wrote:

We can also hope that, in the 1980s, Canadians gained a deeper understanding of the faultlines running through their society, and that they will avoid measures that widen them, thereby concentrating on making new arrangements and re-forming old ones, so that what the rest of the world rightly believes to be a successful experiment in managing diversity will endure and prosper (Simpson 1993:368).

The beginning of formal French-English relations in Canada stretches back to the British conquest of the French on the Plains of Abraham in 1759, an event that remains a dark page in French-Canadian history. In 1760, the French Canadians watched the remnants of the French army and the French élite board ships to return to France. The French Canadians had no thought of leaving as they were born in the New World, but what would happen to them under British military rule? Would they, like their Acadian brethren, be deported to other British colonies? Britain did not need to take such drastic action as there was no French threat to Britain's North American possessions. In the 1763 Treaty of Paris, France ceded New France to Britain, which 'forced' French Canadians to become a defensive minority in their own homeland. While the English lived in cities in Québec and dominated the Québec economy, French Canadians lived mostly in rural areas and managed to successfully maintain their culture within a British North America, an achievement that was made possible because of two factors. First, the French Canadians were a large homogeneous population that occupied a contiguous geographic area. Second, Britain forged a close relationship with the former élites of New France (the Roman Catholic clergy and the landed gentry) to ensure the French Canadians' loyalty because Britain wanted to secure its northern colony against its restless American colonies to the south. This relationship between Britain and the French Canadians would be strengthened with the Québec Act of 1774.

The Québec Act, 1774

With the Québec Act of 1774, the individuality and separateness of Québec was recognized, thus ensuring its unique place in British North America. This act is sometimes described as the Magna Carta for French Canada.[5] Its main provisions ensured the continuation of the aristocratic landholding system (seigneury) and guaranteed religious freedom for the colony's Roman Catholic majority. This gave the most powerful people in New France a good reason to support the new rulers. The Roman Catholic Church was placed in a particularly strong position. Not only was the Church allowed to collect tithes and dues but its role as the protector of French culture went unchallenged. Therefore, the clergy played an extremely important role in directing and maintaining a rural French-Canadian society, a role further enhanced by the Church's control of the education system. The *habitants* (farmers) were at the bottom of French-Canadian society's hierarchy. They formed the vast majority of the population and continued to cultivate their land on seigneuries, paying their dues to their lord (seigneur) and faithfully obeying the local priest and bishop. The British granted another important concession, namely, that civil suits would be tried under French law. Criminal cases, however, fell under English law.

The seigneurial system formed the basis of rural life in New France and later, Québec. In 1774, there were about 200 seigneuries in the St Lawrence Lowlands. This type of land settlement left its mark on the landscape (the long, narrow landholdings and the vast estate of the seigneur) and on the mentality of rural

French Canadians (close family ties, strong sense of togetherness with neighbouring rural families, and staunch support for the Church). A *habitant*'s landholding, though small, was the key to his family's prosperity and, by bequeathing the farm to his eldest son, ensured the continuation of this rural way of life. In 1854, the *habitant* was allowed to purchase his small plot of land from his seigneur, but the last vestiges of this seigneurial system did not disappear until a century later. Even today, the landscape in the St Lawrence Lowlands shows many signs of this type of landholding.

While the heart of this new British territory was the settled lands of the St Lawrence Lowlands, its full geographic extent was immense. Essentially, the Québec Act of 1774 recognized the geographic extent of former French territories in North America. Québec's territory in 1774 was extended from the Labrador coast to the St Lawrence Lowlands and beyond to the sparsely settled Great Lakes Lowlands and the Indian lands of the Ohio Basin.

The Loyalists

The American War of Independence changed the political landscape of North America. Within the newly formed United States, a number of Americans, known as the **Loyalists**, remained loyal to Britain. Like the French-speaking people in North America, most of these Loyalists were born and raised in the New World. For them, North America was their homeland. During the rebellion, they had sided with the British. They were hated by the American revolutionaries and lost their homes and property. As they were not wel-

come in the new republic, many Loyalists resettled in the remaining British colonies in North America, where Britain offered them land. The majority (about 40,000 Loyalists) settled in the Maritimes, particularly in Nova Scotia. About 5,000 relocated in the forested Appalachian Uplands of the Eastern Townships of Québec. A few thousand, including Indians who had supported Britain, took up land in the Great Lakes Lowlands along the northern shores of Lake Ontario. These Loyalists strengthened Britain's hold on its North American possessions, but those that settled in the major cities of Québec and in the Eastern Townships came from a different cultural world than the local francophone residents. Social and political tensions arose from time to time between the two cultural groups.

Within a few decades, the English-speaking settlers in the Great Lakes grew in number. They felt separate from the rest of Québec and demanded British civil law, British institutions, and an elected assembly. In the Constitutional Act of 1791, Québec was split into Upper and Lower Canada.

The Constitutional Act, 1791

In 1791, the British government passed the Constitutional Act (Canada Act) in an attempt to satisfy the political needs of its French- and English-speaking subjects. These were the main provisions of the act: (1) the British colony of Québec was divided into the provinces of Upper and Lower Canada, with the Ottawa River as the dividing line, except for two seigneuries located just southwest of the Ottawa River; and (2) each province was governed by a British lieutenant-governor appointed by

Britain. From time to time, the lieutenant-governor would consult with his executive council and acknowledge legislation passed by an elected legislative assembly.

In 1791, Lower Canada had a much larger population than Upper Canada. At that time, about 30,000 colonists lived in Upper Canada, most of whom were of Loyalist extraction, with about 10,000 Indians, some of whom had fled northward after the American Revolution. Lower Canada's population was nearly four times larger. It consisted of about 100,000 French Canadians, 10,000 English Canadians, and perhaps as many as 5,000 Indians.

Following the Constitutional Act, Upper and Lower Canada each had an elected assembly, but the real power remained in the hands of the British-appointed lieutenant-governors. In Lower Canada the lieutenant-governor had the support of the Roman Catholic Church, the seigneurs, and the Château Clique. The **Château Clique**, a group consisting mostly of anglophone merchants, controlled most business enterprises and, as they were favoured by the lieutenant-governor, wielded much political power. In Upper Canada the **Family Compact**—a small group of officials who dominated senior bureaucratic positions, the executive and legislative councils, and the judiciary—held similar positions in commercial and political circles. While these two élite groups promoted their own political and financial well-being, the rest of the population grew more and more dissatisfied with blatant political abuses, which included patronage and unpopular policies that favoured these two groups. Attempts to obtain political reforms leading to a more democratic political

system failed. Under these circumstances, social unrest was widespread.

In 1837 rebellions broke out. In Lower Canada Louis-Joseph Papineau led the rebels, while William Lyon Mackenzie headed the rebels in Upper Canada. Both uprisings were quickly suppressed by British troops. The goal of both insurrections was to take control by wresting power from the colonial governments in Toronto and Québec and putting it in the hands of the popularly elected assemblies. In Lower Canada the rebellion was also an expression of Anglo-French animosity. While both uprisings were unsuccessful, the British government nevertheless sent Lord Durham to Canada as governor-general to investigate the rebels' grievances. He recommended a form of responsible government and the union of the two Canadas. Once the two colonies were unified, the next step, according to Durham, would be the assimilation of the French Canadians into British culture. When Durham left in 1838, a second rebellion broke out in Lower Canada, but it was as unsuccessful as the first.

The Act of Union, 1841

In response to Durham's report, in 1841, the two largest colonies in British North America, Upper and Lower Canada, were united into the Province of Canada. This Act of Union gave substance to the geographic and political realities of British North America. The geographic reality was that a large French-speaking population existed in Lower Canada, while an English-speaking population was concentrated in Upper Canada. The political reality was twofold. Both groups had to work together to accomplish their political goals and neither group could achieve all its goals without some form of compromise. When the two cultures were forced to work together in a single legislative assembly, a new beginning to the French/English faultline surfaced.

The new governor, Sir Charles Bagot, was appointed by, and reported to the Colonial Office in London. The governor had the authority to appoint members to a Legislative Council and an Executive Council. The only representative body was a Legislative Assembly. Even though Lower Canada (after union known as Canada East) was slightly larger in economic strength and population size (670,000) than Upper Canada (after union known as Canada West, 480,000), each elected forty-two members to the Legislative Assembly. The vast majority of inhabitants in Canada East and Canada West lived in the rural countryside. For example, in 1841, the principal

towns of Montréal and Toronto had populations of about 40,000 and 15,000 respectively.

Demographic Shifts

Since 1791, the balance of French- and English-speaking inhabitants of British North America had begun to tilt more and more in favour of the English. This demographic shift began after the American Revolution when thousands of Loyalists who sided with Britain relocated in British North America. In 1791, the European population of British North America was about 225,000 (mostly French Canadians). Some 162,000 (72 per cent) lived among the St Lawrence River in what is now the province of Québec. Perhaps as many as 50,000 (22 per cent) lived in Atlantic Canada. The remaining 15,000 (6 per cent) were scattered along the north shores of lakes Erie and Ontario, in what is now part of the province of

Table 3.4
Population by Colony or Province, 1851–1871 (percentages)

Colony/Province	1851	1861	1871
Ontario	41.1	45.2	46.5
Québec	38.5	36.0	34.2
Nova Scotia	12.0	10.7	11.1
New Brunswick	8.4	8.1	8.2
Manitoba			< 0.1
British Columbia			< 0.8
Total per cent	100.0	100.0	100.0
Total population	2,312,919	3,090,561	3,525,761

Source: Wayne W. McVey and W.E. Kalbach, *Canadian Population* (Toronto: Nelson Canada, 1995):38. Reprinted with permission of ITP Nelson.

Ontario. By this time, the Aboriginal population of Upper Canada, Lower Canada, and Atlantic Canada had declined substantially to about 25,000.

Within fifty years, not only had this geographic pattern changed but the balance of demographic power had shifted. French Canadians were no longer the majority in British North America. In 1841, British North America had about 1.5 million inhabitants, 45 per cent of them located in Lower Canada, about 33 per cent in Upper Canada, and 12 per cent in the Atlantic colonies. Over the next thirty years, the balance continued to swing in favour of English-speaking regions, thanks to massive immigration from the British Isles. By 1871, Québec's population was only 34 per cent of Canada's population (Table 3.4).

As the country expanded its boundaries and more land was settled, Canada's English-speaking population grew, while Québec's French-speaking population diminished in proportion. Manitoba joined Confederation in 1870 with a population of almost 12,000, which was comprised largely of French- and English-speaking Métis. In 1871, British Columbia, with an estimated population of 28,000 British subjects, became a member of the Dominion of Canada. Beyond these provinces, Indian tribes and Inuit peoples inhabited the land. The total Aboriginal population of all territory that would eventually become Canada was about 100,000 at the time of Confederation. In the subsequent decades, these new lands would be settled by Canadians, Europeans, and Americans. With few exceptions, English became the dominant language of these settlers. Manitoba was one such

exception, but when the English-speaking majority gained control of the government and the public institutions, French-speaking residents found it difficult to maintain their culture and language.

Strained Relations

During these formative years, the Dominion underwent several events that seriously strained relations between its two founding peoples. Two cultures, French and English, were in firm opposition to each other. In the settling of the West, these two cultures clashed over language and religious rights. Four events illustrate the intensity of this power struggle:

- The Red River Rebellion, 1869–70
- The North-West Rebellion, 1885
- The Québec Jesuits' Estates Act, 1888
- The Manitoba Schools Question, 1890

The Red River Rebellion, 1869–70

The Red River Rebellion took place in 1869. In the previous year, the British government passed the Rupert's Land Act, which transferred the Hudson's Bay Company lands to the government of Canada. At first, nothing changed for the Aboriginal peoples inhabiting the vast grasslands west of the Great Lakes, but Ottawa's plan was to settle these lands. With the arrival of surveyors and then settlers, the world of the hunters and trappers soon disappeared. The first to sense this threat were the French-speaking Métis who lived near the Red River. They became alarmed by the arrival of the land survey teams from Ottawa. The Métis of Assiniboia feared that they might lose their culture, religion, and freedom to hunt buffalo on the open Prairies, so they reacted swiftly. The Métis rebellion, led by Louis Riel, soon became a national issue, reopening differences between English, Protestant Ontario and French, Roman Catholic Québec.[6] Québec considered Riel a French-Canadian hero who was defending the Métis, a people of mixed blood who spoke French and followed the Catholic religion. Ontario, on the other hand, considered Riel a traitor and a murderer. For Canada, the larger issue was the place of French Canadians in the West. A compromise was achieved in the Manitoba Act of 1870. Accordingly, the District of Assiniboia became the province of Manitoba. Under this act, land was set aside for the Métis, and the elected legislative assembly of Manitoba provided a balance between the two ethnic groups with twelve English and twelve French electoral districts. Equally important, Manitoba had two official languages (French and English) and two religious school systems (Catholic and Protestant) financed by public funds.

The North-West Rebellion, 1885

During the 1870s, many Ontarians settled in Manitoba while some Métis sought a new home on the open Prairie. Seeking to remain hunters, one group settled along the South Saskatchewan River where they established a Métis colony around Batoche. Batoche, located about 60 km northeast of what is now Saskatoon, became the new centre of the French-speaking Métis in Western Canada. As settlers spread into Saskatchewan, the Métis again feared for their future. In 1884, a party of Métis went to Montana where Louis Riel, their old leader, was living and pleaded with him to

return to Batoche and lead them again. Convinced of his destiny, Riel accepted this challenge. Late in 1884, Riel sent a petition to Ottawa with various demands for all the inhabitants of the North-West—Indians, Métis, and Whites. After Ottawa ignored his petition, Riel established a provisional government and began to organize the Métis into armed bands. Ottawa responded by sending a militia to suppress the rebellion. The militia travelled from Ontario to Saskatchewan in eight days, thanks to the new railway. Within a relatively short period, the Canadian militia captured Batoche, took Riel prisoner, and defeated the Indians, who were led by Chief Poundmaker and Big Bear. For Québec, the defeat of the Métis and the subsequent hanging of their leader, Louis Riel, not only represented a defeat for a French presence in the West but also widened the gulf between French and English Canadians.

The Québec Jesuits' Estates Act, 1888

In the nineteenth century religious and linguistic intolerance was widespread. For example, Protestant extremists in Ontario were ready to pounce on any perceived injustice to their cause. The Jesuits' efforts to obtain financial compensation for lands that the British took from them in 1763 and later transferred to Lower Canada proved to be such a case.

The Jesuit estates, which were granted under the French regime and used for schools and missions, were appropriated by Britain after the British Conquest and given to Lower Canada in 1831. In 1838, Catholic bishops petitioned unsuccessfully for the return of the Jesuit estates. After Confederation, the ownership of the estates passed to the Québec government, with which the Jesuits began negotiating in 1871 for financial compensation. However, the archbishop of Québec argued that the money should be divided among Catholic schools rather than given in its entirety to the Jesuits, who wanted to establish a university in Montréal that would compete with Québec's Université Laval. Québec Premier Honoré Mercier asked Pope Leo XIII to act as arbiter in the dispute among the Roman Catholic hierarchy. In 1888, Québec's Legislative Assembly passed the Jesuits' Estates Act, which determined the division of the financial compensation: $160,000 went to the Jesuits, $140,000 went to the Université Laval, and $100,000 went to selected Catholic dioceses.

Ontario's Orange Order, a Protestant fraternal society, opposed the settlement arguing that the arbiter, Pope Leo XIII, was a papist intruder into Canadian affairs, and that public funds should not be used to support Catholic schools. In March 1889, the House of Commons debated the motion to disallow the Québec Jesuits' Estates Act and eventually voted against this motion. Similar anglophone, Protestant, anti-Catholic sentiment surfaced in Manitoba regarding the Manitoba schools question.

The Manitoba Schools Question, 1890

The British North America Act of 1867 established English and French as legislative and judicial languages in federal and Québec institutions. The remaining three provinces (New Brunswick, Nova Scotia, and Ontario) had only English as the official language. The question of French language and religious rights in acquired western territories first arose in Manitoba.

The French/English issue became the focal point for the entry of the Red River Colony (now Manitoba) into Confederation. Local inhabitants—mostly French-speaking, Roman Catholic Métis, and the less numerous English-speaking Protestants—were determined to have some influence over the terms that would include their community as part of Canada. One of their concerns was language rights, which was ultimately resolved when a list of rights drafted by the provisional government became the basis of federal legislation. When the Red River Colony entered Confederation in 1870 as the province of Manitoba, it did so with the assurance that English- and French-language rights, as well as the right to be educated in Protestant or Roman Catholic schools, were protected by provincial legislation.

During the 1870s and 1880s, a large number of Anglo-Protestants settlers, most of whom were from Ontario, moved to Manitoba, causing the proportion of Anglo-Protestants in the population to increase, and the proportion of French and Roman Catholic inhabitants to decrease. This change in the demographics created a stronger Anglo-Protestant culture in Manitoba. In 1890, the provincial government ended public funding of Catholic schools. From Québec's perspective, this legislation led to the most significant loss of French and Catholic rights outside of Québec.

One Country, Two Visions

The greatest challenge to Canadian unity comes from the cultural divide that separates French- and English-speaking Canadians and their respective visions of the country. In the early years of Confederation, events such as the Red River Rebellion, the North-West Rebellion, the Jesuits' Estates Act, and the Manitoba Schools Question widened the French/English faultline. For French Canadians these events demonstrated the 'power' of the English-speaking majority and their unwillingness to accept a vision of Canada as a partnership between the two founding peoples. The root of each vision lies in the history of Canada.

One vision of Canada is based on the principle of two founding peoples. This vision originated in French-Canadian historical experiences and compromises that were necessary for the sharing of political power between the two partners. This vision began with the conquest of New France in 1760, but its true foundation lies in the formation of the Province of Canada in 1841. From 1841 onward, the experience of working together resulted in a Canadian version of cultural dualism.

Henri Bourassa, an outstanding French-Canadian thinker (and Canadian nationalist) at the turn of the nineteenth century, was a strong advocate of cultural dualism. He wrote, 'My native land is all of Canada, a federation of separate races and autonomous provinces. The nation I wish to see grow up is the Canadian nation, made up of French Canadians and English Canadians' (quoted in Bumsted 1998:253). Bourassa argued that a 'double contract' existed within Confederation. Even today, Bourassa's 'double contract' is an essential element in the two founding peoples concept. He based the notion of a double contract on a liberal interpretation of Section 93 of the BNA Act. This section guarantees denomina-

tional schools; Bourassa expanded the interpretation of the religious rights to include cultural rights for French- and English-speaking Canadians. In more practical terms, Bourassa regarded Confederation as a moral contract that guaranteed French/English duality, the preservation of French-speaking Québec, and the protection of the language and religious rights of French-speaking Canadians in other provinces.

From a geopolitical perspective, Canada is a bicultural country. In one part the majority of Canadians speak English, and in another part the majority speak French. For example, French culture predominates in Québec and has a strong position in New Brunswick. In addition to provincial control over culture, two other geopolitical factors ensure the dynamism of French in those provinces. One factor is the large size of Québec's population—the vitality of Québécois culture is one indication of its success. The second factor is the geographic concentration of French-speaking Canadians in Québec and adjacent parts of Ontario and New Brunswick. For instance, in New Brunswick the French-speaking residents, known as Acadiens, constitute over one-third of the population. Federal policies of bilingualism and biculturalism also helped rejuvenate francophone minorities. Before these policies were instituted, assimilation into the much larger English-speaking culture had seriously weakened the position of francophones in all the provinces, except Québec, and in the two territories. While the attraction of joining the dominant anglophone culture remains, financial support from Ottawa for French educational and cultural facilities in the English-speaking provinces has ensured a place for bilingualism in all provinces and territories.

The Royal Commission on Bilingualism and Biculturalism was an attempt to bridge the gap between English and French Canadians. This commission, set up in 1963, examined the issue of cultural dualism; that is, an equal partnership between the two cultural groups. But by the 1960s, English-speaking Canada had changed and it was no longer the British world of the nineteenth century. English-speaking Canada had evolved from a predominantly British population to a more diverse one with several large minority groups who also spoke other languages, namely, German and Ukrainian. Ottawa, in searching for a compromise, announced two policies, bilingualism and multiculturalism.

In the second vision, Canada consists of ten equal provinces—yet this too is complex. On the one hand, this vision represents the simple notion based on provincial powers granted under the British North America Act (Vignette 3.5), which ensured that Canada consists of a union of equal provinces, all of which have the same powers of government. Nonetheless, by assigning provinces, including Québec, powers over education, language, and other cultural matters within their provincial jurisdictions, Québec's French culture was secure from political tampering by the anglophone majority in the rest of Canada. Confederation then provided a form of collective rights for French culture within Québec. Under Canada's federal system, the powers of government are shared between the federal government and ten provincial governments. But are all provinces really equal? As noted earlier,

population size, geographic extent, and financial strength vary considerably, which is reflected in the need for equalization payments (Vignette 3.6).

The vision of ten equal provinces may reflect an English-Canadian nationalism. For some time, English-speaking Canadians have been searching for their cultural identity. Before the First World War, Canadians saw themselves as part of the British Empire. By the end of the Second World War, this perspective began to change. The Maple Leaf flag, adopted by Parliament in 1964, and 'O Canada', the new national anthem approved by Parliament in 1967 and officially adopted in 1980, were signs of this cultural change. While the Québécois culture was flourishing, English-speaking Canadians continued to lean heavily on American culture. Some looked with envy at the cultural accomplishments of the Québécois and wondered aloud if similar achievements in English-speaking Canada were possible.

Compromise

Given the incompatibility of the two visions, two founding peoples versus ten equal provinces, and the historical development of the country, Canadian politicians have had the unenviable task of trying to accommodate demands from different groups—especially French Canadians and Aboriginal peoples—and from different regions without offending other groups or regions. As in the past, politicians have continued to struggle with this Canadian dilemma, but in reality there is no perfect solution, only compromise. With this object in mind the federal government has made many efforts in search of the elusive middle ground.[7] It seems the search for an acceptable compromise between the two opposing visions of Canada will never end. To understand the current struggle for compromise, it is important to understand the political, economic, and cultural developments that have taken place in Québec over the past four decades. These, and their effects on the English/French faultline, are outlined in the following pages.

Resurgence of Québec Nationalism

After the Second World War, Québec broke with its past. A rise of Québec nationalism had begun much earlier but gained political momentum during the **Quiet Revolution**. This development was the result of four major events. The most important was the resurgence of ethnic nationalism; that is, a pride in being a Québécois. The second was Québec's joining the urban/industrial world of North America and the subsequent expansion in the size of its industrial labour force and business class. The third was the removal of the old élite. This reform movement was profoundly anticlerical in its opposition to the entrenched role of the Church in Québec society, particularly the Church's control over education. In many ways, this reform was based on the aspirations of the working and middle classes in the new Québec economy. The fourth was the state's aggressive role in the province's affairs.

With the election of Jean Lesage's Liberal government in 1960, the province moved forcefully in a new direction. It created a more powerful civil service that allowed francophones access to middle and senior positions that were often denied them in the private sector of the Québec economy, which was controlled by English-speaking Quebeckers and American companies. It nationalized the province's electric system, thereby creating the industrial giant known as Hydro-Québec, now a powerful symbol of Québec's revitalized economy and society. In turn, Hydro-Québec built a number of huge energy projects that demonstrated Québec's industrial strength. By 1968, this Crown corporation had constructed one of the largest dams in the world on the Manicouagan River. Called Manic 5, this dam demonstrated Hydro-Québec's engineering and construction capabilities. To Quebeckers, Hydro-Québec was a symbol of Québec's economic liberation from the years of suffocation associated with Maurice Duplessis and his Union Nationale government, which had been closely tied to big businesses owned by English-speaking Canadians and Americans. Clearly, Lesage's political goal of becoming 'maîtres chez nous' (masters in our own house) had materialized with the success of Hydro-Québec, thus sparking a growth in Québec nationalism. Québec's desire for more autonomy in its own affairs intensified with increased confidence. In short, a new society had arisen in Québec, a society that wanted to chart its destiny. Charles Taylor (1993:4) summed up this new feeling as 'a French Canada which, after a couple of centuries of enforced incubation [under London and then Ottawa], was ready to take control once more of its history.' The political question Taylor raised is a simple one: Would this 'control' take place within the framework of Canada's political system or outside it?

Separatism

Separatism grew out of the Quiet Revolution. It is a form of ethnic nationalism that is popular with francophones but unpopular with anglophones and allophones (those whose first language is neither French nor English). In 1967, French President Charles de Gaulle visited Québec and ignited the forces of French-Canadian nationalism with his now famous words, 'Vive le Québec. Vive le Québec libre.' From that moment on, separatism gained support and took on a political form. By the time of the first referendum, the separatists formed a substantial minority within Québec's population, with perhaps as many as 20 per cent dedicated separatists and another 20 per cent strongly dissatisfied with their place within Canada.

Separatism has had two distinct branches— the Front de libération du Québec (FLQ) and the Parti Québécois (PQ). The FLQ was a small fringe group within the separatist movement. It sought political change through revolutionary means, including bombing, kidnapping, and murder. However, the vast majority of separatists sought change through democratic means. The PQ was committed to a democratic solution by means of a referendum followed by negotiations with the rest of Canada. Referendums, the process of referring a political question to the electorate for a direct decision by general vote, are notoriously tricky political instruments, but Québec Premier René Lévesque offered Quebeckers what he thought was a clear choice—the unpopular status quo or a bold new beginning under sovereignty association. **Sovereignty association** would not mean political separation but a new economic association with the rest of Canada.

Prior to the referendum, the PQ vigorously tackled challenging economic problems and critical cultural issues, all of which had three purposes:

- to accelerate the modernization processes that began with the Quiet Revolution
- to promote the Québécois culture
- to demonstrate that a PQ government could run the affairs of an independent state

The provincial government was involved in the marketplace, often through Crown corporations and government assistance for francophone business operations. The provincial government also promoted Québécois culture in a variety of ways. Under the Liberal government of Robert Bourassa, the French language had been declared the sole official language in 1974, but the PQ government went much further with Bill 101 in 1977.[8] This bill made it necessary for most Quebeckers, regardless of background or preference, to be educated in French-language schools, and was a key measure in ensuring the supremacy of the French language in the province. Among its many goals, Bill 101 was designed to ensure that the children of new immigrants went to French schools and thus guaranteed that the French-speaking population of Québec would continue to grow.

In 1980, Québec voters rejected sovereignty association in a referendum. Almost 60 per cent voted to remain in Canada, which suggests that just over half of the francophone voters sided with the *Non* side, along with almost all the English-speaking residents. The rest of Canada responded with a collective sigh of relief, but separatism was far from dead. Several political events renewed separatist sentiment. One was the Constitution Act of 1982, which patriated the Constitution and gave Canadians the Charter of Rights and Freedoms. The Trudeau government accomplished this political feat at the cost of poisoning relations with Québec City by including the Charter of Rights and Freedoms in the Constitution. The Charter curtailed the power of the Québec government, and the Constitution was patriated without the approval of the Québec government. There were also the failed attempts of Brian Mulroney's Conservative government to achieve provincial unanimity for constitutional reform. The first attempt was the Meech Lake Accord, a package of constitutional revisions incorporating Québec's 'minimum' demands for political reform. This Accord was agreed to in principal by Ottawa and the ten provinces in 1987, but two provinces, Manitoba and Newfoundland, failed to pass it within the required three years. Quebeckers felt humiliated and rejected by the rest of Canada. In 1991, the Mulroney government attempted a second round of constitutional negotiations called the Charlottetown Accord, which would give Québec distinct society status, the provinces the right to veto, Aboriginal peoples self-government, and the nation reform of the Supreme Court and the Senate. In 1992, the Charlottetown Accord was roundly rejected in a national referendum and only narrowly approved in four provinces (but not Québec).

These three political misadventures revived the spirits of the separatists, led by Jacques Parizeau. His party, the PQ, returned to power in 1994, promising a referendum on sovereignty. While the referendum question referred to a new partnership, Parizeau believed that such an arrangement was impossible and saw only one solution—an independent Québec. In the previous year, the Bloc Québécois (a new federal party representing Québec interests), led by Lucien Bouchard (a former cabinet minister in the Mulroney government), formed the official opposition party in the House of Commons. The Québec public expressed their dissatisfaction with Ottawa by rejecting traditional political parties. This left Québec federalists in a vulnerable position where they grew steadily weaker and more disorganized. The federal Liberals could offer little help. In fact, Jean Chrétien, himself a Quebecker, was extremely disliked by many in Québec for a variety of reasons, including his role in the patriation of the Constitution in 1982. He had become a 'tête de turc' (a scapegoat for federal policies), a symbol of those Quebeckers who, as federal ministers, put Québec 'in its place'.

The results of the 1995 referendum vote in Québec were extremely close. 'No—by a Whisker' screamed the headlines of *The Globe and Mail* on the morning after the referendum of 30 October 1995. Québec came within 40,000 votes of approving the separatist dream of becoming an independent state (Vignette 3.7).

Moving Forward

The 1995 referendum was a low point in French/English relations, and its after-effects

Vignette 3.7

No—by a Whisker
The Results of the 30 October 1995 Referendum

The Question: 'Do you agree that Québec should become sovereign, after having made a formal offer to Canada for a new Economic and Political Partnership within the scope of the Bill respecting the future of Québec and of the agreement signed on June 12, 1995?'

The Answer (at 10:30 p.m. Eastern Time, 21,907 of 22,427 polls):

	Number	Per cent
No	2,294,162	49.5
Yes	2,254,496	48.7
Rejected	83,340	1.8
Total	4,631,998	100.0

Source: *The Globe and Mail*, 31 October 1995:A1.

were many and varied. English Canada, dazed by the outcome, attempted to respond. Ottawa reacted almost immediately after the October referendum by passing a unilateral declaration that recognized Québec as a distinct society. The leader of the separatist forces, Jacques Parizeau, had shocked all Quebeckers on the night of the referendum when he blamed the *Oui* side's loss on 'money and the ethnic vote'. In an electrifying moment, the dark side of ethnic nationalism had been revealed. Other centrifugal forces were released too, including the partitionists, who argued that 'if Canada is divisible, so is Québec'. The Cree in northern Québec threatened secession, and partitionists pressured dozens of municipalities around Montréal and Hull to declare their allegiance to Canada.

By 1996, the federal government had decided to take a hard line with Québec, which included having the Supreme Court of Canada determine the conditions of separation. In the same year, the premiers attempted to address the unity issue. In a much more conciliatory manner, they announced the Calgary Declaration: 'the unique character of Québec society with its French-speaking majority, its culture and its tradition of civil law is fundamental to the well-being of Canada'. In a typically Canadian decision, the premiers added to their Declaration that 'any power conferred to one province in the future must be available to all'. This Declaration was the third attempt at reconciliation with Québec since the patriation of the Constitution in 1982.[9] The next step was for each provincial government to pass the appropriate legislation, giving this declaration legal status. By

July 1998, all provinces (except Québec) and territories had passed this resolution in their legislatures.

Changes are taking place in Québec too. Like other Canadians, Quebeckers have grown weary of referendums, political bickering, and the resulting unsettling effect on the national and Québec economies. After a decade of weak economic performance, Québec and the rest of Canada are mainly concerned with economic matters, particularly high levels of unemployment and insufficient funds for education and health. With Jean Charest as the leader of the Québec government opposition, the future of federalism in Québec may brighten: it may be possible to redirect Québec's energies to seeking a political solution within Canada. Jean Charest may take a page from history by using the theme of George-Étienne Cartier, 'Avant tout, soyons Canadiens' (Before all else, let's be Canadiens).

Towards the Future

History and geography explain the basis of the French/English relationship. This dynamic relationship goes to the heart of the nation. Geography compels Canadians to recognize that Québec represents a distinct region of Canada in which a different language and culture dominate. History teaches Canadians that compromise leads to national unity, while conflicts drive a wedge between the two founding peoples of Canada.

Canada's history reflects over 200 years of French/English relations. Through conflicts and compromises, these innumerable interactions, both large and small, have shaped the essential components of the national character, namely, the capacity and willingness to find solutions to complex questions. Both Québec and the rest of Canada are different places and those changes have shaped the French/English faultline and the attempts at compromise in this relationship.

On the surface, reconciliation seems an impossible task, but political realities demand some form of compromise or at least a willingness to search for a solution. Perhaps Paul Villeneuve (1993:104) was correct when he observed that 'for Canada, survival lies in the travelling toward an identity'. For most Canadians, this implies a recognition of the deep divide between French and English Canada but also an acceptance of the importance of each vision. Canadians have also learned that the political gains achieved by having the dominant English-Canadian society impose its will over the minority French-Canadian society are short-term and will eventually weaken national unity.

Summary

Canada is both an old and a young country. The first people arrived in Canada about 30,000 years ago. Much later, Europeans reached its Atlantic shores and England and France established colonies. In 1760, Britain gained control of New France, thus ending France's dream of an empire in the New World. As a result French Canadians' world changed. Their future was uncertain under British rule. The geographic reality of British North America prevailed, however, and in 1774, the Québec Act granted the French-speaking population its basic rights. Both the Act of Union in 1841 and the British North America Act of 1867 recognized Québec's French heritage.

But the real story of Canada begins in 1867. From the early days of Confederation to 1949, Canada grew from a small country of four provinces to the second largest political state in the world. Its territorial expansion only ceased in 1949 when Newfoundland joined Canada. Canada's vast territorial extent is divided into a series of regions. While envied by most peoples of the world, Canada's vast size, physical structure, and human geography have often led to regional tensions, divides, or faultlines.

Throughout its history, Canada has struggled to overcome its difficult geography in order to bind the country together and balance regional differences. This struggle is one element that shapes the Canadian identity. The federal government has a difficult role to play in this struggle, and disagreements with various regions and groups of people are common. Occasionally, tensions turn into bitter and ongoing disputes. These disputes, or faultlines, often take three forms: (1) Aboriginal/non-Aboriginal tensions, (2) centralists/decentralist tensions, and (3) French/English tensions. Of these three forces, the French/English divide has the most serious implications for Canada's future as a nation. If the next Québec referendum points conclusively to sovereignty, a reorganization of Canada's territorial boundaries becomes a real possibility. An important element of Canada's identity is the willingness to compromise, which is the only way the country can make any progress towards resolving the most divisive issue facing Canada: French/English differences over the nature of Confederation.

Notes

1. Although national, provincial, territorial, and municipal governments exist in Canada, only the federal and provincial governments have powers that no other level of governments can usurp.

2. A group of Irish Americans, known as Fenians, was struggling for Irish independence. They believed that attacking British possessions in North America would advance the cause of a free Ireland. Between 1866 and 1870, the Fenians launched several raids across the border into Canada. The United States did not encourage these raids and eventually forced the Fenians to disband. By the end of the American Civil War, Anglo-American relations again were strained because of Britain's tacit support for the Confederacy in the American Civil War. For that reason, the United States allowed the Reciprocity Treaty to lapse in 1864. This treaty, a free trade agreement between British North America and the United States, began in 1854; the subsequent ten years were prosperous ones for British North America. Its loss forced the Province of Canada to seek an alternative economic union with the other British colonies in North America.

3. The Manitoba Act of 1870 recognized the legal status of farms and other lands occupied by the

Métis as 'fee simple' private property. As well, the act provided that 1.4 million acres (566,580 ha) be reserved for the children of Métis. The land was allocated to these Métis in 240-acre (97-ha) parcels, plus 160 acres (65 ha) in 'scrip' for each adult head of a family. These lands were distributed after 1875, but much of the 'scrip' land was sold and then occupied by non-Métis. For more on this subject, see Tough (1996:Chapter 6).

4.　Pontiac, the Ottawa chief in the Ohio Valley, led a successful uprising against the British in 1763. By capturing the forts in the Ohio Territory, he exposed Britain's precarious hold on this region, which they had just obtained from the French. However, Pontiac and his followers could not hold these forts against the British because they could no longer obtain ammunition and muskets from the French. Pontiac concluded that his best move would be to make peace with Britain. Britain also came to that same conclusion, though for other reasons. Without the help of Pontiac and the other chiefs in this region, Britain would lose control over these lands. Britain therefore had to form an alliance with them. With that objective in mind, George III announced an important concession to these Indians in the Royal Proclamation of 1763, namely, that the king recognized these Indians as valued allies and that the land they used to hunt and trap was 'Indian land' within the British Empire.

5.　In 1215, King John of England was forced to sign the Magna Carta. In this charter, he promised to stop interfering with the Church and the law and to consult regularly with the country's leaders before collecting new taxes.

6.　Louis Riel was the leading Métis leader in the late-nineteenth century. This controversial figure is considered both a Father of Confederation and a traitor to the country. Riel, who was born in the Red River Colony in 1844, studied for the priest-

hood at the Collège de Montréal. The founder of Manitoba and the central figure in both the Red River Rebellion (1869–70) and the North-West Rebellion (1885), he was captured shortly after the Battle of Batoche, where the Métis forces were defeated. After a trial in Regina, the jury found Riel guilty, but recommended clemency. Appeals were made to Manitoba's Court of Queen's Bench and to the Judicial Committee of the Privy Council. Both appeals were dismissed. A final appeal went to the federal cabinet, but the government of John A. Macdonald wanted Riel executed. Riel was hanged in Regina on 16 November 1885. His body was interred in the cemetery at the Cathedral of St Boniface in Manitoba.

Riel's execution has had a lasting effect on Canada. In Québec, French Canadians felt betrayed by the Conservative government and federalism. Riel's execution was proof for Canada's French-speaking population that they could not count on the federal government to look after French-Canadian interests. It was also a blow against a francophone presence in the West. In Ontario the hanging of Louis Riel satisfied the anti-Catholic and anti-French majority. For the Orange Order (the Protestant fraternal society that blamed Riel for the death of one of their members, Thomas Scott, who was executed by a Métis firing squad during the Red River Rebellion), Riel's execution was long overdue. In the West, Riel's hanging resulted in the marginalization of both the Métis and Indian tribes, especially those who participated in the uprising.

7.　A recent example is the political fallout from the Québec referendum. Prime Minister Chrétien sought to fulfil his verbal promises made in the closing days before the vote on the 1995 referendum. In a House of Commons resolution, the federal government proposed three concessions to Québec: (1) a veto over constitutional changes, (2) recognition of Québec's distinct society status, and (3) devolution of federal powers to Québec. In the case of the veto, Ottawa was prepared to 'lend' its Constitution veto to Québec, Ontario, Atlantic Canada, and the four western provinces. Not only was the federal government committing itself to seeking permission from the four regions before putting its stamp of approval on any constitutional change, it was also recognizing that Canada consisted of four major regions. The premiers of Alberta and British Columbia reacted negatively to that concept of regionalism. British Columbians in particular saw this arrangement as another example of Ottawa's failure to recognize the West Coast as a 'distinct and powerful' part of Canada. The federal government retreated from this issue and quickly amended its resolution to extend the veto to British Columbia. In December 1995, this resolution passed in both the House of Commons and the Senate. It then became the law of the land that Canada consists of five major regions!

8.   In 1974, the Liberal government passed Bill 22 (Loi sur la langue officielle), which made French the language of government and the workplace. English was no longer an official language in Québec. In 1977, the Parti Québecois government introduced a much stronger language measure in the form of Bill 101 (Charte de la langue française). This legislation eliminated English as one of the official languages of Québec, and required the children of all newcomers to Québec to be educated in French. Four years later, Bill 178 required all commercial signs to use only French. The French language has made modest gains outside of Québec.

In 1969, New Brunswick passed an Official Languages Act, which gave equal status, rights, and privileges to English and French, and Ottawa approved the Official Languages Act, which declared the equal status of English and French in Parliament and in the Canadian public service.

9.   Separatists argue that Québec does not have enough powers; that is, Québec is subordinate to Ottawa. For separatists, the solution lies in independence, whether achieved through Parizeau's 'chicken-plucking' strategy or Bouchard's 'winning' referendum strategy. A federalist counterargument is that, as a member of a federation, Québec has many 'exclusive' powers, such as power over language and education. However, circumstances may force Ottawa to make a decision that adversely affects some provinces while favouring others. In 1982, the patriation of the British North America Act as Canada's Constitution was such a decision. The Constitution differs from the British North America Act in several ways, but without a doubt the most important difference is the Charter of Rights and Freedoms. These rights and freedoms strengthen the rights of individuals and weaken collective rights. Prime Minister Trudeau, who conceived of society as an agglomeration of individuals (not collectivities), whose rights accrued to them as individuals, saw the Charter as protecting individuals from governments that try to suppress individual rights. Then too there is the Supreme Court of Canada's new role, which can challenge the authority of provincial governments.

Key Terms

Aboriginal peoples
    All Canadians whose ancestors lived in Canada before the arrival of Europeans; includes Indian, Métis, and Inuit.

Château Clique
    The political élite of Lower Canada; composed of an alliance of officials and merchants who had considerable political influence with the

British-appointed governor; similar to the Family Compact in Upper Canada.

comprehensive land-claim agreements

Agreements based on territory claimed by Aboriginal peoples that were never ceded or surrendered by treaty. Such agreements extinguish the Aboriginal land claim to vast areas in exchange for a relatively small amount of land, capital, and the organizational structure to manage their lands and capital.

culture areas

A region within which the population has a common set of attitudes, economic and social practices, and values.

Family Compact

A group of officials who dominated senior bureaucratic positions, the executive and legislative councils, and the judiciary in Upper Canada.

Inuit

People who are descended from the Thule. The Thule migrated into Canada's Arctic from Asia about 1,000 years ago. The Inuit do not fall under the Indian Act, but are identified as an Aboriginal people under the 1982 Constitution.

Loyalists

Colonists who supported the British during the American Revolution. About 40,000 American colonists who were loyal to Britain resettled in Canada, especially in Nova Scotia and Québec.

Métis

People who have a mixed biological and cultural heritage, usually either French-Indian or English/Scottish-Indian. This 'mixing' between Indians and Europeans took place during the fur trade and continues today. Originally the term was more narrowly applied to French-Indian people who settled in the Red River area and who developed a distinct hunting economy and society based on the French language and the Roman Catholic religion.

non-status Indians

Aboriginal peoples who are not registered as Indians under the Indian Act.

Quiet Revolution

A period in Québec during the Liberal government of Jean Lesage (1960 to 1966) that was characterized by social, economic, and educational reforms, and by the rebirth of pride and self-confidence among the French-speaking members of Québec society, which led to a resurgence of francophone ethnic nationalism. During this time, the separatist movement gained strength.

reserve

Under the Indian Act, reserves are defined as lands 'held by her Majesty for the use and benefit of the bands for which they were set apart; and subject to this Act and to the terms of any treaty or surrender'.

sovereignty association

A concept designed by the Parti Québecois under the Lévesque government and employed in the 1980 referendum. This concept is based on the vision of Canada as consisting of two 'equal' peoples. Sovereignty association calls for a partnership with Canada based on an economic association.

specific land-claim agreements

Agreements designed to rectify a shortfall in the amount of land received by a band under a numbered treaty.

status Indians

Aboriginal peoples who are registered as Indians under the Indian Act.

treaty Indians

Aboriginal peoples who are descendants of Indians who signed a numbered treaty and who benefit from the rights described in each treaty. All treaty Indians are status Indians, but not all status Indians are treaty Indians.

References

Bibliography

Anderson, Robert B., and Robert M. Bone. 1995. 'First Nations Economic Development: A Contingency Perspective'. *The Canadian Geographer* 39, no. 2:120–30.

Bumsted, J.M. 1998. *A History of the Canadian Peoples*. Toronto: Oxford University Press.

Cook, R. 1993. *The Voyages of Jacques Cartier*. Toronto: University of Toronto Press.

Globe and Mail. 1995. 'No—by a Whisker' (31 October):A1.

McVey, Wayne W., and W.E. Kalbach. 1995. *Canadian Population*. Toronto: Nelson Canada.

Marsh, James H., ed. 1988. *The Canadian Encyclopedia*, 2nd edn. Edmonton: Hurtig Publishers.

Nemni, Max. 1994. 'The Case Against Quebec Nationalism'. *The American Review of Canadian Studies* 24, no. 2:171–96.

Simpson, Jeffrey. 1993. *Faultlines: Struggling for a Canadian Vision*. Toronto: HarperCollins.

Smith, P.J. 1982. 'Alberta Since 1945: The Maturing Settlement System'. In *Heartland and Hinterland: A Regional Geography of Canada,* edited by L.D. McCann. Scarborough: Prentice-Hall.

Statistics Canada. 1998. The Daily—1996 Census: Ethnic Origin, Visible Minorities, 17 February 1998 [online database], Ottawa. Searched 16 July 1998; <URL:http://www.statcan.ca/Daily/English/>:21 pp.

Taylor, Charles. 1993. *Reconciling the Solitudes*. Montréal–Kingston: McGill-Queen's University Press.

Tough, Frank. 1996. *As Their Natural Resources Fail: Native Peoples and the Economic History of Northern Manitoba, 1870–1930*. Vancouver: University of British Columbia Press.

Villeneuve, Paul. 1993. 'Allocution présidentielle: L'invention de l'avenir au nord de l'amérique'. *Le géographe canadien* 37, no. 2:98–104.

Further Reading

Coates, Ken. 1992. *Aboriginal Land Claims in Canada: A Regional Perspective*. Toronto: Copp Clark Pitman.

Cook, R. 1969. *Provincial Autonomy, Minority Rights and the Compact Theory, 1867–1921*. Studies of the Royal Commission on Bilingualism and Biculturalism, no. 4. Ottawa: Queen's Printer.

Fremlin, Gerald, ed. 1974. *The National Atlas of Canada*. Ottawa: Macmillan.

Garreau, Joel. 1981. *The Nine Nations of North America*. Boston: Houghton Mifflin.

Gentilcore, R. Louis. 1993. *Historical Atlas of Canada, Volume II: The Land Transformed 1800–1891*. Toronto: University of Toronto Press.

Harris, R. Cole, ed. 1987. *Historical Atlas of Canada, Volume I: From the Beginning to 1800*. Toronto: University of Toronto Press.

———, and John Warkentin. 1991. *Canada Before Confederation: A Study in Historical Geography*. Ottawa: Carleton University Press.

Kerr, Donald, and Deryck W. Holdsworth, eds. 1990. *Historical Atlas of Canada: Volume III:*

Addressing the Twentieth Century 1891–1961. Toronto: University of Toronto Press.

Mitchell, Robert D., and Paul A. Groves. 1987. *North America: The Historical Geography of a Changing Continent*. Totowa: Rowman & Littlefield.

Morton, D. 1983. *A Short History of Canada*. Edmonton: Hurtig.

O'Handley, Kathryn, ed. 1994. *Canadian Parliamentary Guide*. Toronto: Globe and Mail Publishing.

Pielou, E.C. 1991. *After the Ice Age: The Return of Life to Glaciated North America*. Chicago: University of Chicago Press.

Warkentin, John, ed. 1968. *Canada: A Geographical Interpretation*. Toronto: Methuen.

Wynn, G. 1990. *People, Places, Patterns, Processes: Geographic Perspectives on the Canadian Past*. Toronto: Copp Clark Pitman.

Chapter 4

Overview

With a slow-growing population, Canada must continue to rely on immigration for much of its population increase. As more immigrants come from non-European nations, the human face of Canada is changing, making Canada a pluralistic country where multi-culturalism is supported. Internal forces continue to exert pressure on national unity, particularly on the Aboriginal/non-Aboriginal, centralist/decentralist, and French/English tensions.

Another force having an impact on Canada's peoples is the transformation of Canada's economy from an industrial to a postindustrial one. As the economy increasingly shifts to high-technology industries, the country's economic structure and east-west alignment are changing. Both changes are driven by the liberalization of world trade and by the creation of a continental economy in North America. As a result, the 'old' core/periphery structure is being challenged. The net result is high economic growth in more favoured regions, but at the cost of widening the economic gap between the favoured and less favoured regions.

Objectives

- Describe Canada's current demographic, economic, and social changes.
- Relate these changes to the three major internal social tensions and external economic forces, and to the core/periphery model, the demographic transition theory, and the stages of development concept.
- Argue that the national version of the core/periphery model has been replaced by a continental version.
- Recognize that demographic and economic changes are not occurring equally across Canada's six geographic regions.
- Examine the demographic shifts within Canada's three faultlines.

Canada's Human Face

Introduction

Canada is home to about 30 million people. This population influences the economic and social characteristics of the country's different regions. **Demography** is the statistical study of the size, distribution, and composition of human populations. Using demography, we can discern differences in the populations of Canada's regions. For example, the Territorial North has an extremely scattered population; Atlantic Canada has a much higher rate of unemployment than other regions; and Québec has the highest percentage of French-speaking Canadians of all six geographic regions.

In this chapter we first focus our attention on economic and social changes. Next, we examine Canada's population size, density, and distribution, and analyze its demographic features, including population increase, natural growth, age/sex structure, and age dependency. Cultural characteristics of Canadians (as measured by ethnicity, multiculturalism, language, and religion) are then investigated. The importance of Canada's Constitution, with its emphasis on bilingual, pluralistic, and civic nationalism, becomes readily apparent in this discussion. The chapter also explores eco-nomic issues affecting Canada's human geography such as labour force trends, unemployment, and earnings of Canadians. Finally, it looks at current population trends from the perspective of the three faultlines discussed in Chapter 3.

Economic and Social Change

Like other modern industrial countries, Canada has been transformed from an agrarian to an industrial economy. Changes in Canada's industrial society have been especially reflected in its economic structure. By structure, we mean the relative share of primary, secondary, tertiary, and quaternary economic activities in Canada's economy at different times (Table 4.1). In the early-nineteenth century Canada's economy was agrarian, meaning that most people were engaged in primary activities (see Vignette 4.1). As Canada developed an industrial base in the late-nineteenth century, many of its workers were involved in secondary activities. However, this workforce was concentrated in major cities, especially those in the manufacturing belt of Central Canada. The rest of the country was involved either in agriculture or resource development, such as forestry or mining. This created a geographic

Table 4.1

Types of Economic Structure

Economic Activity	Characteristics
Primary	Economic activities that are concerned with natural resources of any kind, such as fishing, forestry, mining, and trapping.
Secondary	Economic activities that process, transform, fabricate, or assemble the raw materials derived from primary activities, or that reassemble, refinish, or package manufactured goods. Examples would include an automobile factory, a meat-packing plant, and a pulp mill.
Tertiary	Economic activities that involve the sale and exchange of goods and services, including professional services, retail sales, and education.
Quaternary	Economic activities that deal with the handling and processing of knowledge and information that lead to decision making by companies and governments. Such activities are often found in research and development units.

Vignette 4.1

Stages of Development

Technological change and economic development have led to major changes in the economic structure of industrial states. These changes are sometimes called stages of development. Often modern industrial nations have gone through three stages of development: preindustrial, industrial, and postindustrial. In the **preindustrial stage, primary activities**, led by agriculture, dominate the economy and occupy most of the labour force. The **industrial stage** occurs when manufacturing activities surpass those of the primary sector of the economy. In this stage, most of the workers are involved in **secondary activities** (such as processing raw materials and manufacturing goods) and **tertiary activities** (such as providing a variety of services to the public). The last stage is the **postindustrial stage**, in which the economy is dominated by tertiary and **quaternary activities**. Table 4.1 provides a fuller description of the types of economic structures.

division in Canada's economic structure, which took on the appearance of an industrial core and a resource hinterland. After the Second World War, the economic structure of Canada, like other modern industrial states, changed again. This time it shifted from a manufacturing-dominated economy to a service-oriented one, marking Canada's entrance into the postindustrial world. This phenomenon is also evident in each region of Canada, though the proportion of manufacturing activity remains concentrated in Central Canada.

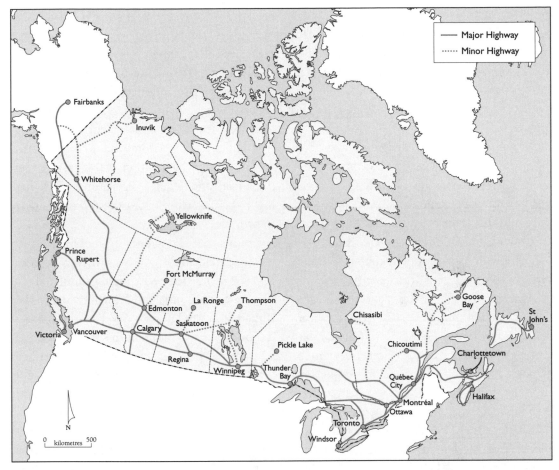

Figure 4.1 Canada's highway system. The highway system reveals three characteristics: its east-west alignment of main highways, its north-south alignment of minor highways, and the paucity of highways in the Territorial North and the Canadian Shield.

Other important changes, particularly related to the labour force, included an increase in the number of female workers and high unemployment rates.

What has made Canada's transformation different from that of other countries is its geography (vast size, varied physical geography, and a highly concentrated population) and its proximity to the United States, the most powerful nation in the world. For instance, Canada's vast size required enormous investment in its transportation system to link the various parts of the country. In 1885, the construction of the Canadian Pacific Railway was completed, but it had exhausted the resources of the private company, forcing it to rely heavily on the federal government for financial support. The Trans-Canada Highway, the next attempt at building a national transportation system, was completed in 1962.

From that point onwards, the highway system grew. By 1979, the Depster Highway had spread northward to the Arctic Ocean. Today, the results of this investment in Canada's transportation system can be seen in the network of highways that crisscross the country (Figure 4.1).

Having the US as a neighbour and trading partner also affected Canada's economic transformation, resulting in close economic ties between the two countries. Before the Auto Pact and the North American Free Trade Agreement, the US's impact on Canada's economy took the form of American branch plants. Both these agreements (discussed in detail in this chapter) called for the economic integration of North America. While there have been benefits from such close ties, Canada has become heavily dependent on the performance of the American economy. Glen Norcliffe, one of Canada's leading industrial geographers, predicts problems stemming from such a dependency: 'Canadian trade in goods and services under an export-led continentalist option promises chronic deficits, progressive subordination to US interests, and increasing pressure for economic and political integration' (Norcliffe 1996:44).

Social changes have also transformed Canadian society. Two significant events made Canada a pluralistic society. One was the change in Canada's immigration policy, which changed immigrant selection to one based on merit instead of race (preferred nationalities). The second was the replacement of the British North America Act with the Canadian Constitution in 1982. The Constitution, with its entrenched Charter of Rights and Freedoms, declared a form of citizenship based on the individual rights of all Canadian citizens regardless of sex, religion, race, ethnic origin, cultural background, or place of birth. The Constitution also reflected the principle that collective and individual rights would be protected by Canada's legal system. This was a substantial change from the British North America Act, which assigned this task to the provincial and federal legislatures. Together these changes represented the birth of a new Canada.

Other social changes have taken place within Canadian society, most driven by economic factors that were often stimulated by the new economic relationship with the United States and the rest of the world. Urbanization is one indication of this transformation into an industrial society. As Bourne and Ley point out, cities 'serve as both mirrors and moulders of a nation's society, culture, and politics' (Bourne and Ley 1993:3). The measure of urbanization is revealed in the shift of Canada's population from rural to urban areas that began long ago, but accelerated after the Second World War. It was slower to start in Québec, but once it started, there was no turning back. This demographic change had profound cultural and economic implications, especially in Québec, where it provided the foundation for the Quiet Revolution; that is, there was a shift from a conservative rural population dominated by the Catholic clergy to a liberal urban population that looked to the provincial government for its economic and social well-being. Canada is now a highly urbanized society with three cities—Toronto, Montréal, and Vancouver—that have popula-

As Canada's largest metropolitan area, Toronto exerts an economic influence that extends across the country. (*Corel Photos*)

Montréal, located at the confluence of the St Lawrence and Ottawa rivers in southwestern Québec, is a major industrial, commercial, and financial centre. (*Corel Photos*)

Vancouver, Canada's third-largest city, is bordered by Burrard Inlet, the Strait of Georgia, and the Fraser River. BC Place and the downtown core are in the foreground, while Stanley Park, West Vancouver, and North Vancouver appear in the distance. (*Corel Photos*)

tions exceeding 1 million. The concentration of immigrants in Canada's major cities, particularly the three major ones, has altered these cities' social character. The impact of immigrants, especially visible minorities, is much less evident outside these core cities.

Change, both economic and social, continues to affect Canada today. Indeed, one overriding characteristic of all modern industrial societies is rapid change. Technological advances often trigger such changes. Computer technology, for instance, links the world community together instantaneously, thereby overcoming the obstacle of distance. This innovative technology is only one factor in the recent modification of Canada's economic structure. Another factor is the openness of Canada's economy to external forces as a result of international agreements such as the North American Free Trade Agreement. (For a more detailed discussion of Canada's open economy, see Norcliffe 1996:21–47.) However, with change often comes challenge. If high unemployment, for instance, becomes a permanent feature of Canada's postindustrial society, it will have a profound impact, especially in the weaker hinterland regions of the country. Nevertheless, change, as reflected in social and economic characteristics, will remain an important feature of Canadian society.

Canada's Population

Population is a significant factor of social and economic change. Such changes are often reflected in, and affected by the country's population. For instance, some of the significant population trends being experienced today in Canada include a decline in the rate of natural increase, the aging of the population, and the high fertility rate among Aboriginal peoples. Three measures of Canada's population—size, density, and distribution—are outlined below.

Population Size

Since Confederation, Canada's population size has increased from 3.4 million to 30 million. What factors account for this population growth? Three factors predominate: natural increase, population gained from territorial expansion, and immigration. This remarkable population growth has transformed Canada into a medium-size country. For instance, approximately three-quarters of the nations of the world have smaller populations than Canada. Canadians, however, do not measure their population size against all the countries of the world but against the United States, which is far more populous than Canada. Canadians also take into consideration the geographic size of their country. As the second largest country in the world by geographic area, Canada's population density is one of the lowest in the world. Based on these comparisons, Canada's population seems small.

Population Density

Population density is determined by dividing the total number of people by the total land area. Canada has a population density of three people per square kilometre, which means that Canada has a low population density (but

Geographic Region	Population (000)	Land Area (000 km²)	Population Density
Territorial North	95.2	3778.2	0.01
Atlantic Canada	2,333.8	501.7	1.9
Western Canada	4,801.0	1756.0	2.7
British Columbia	3,724.5	892.7	4.2
Québec	7,138.8	1357.8	5.3
Ontario	10,753.6	916.7	11.7
Canada	28,846.9	9203.1	3.1

Table 4.2
Population Density, 1996

Sources: Statistics Canada, 1997c:11; 1996b:22.

not the lowest in the world). Australia has a slightly lower population density with two people per square kilometre. All other countries of the world have higher population densities. The United States, for example, has twenty-nine people per square kilometre, while Bangladesh, the most densely populated country in the world, has 900 people per square kilometre.

The explanation for these variations is simple. Land varies greatly in its capacity to support human settlement. Most land in Canada lies beyond the northern limits of agriculture. For example, land in the Territorial North has a low capacity to support human life. Wildlife constitutes most of the food. The region's population density is only 0.01 people per square kilometre (Table 4.2). On the other hand, Ontario has a much more richly endowed natural environment. Consequently, its population density is the highest in Canada.

Population Distribution

Population distribution is the dispersal of people within a geographic area. Canada's population is extremely unevenly distributed across the country. In fact, few nations have so much of their population concentrated in one area, while the rest of the country is almost vacant. A pattern of settlement at its southern border characterizes Canada's population distribution. In 1967, an American geography text described Canada's population as if it were 'drawn by a magnet toward the giant-neighbor on the south, for they [Canada's inhabitants] are strikingly concentrated along the United States border (Trewartha, Robin-

son, and Hammond 1967:542).' Others view this same distribution as consisting of a population core and a population hinterland. The population core is sometimes described as Canada's national **ecumene** (inhabited area).

In the 1990s, the greatest population increases occurred in Canada's national ecumene, particularly in its cities. Most of the increase was due to rural-to-urban migration, while some was due to people leaving have-not provinces for large urban centres in other, more prosperous provinces. But most of the increase has been attributed to the influx of immigrants to Toronto, Montréal, and Vancouver.

Population by Geographic Region

Among the six geographic regions of Canada, there are notable differences with respect to population patterns. For instance, the country's uneven population distribution is illustrated in the variation between each region's percentage of Canada's population (Table 4.3). The smallest population is in the Territorial North. In 1996, the Territorial North had fewer people (only 0.3 per cent of Canada's population) than Peterborough, Ontario. At the other extreme, Ontario (with 37 per cent of the national population) has the largest population. Québec takes second place, with 25 per cent of Canada's population. Together, Ontario and Québec command a dominating demographic position within Canada. In combination, the remaining regions have only 38 per cent of Canada's population.

Table 4.3

Population by Geographic Region, 1996

Geographic Region	Population (per cent)	Population (000)
Territorial North	0.3	95.2
Atlantic Canada	8.1	2,333.8
British Columbia	12.9	3,724.5
Western Canda	16.6	4,801.0
Québec	24.8	7,138.8
Ontario	37.3	10,753.6
Canada	100.0	28,846.8

Source: Adapted from Statistics Canada, 1997c:11.

Table 4.4

Population Zones, 1996

Zone	Population Size (millions)	Percentage of Total Population	Major City	Population of Major City
1	17.0	59	Toronto	4,263,757
2	11.5	40	Vancouver	1,831,665
3	0.3	1	Fort McMurray	35,213
4	<0.1	<1	Labrador City	8,455

Source: Based on Statistics Canada, 1997c:tables 4 and 10.

Population Zones

Population zones provide another picture of the spatial patterns of population in Canada (Figure 4.2). Four population zones reinforce the image of a highly concentrated population surrounded by a more dispersed population (Table 4.4).

The first zone is the principal population core within the national ecumene, a densely populated area in the Great Lakes–St Lawrence Lowlands. It has about 17 million people (nearly 60 per cent of the national population) and almost three-quarters of Canada's metropolitan cities. For example, Toronto, Montréal,

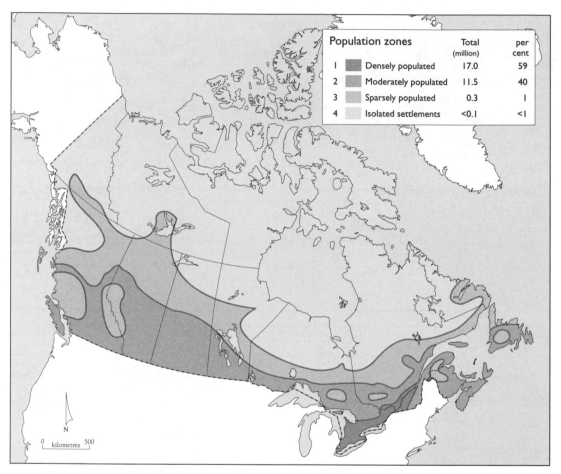

Population zones		Total (million)	per cent
1	Densely populated	17.0	59
2	Moderately populated	11.5	40
3	Sparsely populated	0.3	1
4	Isolated settlements	<0.1	<1

Figure 4.2 Canada's population zones. Canada's population is heavily concentrated in southern Ontario and southern Québec, where a favourable physical geography and an advantageous geographic location have resulted in a dense population. A secondary belt of population spans a southern strip of Canada. Together the densely and moderately populated zones account for 99 per cent of Canada's population.

Ottawa–Hull, Québec City, Hamilton, and London are located in this population core.

The second zone forms a secondary population core within the national ecumene. It stretches across southern Canada, its outer limits corresponding to those areas that are associated with agriculture. About 11.5 million Canadians (over one-third of the country's total population) live in this moderately populated zone. Approximately one-quarter of the major cities are within this zone, including Vancouver, Edmonton, Calgary, and Winnipeg.

The third zone corresponds with, and contains the geographic limits of the commercial forest area. Approximately 1 per cent of all Canadians (about 300,000) live in this zone.

None of Canada's major cities are in this zone. The populations of the largest cities range between 10,000 and 35,000. Most of these cities are administrative and service centres or resource towns. Whitehorse and Yellowknife, as the capital cities of Yukon and the Northwest Territories, are administrative centres. Since they also provide most of the service functions in the Territorial North, these two cities are also **regional service centres**. **Resource towns** are centres that have a single industry, such as mining. Most resource towns are small, with the exception of Fort McMurray. It has the largest population of all cities in the third zone. In 1996, Fort McMurray had a population exceeding 35,000.

The fourth zone is the largest population zone in *area* but, with less than 100,000 people, it is by far the smallest zone by *population*. It lies in the northern boreal forest and the tundra lands of the Arctic. From a resource perspective, it is the least productive area in

A copper smelter at Flin Flon, Manitoba, a resource town in the Canadian Shield. (*Courtesy R.M. Bone*)

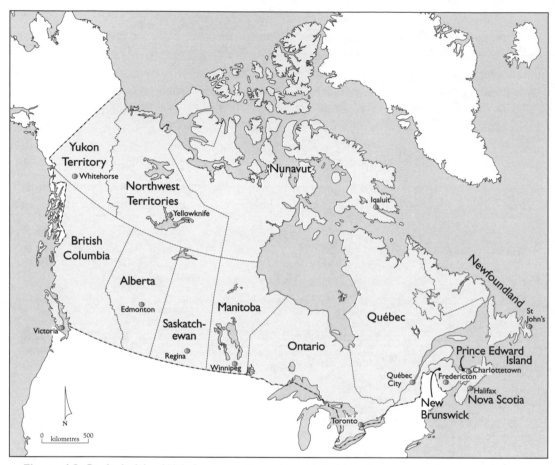

Figure 4.3 Capital cities. With four exceptions, the capital cities of the ten provinces and three territories are the largest urban centres in each political unit. The largest city in New Brunswick is Saint John; in Québec, it is Montréal; in Saskatchewan, it is Saskatoon; and in British Columbia it is Vancouver.

the country. Most commercial activities centre on exploitation of non-renewable resources, including mineral bodies and petroleum deposits. Unlike in the other zones in Canada, Aboriginal peoples are the majority in this zone. Many continue to rely on wildlife for food. Urban centres are small, most with populations under 5,000. Single-industry resource towns are the largest urban centres. The iron-mining town of Labrador City has

the largest population with just over 8,000 inhabitants.

Urban Population

As defined in the census of 1996, all people who live in either incorporated cities, towns, and villages, or unincorporated places with a population of 1,000 or more and a population density of at least 400 per square kilometre, form part of the **urban population**. By 1996, more than three out of every four Canadians were living in an urban centre with a population of at least 1,000 people. However, this was not always the case. Over the last 100 years, a massive rural-urban migration has taken place. While this migration has slowed since the 1970s, the percentage of Canadians living in urban centres, both large and small, has continued to increase in most regions (Table 4.5).

The process of urbanization has had a profound impact on the geography of population. In 1901, for example, approximately 35 per cent of all Canadians lived in an urban setting (Table 4.5). By 1961, the percentage had doubled to 70 per cent. In the most recent census (1996), urban dwellers constituted close to 80 per cent of the Canadian population. By the turn of the century, it is expected that four out of five Canadians will be living in an urban environment.

The rate of urbanization varies across Canada. The highest level of urban dwellers is in the industrial heartland of Canada. In 1996, Ontario, with 83 per cent of its population living in urban places, led all other geographic regions (Table 4.5). Statistics Canada classifies the largest cities in Canada as census metropolitan areas (CMAs). A **census metropolitan area** is a very large urban area of at least 100,000 people. Adjacent urban and rural areas have a high degree of economic and social integration with a CMA. Toronto, for instance, has a population of over 4 million, which is roughly 14 per cent of Canada's population (Table 4.6). In terms of **population strength**, Toronto is more 'powerful' than

Table 4.5
Percentage of Urban Population by Region,* 1901–1996

Region	1901	1921	1941	1961	1981	1991	1996
Ontario	40.3	58.8	67.5	77.3	81.7	81.8	83.3
British Columbia	46.4	50.9	64.0	72.6	78.0	80.4	82.1
Québec	36.1	51.8	61.2	74.3	77.6	77.6	78.0
Western Canada	19.3	28.7	32.4	57.6	71.4	74.4	74.4
Atlantic Canada	24.5	38.8	44.1	50.1	54.9	54.1	52.8
Canada	34.9	47.4	55.7	70.2	76.2	77.2	77.9

*Newfoundland was not included in Atlantic Canada's figures until 1961.
While comparable statistics are not available for the Territorial North, two observations are possible: (1) prior to the 1950s, few people in this region lived in settlements, and (2) by 1996, only three communities, Whitehorse (20,075), Yellowknife (17,275), and Iqaluit (4,220), had populations exceeding 4,000.

Sources: McVey and Kalbach, 1995:149; Statistics Canada, 1997c:Table 15.

Table 4.6
Population and Percentage Increase
for Census Metropolitcan Areas, 1991–1996

Rank	Census Metropolitan Area	1996	Per cent Increase 1991–6
1	Toronto	4,263,759	9.4
2	Montréal	3,326,510	3.7
3	Vancouver	1,831,665	14.3
4	Ottawa–Hull	1,010,498	7.3
5	Edmonton	882,597	2.6
6	Calgary	821,628	9.0
7	Québec City	671,889	4.1
8	Winnipeg	667,209	1.0
9	Hamilton	624,360	4.1
10	London	398,616	4.5
11	Kitchener	382,940	7.4
12	St Catharines–Niagara	372,406	2.2
13	Halifax	332,518	3.7
14	Victoria	304,287	5.7
15	Windsor	278,685	6.3
16	Oshawa	268,773	11.9
17	Saskatoon	219,056	3.8
18	Regina	193,652	1.0
19	St John's	174,051	1.3
20	Sudbury	160,488	1.8
21	Chicoutimi–Jonquière	160,454	-0.3
22	Sherbrooke	147,384	4.7
23	Trois-Rivières	139,956	2.7
24	Saint John	125,705	-0.1
25	Thunder Bay	125,562	0.5

Source: Adapted from Statistics Canada, 1997c:12.

three geographic regions (the Territorial North with 0.3 per cent of Canada's population, Atlantic Canada with 8.1 per cent, and British Columbia with 12.9 per cent) (see Table 4.3). In addition to Toronto, Ontario has several census metropolitan areas, including Ottawa and Hamilton.

In 1996, British Columbia had the second highest percentage of urban population at 82 per cent (Table 4.5), most of which is in Vancouver. In 1996, nearly 2 million Canadians called this West Coast city home (Table 4.6). Québec is in third place with 78 per cent of its population living in urban areas, particularly Montréal and Québec City. With a population of over 3.3 million, Montréal has most of the province's urban population, while Québec City has a population close to 675,000. Western Canada is close behind Québec, with nearly 75 per cent of its population in urban centres like Edmonton, Calgary, and Winnipeg. Edmonton has a population of over 882,000, Calgary has almost 822,000, and Winnipeg over 667,000. Atlantic Canada, with only 52 per cent of its population living in villages, towns, and cities, has the smallest urban popu-

lation. Its major city, Halifax, has a population of about 330,000.

In this discussion of urban population, the Territorial North region has not been included in the comparison because to do so would be unfair and misleading. The Territorial North has not followed the same settlement patterns as the other geographic regions

The Eastern Townships, a cluster of small urban centres located east of Montréal, are at the heart of Québec's Appalachian subregion. (*Corel Photos*)

An Inuit fisher mends a net. The Aboriginal majority in the Territorial North has shaped a settlement pattern that is unique in Canada. (*Corel Photos*)

has had a greater impact on settlement life than in southern Canada. For example, prior to the 1950s, Aboriginal peoples lived on the land as hunters and trappers. Since they moved from place to place in search of game, Aboriginal peoples did not form permanent settlements. After the 1950s, however, Aboriginal peoples in the north relocated into settlements, marking a change in their lifestyle from nomadic to sedentary. Still, most settlements in the Territorial North have fewer than 1,000 inhabitants. For reasons of geography and culture then, the urban population of the Territorial North has followed a different historical path than the rest of Canada.

Census Metropolitan Areas

The emergence of CMAs across Canada is the latest outcome of urbanization (Table 4.6). CMAs have a minimum population of 100,000. The proportion of Canada's population residing in metropolitan areas has increased from 30.3 per cent in 1931 to 61.9 per cent in 1996. In 1931, there were ten CMAs: Halifax, Hamilton, Montréal, Ottawa, Québec, Saint John, Toronto, Vancouver, Windsor, and Winnipeg. By 1996, there were twenty-five metro-

because of cultural and physical differences. For instance, the Territorial North has a cold environment. Consequently, farming, which is the basis of Canada's rural populations in the other regions, is virtually non-existent in the northern region. Furthermore, much of the Territorial North has a majority of Aboriginal inhabitants and therefore Aboriginal culture

politan centres (Table 4.6). Beyond their demographic importance, metropolitan cities serve as centres for businesses and governments. Toronto is recognized as the corporate and financial headquarters of Canada, along with Montréal, Calgary, and Vancouver.

Since 1951, the populations of the six largest CMAS have increased remarkably. The greatest increase in population occurred in Toronto. During this time, Toronto surpassed Montréal as Canada's largest metropolitan centre. By 1996, Toronto had a population exceeding 4.2 million, while Montréal's was about 3.3 million. Of the six largest metropolitan centres, the greatest rate of increase took place in Vancouver. From 1991 to 1996, Vancouver's increase was a phenomenal 14.3 per cent (Table 4.6).

A common problem facing all CMAS is urban sprawl. As cities grow, they spread laterally, often overtaking and occupying agricultural land. Since land values for urban land are much higher than for farmland, the marketplace encourages urban expansion into the rural countryside. The downside is the loss of scarce productive farmland. Urban sprawl is most acute in the largest urban centres; that is, in Toronto, Montréal, and Vancouver.

Population Change

Population change has three components: births, deaths, and migration. **Population increase** is the sum of natural increase and net migration over a given period. The term **population growth** is used when this increase is expressed as a rate; that is, as a percentage change over time. **Natural increase** is the difference between the **crude birth rate** (CBR) and the **crude death rate** (CDR). CBR is the number of live births per 1,000 people in a given year. CDR is the number of deaths per 1,000 people in a given year. **Net migration** is the difference between in- and out-migration. Canada has a large influx of migrants, but a number of Canadians also leave the country, often for the United States. While reasons for out-migration vary, many Canadians move to the United States because of job opportunities and/or because of a more temperate climate.

Since 1851, Canada has enjoyed continuous population growth (Table 4.7). At first, high rates of natural increase and high levels of immigration propelled this growth. When Canada became an industrial country, its rate of natural increase slowed as parents opted for smaller families. After the Second World War, however, the baby boom occurred. During the 1950s and 1960s, the rate of natural increase rose sharply and the average family size increased, accounting for almost 70 per cent of Canada's population growth. Combined with high levels of immigration, Canada's rate of growth reached record levels when the average annual rate of increase soared over 2 per cent per year during the 1950s (Vignette 4.2). Since Confederation, this rate was exceeded only once—in 1911 when it reached 3 per cent (Table 4.7). Since the baby boom, population growth has slowed. By the 1980s, natural increase made up just over 50 per cent of Canada's population growth, and in the early 1990s, this figure dropped below 50 per cent. Natural increase made up only 45 per cent of Canada's population growth in 1995. That year, Canada's population rose by 376,000,

Table 4.7
Population Increase, 1851–2001

Year	Population (000)	Percentage Change	Average Annual Rate
1851	2,436.3	—	—
1861	3,229.6	32.6	2.9
1871	3,689.3	14.2	1.3
1881	4,324.8	17.2	1.6
1891	4,833.2	11.8	1.1
1901	5,371.3	11.1	1.1
1911	7,206.6	34.2	3.0
1921	8,787.9	21.9	2.0
1931	10,376.8	18.1	1.7
1941	11,506.7	10.9	1.0
1951	14,009.4	21.8	1.7
1956	16,080.8	14.8	2.8
1961	18,238.2	13.4	2.5
1966	20,014.9	9.7	1.9
1971	21,568.3	7.8	1.5
1976	22,992.6	6.6	1.3
1981	24,343.2	5.9	1.2
1986	25,309.3	4.0	0.8
1991	27,296.9	7.9	1.6
1996	28,846.8	5.7	1.1
2001	30,664.0*	6.3*	1.3*

*Statistics Canada estimate based on series 3 projection

Sources: McVey and Kalbach, 1995:42; Statistics Canada, 1997c:11.

Vignette 4.2

The Baby Boom Effect

After Confederation, fertility (also known as the crude birth rate) and mortality (also known as the crude death rate) declined until the late 1940s when the birth rate suddenly increased. Demographers refer to this aberration as the baby boom. By the late 1960s, however, fertility rates decreased sharply. This long-term downward trend is associated with industrialized countries and described in the demographic transition theory. The end of the baby boom is sometimes described as the baby bust.

The baby boom lasted for almost twenty years. During that time of high fertility, there were almost 10 million births, which created a bulge in the age structure of Canadian society that continues to have both economic and social implications. Thus, while this demo-graphic phenomenon was short-lived, it has left its mark on Canadian society (Foot 1996). As consumers of goods and services, baby boomers have had a decided impact on the economy as they move through their life cycle. Companies have geared their products to meet the strong demand for goods and services created by baby boomers. In the early 1950s, the emphasis was on baby products and larger houses. In the 1960s, a similar age-related pressure was exerted on school facilities, creating a demand for more schools and teachers. As the baby boomers approach old age, the demand for health-care services is expected to rise. Companies are already targeting their advertisements at the growing number of retirees from the baby boom generation.

with natural increase accounting for 167,000 and immigration for 209,000 (Table 4.8; Statistics Canada 1996a:Table 7). In the first half of the 1990s, the annual rate of population growth fell to 1.6 per cent, dropping to 1.1 per cent by 1996.

Natural Increase

In the past, most of Canada's population growth was due to natural increase, for most Canadians lived in rural settings where large families were the rule. As most available farm-land became occupied, children of farm parents had no choice but to seek their fortunes elsewhere. Many went to the cities in search of work. Fertility rates were extremely high, while mortality rates were low, allowing for a high rate of natural increase (Table 4.8). The **fertility rate** is the same as the crude birth rate (number of live births in a year per 1,000 people), while the **mortality rate** is the same as the crude death rate (the number of deaths per 1,000 people in a given year). In the 1870s, natural increase exceeded 2 per cent per year. At that rate, Canada's population would double every thirty to thirty-five years. Like many social phenomena, such rapid growth failed to sustain itself because circumstances changed dramatically. As Canada became an industrial country, fertility rates declined. Except for a short time after the Second World War, the fertility rates declined

Table 4.8

Canada's Rate of Natural Increase, 1851–1995

Year	Crude Birth Rate	Crude Death Rate	Natural Increase (per cent)	Natural Increase (000)
1851	45	20	2.5	61
1861	45	20	2.5	81
1871	42	20	2.2	81
1881	40	19	2.1	90
1891	38	18	2.0	97
1901	35	16	1.9	97
1911	32	14	1.8	129
1921	29.3	11.6	1.8	160
1931	23.2	10.2	1.3	138
1941	22.4	10.1	1.2	145
1951	27.2	9.0	1.6	255
1961	26.1	7.7	1.8	335
1971	16.8	7.3	1.0	205
1981	15.2	7.0	0.8	200
1991	14.3	7.0	0.7	207
1995	12.8	7.1	0.5	167

Sources: Adapted from Statistics Canada, 1997d: Table 1B; McVey and Kalbach, 1995:268, 270.

steadily from about forty-five births per 1,000 people per year to approximately thirteen births per 1,000 people. Over the same period, the mortality rate dropped from about twenty deaths per 1,000 people per year to about seven. While there is no single explanation for these changes, medical advances and the development of public health-care systems across the country account for most of the sharp reduc-

tion in the mortality rate. The decline in the fertility rate is more complex. A series of social and economic changes caused parents to opt for smaller families. Among these changes, two stand out: the shift of people from rural areas to towns and cities, and the sharp increase in the number of women in the labour force.

By the 1990s, fertility rates were at an all-time low. The shift from high to low fertility rates in Canada follows the fertility patterns of most industrial countries. In recent decades, Canada has had a low rate of natural increase. Since 1970, the rate of natural increase has remained below 1 per cent per year. By 1995, it was 0.5 per cent, and every indication pointed to a further decline (Table 4.8). These trends in birth and death rates, which are consistent with the experience of other industrialized countries, are described in the demographic transition theory.

The Demographic Transition Theory

The **demographic transition theory** is the most widely accepted theory that describes population change in industrial societies. This theory is based on the assumption that changes in birth and death rates occur as a society moves from a preindustrial to an industrial economy. These demographic changes occur in four phases, each of which has a distinct set of vital rates that coincide with the phases in the process of industrialization (Table 4.9).

A cursory examination of Canada's birth and death rates over the last 150 years reveals strong similarities to the early industrial, late industrial, and postindustrial phases. In the nineteenth century high birth rates and lower death rates were common. These vital rates are similar to those in the second phase of the demographic transition theory. Then Canada entered the late industrial phase, characterized by low death rates and a declining birth rate. By the 1980s, birth rates had declined, resulting in little natural increase. Such rates characterize the postindustrial phase of the demographic transition theory. Based on a more precise measure of fertility, demographers argue that Canada's rate of natural increase has fallen below its replacement level (Vignette 4.3).

Table 4.9

Phases in the Demographic Transition Theory

Phase	Birth and Death Rates	Rate of Natural Increase
Preindustrial	High rates often exceed forty births and deaths per 1,000 people.	Little or no natural increase, though population fluctuations caused by a temporary rise in the death rate do occur.
Early industrial	Falling death rates and a continuation of high birth rates.	Extremely high rates of natural increase.
Late industrial	Falling birth rates and low death rates.	Extremely high rates of natural increase.
Postindustrial	Low rates often below ten births and deaths per 1,000 people.	Little or no natural increase, though population fluctuations caused by a temporary rise in the birth rate do occur.

Vignette 4.3

The Concept of Replacement Fertility

The concept of replacement fertility refers to the level of fertility at which women have enough daughters to replace themselves. If women have an average of 2.1 births in their lifetime, then each woman, on average, will have given birth to a daughter and a son. The number 2.1 was determined to represent the minimum level of replacement fertility because, on average, slightly more boys than girls are born. In 1961, the Canadian fertility rate was 3.8 births per woman of childbearing age (15–49). By 1991, it had dropped below replacement level to 1.8 births per woman of childbearing age.

Age and Sex Structure

Age and sex are the most basic characteristics of a population. Every region in Canada has a different age and sex composition and this structure can have considerable impact on its demographic and socio-economic behaviour. Some regions have populations that are younger, others that are older than the national average. In rural areas of Western Canada, in the hinterlands of Ontario and Québec, and in Atlantic Canada, populations are older than the national average; that is, a high proportion of their population is over the age of fifteen. The explanation for this age difference is twofold: (1) new Canadians tend to settle in large urban centres and most of these immigrants are young; (2) young families and individuals born in hinterlands tend to relocate to large urban centres.

Since Confederation, Canada's population has grown older. In 1901, for example, nearly 35 per cent of the nation's population was under the age of fifteen. By 1996, however, this figure had dropped to 20 per cent. The drop in fertility rates is the main reason for this decline. Only the Territorial North still has a young population. In 1996, approximately 30 per cent of the Territorial North's population was under the age of fifteen. Again, the explanation lies in the higher fertility rate in this region compared to the national fertility figure. High fertility rates in the North are attributed to Aboriginal peoples, who constitute almost 40 per cent of the northern population.

Since the probability of a woman giving birth to a female or male baby is almost the same, the gender structure of a population should be equal (Vignette 4.3). But other factors, such as migration, can create an imbalance. Therefore, some regions have populations that have a higher proportion of males to females, while others have more females. In demographic terms, **sex ratio** is the number of males divided by the number of females multiplied by 100. Accordingly, the sex ratio indicates the number of males for every 100 females in a population. A sex ratio of 100 represents perfect balance between the sexes, while a sex ratio greater than 100 would indicate more males than females. From 1871 to 1971, Canada's sex ratio exceeded 100. In 1911, it was nearly 113! This imbalance was due to the large number of male immigrants coming to Canada in the first decades of the twentieth century. Since 1981, the sex ratio has remained just below 100, which reflects two factors: (1) the longer life span of females compared to males, and (2) the higher mortality rates for males than females.

Within the geographic regions of Canada, only the Territorial North has a sex ratio exceeding 100. The high proportion of males to females in the North reflects its frontier nature; that is, the tendency for more male migrants to move to the North for employment in the resource industry. Capital cities often have a higher proportion of females to males. This phenomenon is caused by the in-migration of large numbers of young female adults who are attracted to the job opportunities within the public sector.

Age Dependency

Canadian society is aging rapidly. For instance, the number of people age sixty-five and over has increased sharply. In 1961, there were 1.4 million senior Canadians, but by 1991, this number had reached 3.2 million. This trend, however, began long ago in Canada's history. Since Confederation, Canada's demographic structure has changed from a youthful one in 1867 to a mature one in 1996. Social and economic implications of such shifts are far-reaching. One such impact is measured by age dependency, while another by old-age dependency.

The **age dependency ratio** is the ratio of persons in the 'dependent' age groups (under fifteen and over sixty-four years) to those in the 'economically productive' age group (between fifteen and sixty-four years). The assumption is that productive members of society are those between the ages of fifteen and sixty-four, while unproductive members are either too young (under fifteen) or too old (over sixty-four) to make an economic contribution. The purpose of the age dependency ratio is to compare the number of dependants with the number of economically productive members of society. While the assumptions underlying this measure are not perfect, the ratio provides an approximation of changes in the age structure of a population over time.

In Canada the age dependency ratio declined substantially from 1961 to 1996. In 1961, the ratio recorded seventy dependants per 100 people of working age. By 1996, it had declined to about forty-eight per 100, which is where it remains today. This decline suggests that the costs borne by the working population have diminished. However, this decline has been significantly affected by the large number of baby boomers who have entered the workforce (leaving the under fifteen group) over this period and, to a lesser degree, by immigrants to Canada. For instance, the percentage of Canadians under fifteen years fell from 34 per cent in 1961 to 20 per cent in 1996. As Canada enters the next century, baby boomers and immigrants will increasingly move from being productive to dependent members of society.

The **old-age dependency ratio** measures only the number of people over sixty-four as dependants and compares it to the number of those in the productive age group. Given Canada's aging population, age dependency is an important issue. From 1961 to 1991, the old-age dependency ratio increased from four-teen elderly Canadians per 100 people of working age to eighteen per 100. By 1996, this figure had reached twenty per 100. This form of dependency occurs because, with an aging population, the proportion of people actively participating in the labour force decreases and the cost of supporting the elderly increases. The burden of paying for public services, such as health services, becomes greater for those still in the workforce.

Population Trends and Canadian Society

Canada has evolved from a dualistic society that was characterized by French and English settlements into a pluralistic society character-ized by a multicultural population. A plural-istic society, by definition, is composed of many groups of people of different cultural backgrounds. People who speak different lan-guages, hold different religious beliefs, and belong to different ethnic groups have trans-formed the cultural composition of Canadian society. Much of this social change stems from immigration, as well as from Aboriginal peoples' greater participation in the daily affairs of Canada.

While Indians played a key role in the development of British North America, by the time of Confederation they were pushed to the margins of Canadian society, hardly noticed by the average citizen and certainly not by politi-cians. In today's world, Aboriginal issues have moved from the margins of political concern to the forefront, and Aboriginal peoples are playing a more significant role in the multicul-tural nature of Canada.

All of these social changes are taking place within a new Canada, one based on civic cit-izenship as defined in the Constitution Act (1982). The role of immigration in redefining Canadian society, and the nature and signific-ance of ethnicity, religion, and multicultural-ism are explored in the following sections.

Waves of Immigration

Immigrants have been an important compon-ent of society throughout Canada's history. For that reason, Canada is often described as a country of immigrants. These newcomers be-came the farmers and workers who built the nation. Some historians venture to say that without immigrants in the late-nineteenth century, Canada would have been settled by Americans and, like the Oregon Territory, become part of the United States.

People have various reasons for moving from one place to another. 'Push' factors (such as wars, natural disasters, a declining local economy, limited job prospects, or political oppression) are unfavourable aspects of an environment that prompt people to relocate. 'Pull' factors (such as better career opportun-ities, proximity to family and friends, or a more favourable climate or neighbourhood) may make another location seem more desirable to inhabit. Whatever the reasons for relocating, successive waves of immigrants have not only increased the Canadian population but have also changed its ethnic composition and stimu-lated its industrial development by bringing labour, capital, and innovative ideas to Canada. They have also changed the human face of this country. For example, at the time of Confederation only 1.5 per cent of the Canadian population claimed non-European origin. By 1996, Canadians of non-European origin comprised 15 per cent of the total population (Statistics Canada 1998c). In 2000, this figure could reach 20 per cent.

During the nineteenth century, the annual influx of immigrants to Canada was usually less than 50,000 people. With the opening up of land in the Prairies, the annual number of immigrants quickly surpassed 50,000, reaching a record number of 331,000 in 1911. Until the 1950s, the annual number of new Canadians fluctuated, but rarely exceeded 100,000. For example, during the Great Depression of the 1930s, the level of immigration fell well below 50,000 people per year. Since the Second World War, the number has grown and is now about 250,000 per year.

The origins of Canadian immigrants has changed over time. More and more immigrants, with different languages and religions, are coming from non-English-speaking countries. In recent years, a small number of French-speaking immigrants have come to Canada, many of them settling in Québec. For over a century after 1760, the British authorities did not encourage French-speaking immigrants to come to British North America. Instead, Britain wanted to assimilate French Canadians into a British society by settling English-speaking immigrants in Québec.

In the first half of the nineteenth century, nearly 1 million immigrants came from Britain. All were English-speaking and mostly Protestant. By 1840, most English-speaking immigrants came from famine-stricken Ireland. Many were Irish Catholics. Hundreds of thousands settled in the British colonies of Nova Scotia, New Brunswick, Prince Edward Island, Lower Canada, and Upper Canada. After American settlers from New England occupied most of the Ohio Basin, some Americans moved northward to Upper Canada where land was still available. Again, most American immigrants were English-speaking Protestants.

The opening of the Canadian West brought another surge of immigrants to Canada. By the late-nineteenth century, the Great Plains in the United States had been settled, which left the Canadian Prairies as the only natural area suitable for agricultural settlement in North America. Between 2 and 3 million homesteaders moved to the Canadian Prairies. Some came from Ontario and Atlantic Canada, but most were from foreign countries, especially Europe and the United States. Over 1 million immigrants came from Europe, including Britain, Germany, Scandinavia, Central Europe, Russia, and the Ukraine. Even today, the rural landscape in Western Canada reflects this ethnic influx with distinctive place names (Humboldt), church architecture (Greek Orthodox), and even land holdings (Hutterite colonies). By the 1920s, Canada's agricultural lands were occupied, causing the pattern of immigrant settlement to change once again. Most immigrants were now drawn to the cities to seek work in factories, or to the hinterland to work in the forests and mines.

Recent Immigration

In 1967, a change in immigration regulations heralded a new era. Immigration laws were relaxed to encourage immigration from all over the world. In one gesture, the new regulations removed the 'preferred nationalities' clause, which favoured immigrants from the British Isles and other European countries, and replaced it with a universal point system for assessing a candidate's suitability for entry into Canada regardless of geographic origin or ethnic background. Since 1967, many immigrants have come from Asia, Africa, the Caribbean, and Latin America. Between 1991 and 1996, just over 1 million people settled in Canada (Statistics Canada 1997a), with most new Canadians coming from Asia (57 per cent).

This non-European stream of immigrants has caused a dramatic shift in the ethnic and religious composition of Canada's population. While in 1967, the major sources of immigrants were Britain, Italy, and the United States; by 1996, Hong Kong, India, and the Philippines were the top three sources of immigrants (Table 4.10). This shift in the origins of immigrants—from Europe to Asia and other countries—has been the result of Canada's more liberal immigration policy.

The impact of this new immigration policy is illustrated in the following statistics: before 1961, 10 per cent of immigrants were born outside Europe, whereas by the 1980s, this number had increased to 75 per cent (Badets and Chui 1994:13). The trend is also revealed in Table 4.11. As a result of this more open immigration policy, Canada has a more diverse society that includes non-European languages and non-Christian religions. Canada also has as a result a faster-growing population. For instance, in the 1980s, the number of immigrants almost matched the number of births in Canada. By the 1990s, the number of immigrants exceeded the number of births.

Canada also has a policy of allowing refugees into the country; that is, people who are fleeing the turmoils of war or political oppression are allowed into Canada if they declare themselves as refugees and if their claim is verified by Canadian officials. Refugees constitute about 10 per cent of the total number of immigrants.

Table 4.10
The Newcomers in 1967 and 1996/7

Rank 1967	Country	Number	Rank 1996/7	Country	Number
1	Britain	62,420	1	Hong Kong	29,623
2	Italy	30,055	2	India	20,683
3	US	19,038	3	Philippines	11,355

Sources: Statistics Canada, 1996a: Table 7; 1998b:1.

Table 4.11
Immigration by Country, 1971–5, 1986–90, 1990–5

Origin	1971–5 Number	1971–5 Per cent	1986–90 Number	1986–90 Per cent	1990–5 Number	1990–5 Per cent
Asia	186,637	22.4	395,015	48.2	668,300	57.4
Europe	336,799	40.4	205,011	25.0	219,120	18.8
Caribbean	80,792	9.7	51,638	6.3	*	*
Africa	39,773	4.8	48,289	5.9	81,222	7.0
South America	46,222	5.5	42,325	5.2	151,000*	13.0*
United States	118,922	14.3	34,794	4.2	32,915	2.8
Oceania	7,272	0.9	5,681	0.7	6,357	0.5
Australasia	12,493	1.5	3,882	0.5	5,434	0.5
Total	834,452	100.0	819,477	100.0	1,164,348	100.0

* For 1990–5, South America included people from the Caribbean.

Source: Adapted from Statistics Canada, 1996a: Table 7.

Regional Patterns of Immigration

Immigrants are not evenly dispersed across Canada. The vast majority live in Ontario (over half of all immigrants reside here), British Columbia, Québec, and Alberta. Immigrants within these four provinces have especially been drawn to the major cities. Relatively few live in small towns or rural settings. In 1996, over half the number of immigrants resided in Toronto, Montréal, and Vancouver, resulting in large ethnic concentrations not found in smaller urban centres. In 1996, about 1.8 million immigrants lived in the census metropolitan area of Toronto, making up nearly 42 per cent of its population. Immigrants constitute at least 20 per cent of the population in Vancouver, Hamilton, Kitch-ener, Windsor, and Calgary. The figure for Montréal is about 18 per cent. In sharp contrast, the percentage of immigrants in many cities in Atlantic Canada and Québec—including the census metropolitan areas of Saint John, Sherbrooke, St John's, Québec City, Trois-Rivières, and Chicoutimi–Jonquière—fell below 5 per cent.

New Canadians are attracted to the metropolitan centres mostly because of perceived economic opportunities and because they have a social network of friends and relatives in these cities. Since many new Canadians are from non-European countries, they have dramatically changed the ethnic character of these cities, particularly Toronto. Between 1991 and 1996, for example, 83 per cent of immigrants settling in Toronto were from non-European (especially Asian) countries (Statistics Canada 1998b). New Canadians have therefore diversified the range of religious and social institutions in the large urban centres. The rest of Canada has not received many immigrants and so has not experienced the same cultural impact.

The Importance of Immigration

One measure of the importance of immigration to Canada is the proportion of immigrants in the national population. In 1996, immigrants constituted 17.4 per cent of Canada's population, up from 16.1 per cent in 1991. By 2001, this figure may exceed 20 per cent. The reason for this increase is twofold: the rate of natural increase has declined, and Ottawa's quota for landed immigrants has increased.

Over the last twenty-five years, the annual number of immigrants has increased from about 122,000 in 1971 to a peak of 256,000 in 1993. Over the same period, the rate of natural increase has remained fairly stable, with about 200,000 births in Canada each year. In 1991, the number of births in Canada reached 207,000, but has since begun declining, falling to 170,000 by 1995. On the other hand, the influx of immigrants has remained above 200,000. For example, 207,000 immigrants to Canada arrived in 1995 (Statistics Canada 1996a:Table 7). The demographic implications are threefold:

• The increase in immigration will offset the losses in natural increase, thus keeping Canada's population growth stable.

• Immigrants will form a larger proportion of the national population, thus further increasing the cultural diversity of Canada's pluralistic society.

• Canada's increasing diversity will call into question the validity of the vision of two founding peoples.

Ethnicity

Canadian society is composed of many ethnic groups. An **ethnic group** is made up of members of a population who share a culture that is distinct from other groups. **Culture** is the learned collective behaviour of a group of people.

For Canadians born and raised in this country, the connection with their so-called ethnic origins is tenuous at best (Beaujot 1991:297). Place, as cultural geographers insist, plays a critical role in the development of a regional sense of belonging. From this point of view, Canadian-born citizens, regardless of their ethnic or religious backgrounds, are generally more attached to their locality, region, and country than to their ethnic origin (the artificial category devised by Statistics Canada—see Table 4.12). Nevertheless, the ethnic makeup of Canada, as well as its historical development, have had a significant impact on Canadian society.

Within Canada's cultural complexity, the two founding peoples (the French-speaking and the English-speaking) form the major ethnic groups. The ethnic origins of these two groups are the British Isles and France (Table 4.12). The next two major groups are European and Asian. Since 1871, the proportion of

Canadians who originate from the British Isles and France has declined, while the proportion of people who come from other parts of the world has increased.

The dominance of the British and French ethnic groups in Canada weakened in the years before the First World War when a flood of non-British and non-French migrants settled in the Prairies and moved to the towns and cities of industrial Canada. After the Second World War, immigrants from all over the world came to Canada in ever-increasing numbers. Canadian society now consisted of a number of ethnic groups whose members shared a sense of identity based on descent, language, religion, tradition, and other common experiences. Overarching these shared ethnic experiences was a sense of belonging to Canada. Ottawa now promotes multiculturalism, the idea that a cultural identity and a national one are not mutually exclusive. The results of this policy are reflected in Table 4.12.

In Québec, a strong sense of ethnicity exists and continues to shape that province's society and politics. History and geography are the causes of this sense of ethnic nationalism. Nationalism is nurtured by Québec's place, not

Table 4.12
Ethnic Origins of Canada's Population by Percentage: 1871, 1941, 1991, and 1996

Ethnic Origin	1871	1941	1991	1996[a]
Canadian[b]	—	—	—	29.1
British Isles[c]	60.5	49.7	40.7	17.9
French[d]	31.1	30.3	27.8	14.7
European	6.9	17.8	18.7	20.5
Asiatic	—	0.6	6.0	10.8
Other	1.5	1.6	6.8	7.0
Total per cent	100.0	100.0	100.0	100.0
Total no. (000)	3,486	11,507	26,994	18,304

Notes: [a] Single responses (respondents provide only one ethnic origin).
[b] For the first time, Canadian was included as one of the examples on the 1996 Census questionnaire. Of all the provinces, Québec's population reported the highest 'Canadian' response at 38 per cent.
[c] British Isles origin includes responses of English, Irish, Scottish, and Welsh.
[d] French origin includes responses of French or Acadian.

Sources: Statistics Canada, 1993a: Table 1A; 1998e: 15 of 21.

in Canada but in North America. As a tiny French enclave in the English-speaking North American continent, Québec naturally fears for its cultural existence. Its close proximity to the United States underscores its precarious position in North America much more than its 'sheltered' position within Canada. The United States exerts an enormous cultural impact on the world through the mass media and now increasingly through the world of the Internet. Within North America, American culture poses the major threat to French- and English-Canadian cultures. This fear of assimilation is the basis of ethnic nationalism in Québec, a nationalism that is generally expressed by the Québécois in one of two ways: (1) Québec society is distinct within Canada; (2) it is only through independence that Québec will be fully protected and emancipated.

Ethnicity is therefore far more important in Québec than in the rest of Canada. The Québécois view their provincial government not only as their 'government' but also as the protector of their culture and language and as their instrument of economic growth. Strong ethnic feelings account for Québécois' attachment to their province and government, which is much stronger than what other Canadians feel for their provinces and provincial governments.

Language

Language is a key component of ethnicity. Indeed, language represents the most durable link to the past and the tool to maintaining a culture. Canada's two official languages are English and French. Approximately 85 per

cent of Canadians use one of two official languages. The remainder, who speak other languages, eventually choose English and/or French. New Canadians who speak neither official language hold the key to the future balance between English and French. Where they settle and what official language they learn impacts on the French and English societies. From a political perspective, issues surrounding the two official languages are a delicate matter. In fact, this topic is the basis of a population faultline discussed later in this chapter.

There are a number of French organizations across Canada that aim to protect and promote their language and culture. In Atlantic Canada, these organizations are called Acadiens; in Ontario, Franco-Ontariens; in Manitoba, Franco-Manitobiens; in Saskatchewan, Fransaskois; in Alberta, Franco-Albertiens; in British Columbia, Franco-Columbiens; in Yukon, Franco-Yukoniens; and in the Northwest Territories, Franco-TéNois. Their purpose is to provide a social organization that promotes French heritage and language. Under the 1982 Constitution, such organizations receive federal funding to finance their cultural and educational activities. This public support, plus the presence of French language radio and television programs, have resulted in a rejuvenation of these provincial organizations.

Many languages, other than French or English, are spoken across the country, especially in larger cities where new immigrants tend to settle. Without the official language status enjoyed by French and English, however, foreign languages are often difficult to maintain. The survival of these other languages in Canada is bolstered somewhat by continued immigration and aided by the federal government, which provides funds for heritage language classes. However, Canadian-born members of ethnic groups—the second and subsequent generations of immigrants—usually lose the language of their parents and grandparents.

Religion

Religion is a key element of culture. It is not only a durable link to the past but also provides the institutional organization to preserve ethnicity. Religious organizations provide an institutional structure that consolidate people of similar beliefs. One example is the Roman Catholic Church. In the early history of Canada, the Roman Catholic Church helped sustain Catholicism, the French language, and the French-Canadian way of life. While its role has diminished in recent decades, the Church was the dominant cultural force among francophones from the conquest of New France to the Second World War. Not only did it give spiritual direction to French Canadians within Québec but it also organized its parishes to sponsor group immigration to more isolated areas of Québec (such as the Clay Belt in northern Québec), to other provinces, and to New England. In these group migrations, priests provided the leadership for the move and for organizing the new settlement, including its educational, social, and religious structures.

Religious freedom is the right of all Canadians. This right was inscribed in the British North America Act and is contained in the Constitution Act of 1982. In fact, many immigrants have come to Canada seeking religious freedom. The Doukhobors are one such religious group. They are a sect that broke away from the Russian Orthodox Church. As they were persecuted in Russia, they immigrated to Canada in 1898, settling initially in Saskatchewan and later moving to an isolated area in the southern interior of British Columbia, where they could live a communal life, practise their religion, and speak Russian. While most Doukhobors have integrated into Canadian life, a minority remains committed to their religion, a communal lifestyle, and pacifism.

Most Canadians are Christians. In the 1960s, nearly 90 per cent of Canadians declared their religion as Christian (though some may not have been active church members). Roman Catholics were the largest group of Christians, followed by members of the United Church of Canada. At that time, most non-Christians followed Judaism. By the 1990s, Canada's spiritual world had become much more diversified. Now only 70 per cent of Canadians profess to be Christians. While the actual number of Christians has not decreased, the number of Canadians subscribing to non-Christian beliefs has increased, mainly as a result of immigration from non-Christian areas of the world. While Judaism remains an important non-Christian faith in Canada, other religions are expanding their memberships rapidly. Islam, for example, may become the second largest religious group in Canada by the next century. Other important religions in Canada include Hinduism, Buddhism, Sikhism, and Baha'ism. There has also been a revival of the spiritual practices of Aboriginal Canadians.

Religion and Education

Religion has been an important, sometimes controversial, component of education in Canada. Under the British North America Act, education was assigned to the provinces. This decision was largely due to Québec's desire to control education for its French-speaking and Catholic population and thereby ensure the future of the French language in that province. At the time of Confederation, almost all Canadians declared themselves to be Christian, yet the gulf between Protestants and Catholics was so wide that it affected the structure of the education systems in Canada. This religious difference was originally associated with the two founding peoples, French Canadians and English Canadians, but with the influx of more and more European immigrants who belonged to the Roman Catholic Church, the English-speaking community became divided on religious grounds. Religious differences among Christians were more divisive than today (not as vicious as in Northern Ireland today, but nevertheless very unpleasant).

Each province developed its own solution to the question of religion and education. Generally, dual school systems were established—often one was public, usually non-denominational, and one was separate, almost always Catholic. In Québec, public funds were provided for both a Catholic and a Protestant school system. Upon entering Confederation, Ontario, fol-

The Basilica of Notre Dame, Ottawa. The number of Christian churches in Canada attests to this religion's prominent place in the country's history. (*Corel Photos*)

lowed by Alberta and Saskatchewan, funded both a public and a Catholic (separate) school system, while New Brunswick, Nova Scotia, and Prince Edward Island supported only a public system. When British Columbia joined Canada, it also funded only a public school system. However, since 1977, British Columbia provides tax support to private schools, while the Maritime provinces have given public funds to support Catholic schools when there is a local demand.

This type of dual education system in Manitoba led to a controversy that shook the very foundations of the Dominion. In the 1870 Manitoba Act, both the Protestant and Roman Catholic schools were publicly funded. During the next two decades, a significant number of immigrants from Ontario and the United States moved to Manitoba. Most were English-speaking Protestants which led to an imbalance between English Protestants and French Catholics. In 1890, the Manitoba government abolished public funding of Catholic schools. This created much resentment among French Canadians, both in and outside Manitoba, who felt that the federation was not willing or able to protect their religious (and language) interests. It was not until late 1970 that limited public funding was made available to Roman Catholic schools.

The link between religion and education, however, has begun to weaken, reflecting the social changes that have taken place. For example, in 1997, both Québec and Newfoundland announced plans to remove the control of education from the churches and place their school boards under provincial control. Québec seeks to change the organiza-tion of its school boards from religious to linguistic; that is, the province will have English and French school boards. Under the Terms of Union with Canada, Newfoundland has three types of denominational school boards (one Roman Catholic and two Protestant). It now wishes to have a single 'public' system.

Multiculturalism

In 1971, Prime Minister Pierre Trudeau announced a new government policy of multi-culturalism. Multiculturalism recognizes all Canadians as full and equal participants in Canadian society. Multiculturalism is a form of cultural pluralism. The federal government announced its support for multiculturalism following the criticism of the Royal Commission on Bilingualism and Biculturalism by people whose background was neither British nor French. In fact, the federal government was simply responding to a new reality, namely, that Canada had evolved into a more complex society composed of many ethnic groups such as German- and Ukrainian-speaking Canadians. In adopting multiculturalism, in addition to biculturalism, Ottawa was re-cognizing Canada's pluralistic society. The federal government, however, expected all im-migrants to acquire at least one of Canada's official languages (French or English).

The 1988 Canadian Multiculturalism Act's ultimate goal seeks greater human under-standing and stronger bonds among Cana-dians of different cultural backgrounds and ethnic origins. Multiculturalism is a distinct-ively Canadian approach to social equality in nation-building, encouraging respect for cultural diversity. In sum, multiculturalism is the shared vision of people of diverse cultural backgrounds seeking to live together in harmony.

Multiculturalism, however, has its detractors. Some Canadians fear that multiculturalism might increase ethnic group identification at the expense of Canadian social cohesion. Ottawa, however, maintains that multiculturalism binds Canada's diverse populations together and the government sees no conflict between Canadians having two identities—a cultural identity and a national one. With most immigrants settling in Toronto, Montréal, and Vancouver, the impact of new cultures and religions is most visible in those cities. The notion of other ethnic groups within Canadian cities is still a new phenomenon for some communities. How these groups will fit into the larger Canadian society remains uncertain, but they enrich its culture and create more cosmopolitan cities.

For those committed to the concept of two founding peoples, multiculturalism is very unpopular. Francophones see multi-culturalism as a threat to their position within Canada and view the policy as a recognition of a third political force. In 1971, Premier Robert Bourassa of Québec wrote to Prime Minister Trudeau, arguing that multiculturalism contradicted the principle of 'the equality of the two founding peoples'. He further stated that Québec would not be introducing multiculturalism in its jurisdictions but would continue to ensure the well-being of the French language and culture in its territory and beyond (McRoberts 1997:129). Claude Ryan, a leading spokesman for Québec, maintained that multiculturalism 'omits a central fact: the two official languages of Canada, far from existing in the abstract as the subject of juridical definitions, are the expression of two cultures, two peoples, two societies which give Canada its distinctive shape' (McRoberts 1997:129). Multiculturalism, from the perspective of French-Canadian leaders, widened the gulf between French- and English-speaking Canadians.

Canada's Labour Force

The labour force is comprised of the economically productive members of the population. In 1996, the size of the Canadian labour force was approximately 13.5 million, which included employed workers and temporarily unemployed people. As Canada's economy became more complex and interdependent, its labour force adjusted to these changing circumstances by increasing work specialization. Consequently, the type of work and the participation of different members of the population in the work place have changed dramatically. Aspects of the labour force are examined in this section from these perspectives:

- structure of the labour force
- domestic market
- North American market
- profits and unemployment
- earnings of Canadians
- women in the labour force

Structure of the Labour Force

Major shifts in the percentage of workers in the primary, secondary, tertiary, and quaternary sectors are revealed in Table 4.13 (see also

Table 4.13
Major Sectors of the Labour Force by Percentage

Development Stage	Year	Primary	Secondary	Tertiary*
Agricultural	1881	51	29	19
Early industrial	1901	44	28	28
Late industrial	1961	14	32	54
Postindustrial	1991	6	21	73

*Because data for the quaternary sector is not available, percentages for the tertiary sector conventionally include both tertiary and quaternary jobs.

Source: Adapted from McVey and Kalbach, 1995: Table 10.3.

Table 4.1). These shifts mark three fundamental transitions in the nature of the Canadian economy—from agrarian to early industrial, from early industrial to late industrial, and from late industrial to postindustrial (see also section 'Economic and Social Change' at the beginning of this chapter).

Today, Canada's postindustrial economy is characterized by a large portion of the labour force employed in the tertiary sector, a much smaller portion in the secondary sector, and a very small portion in the primary sector. As discussed earlier in the chapter, the primary sector in Canada includes mainly resource activities such as mining, forestry, and commercial fishing; and the secondary sector includes manufacturing or the processing of raw materials. The tertiary sector extends over a wide variety of service functions such as commerce, education, health, transportation, and communications. Tertiary functions are often divided into public and private subsectors. With the growth of activities in this sector, many scholars have split it into two parts,

adding the quaternary sector to include decision-making occupations such as research and senior management.

The tertiary sector has low- and high-skilled jobs. The postindustrial economy is marked by the growth of highly innovative and high-technology industries such as telecommunications. Many of these jobs fall into the quaternary sector. Professional and technical employees in these businesses earn high wages and look forward to a secure future. Such jobs provide stability for the individual, his or her family, and the larger community. On the other hand, semi-skilled employees have low-paying jobs in retail stores, fast-food outlets, and service stations. These jobs belong to the tertiary sector. Often these workers have no pension plan or job security. Clearly, they represent the low end of the service industry. Furthermore, such jobs are not restricted to those with limited education. Even university graduates who are employed by universities as sessional lecturers fall into the low end of the service industry.

Domestic Market

As a modern industrial country, Canada has a small population and therefore a small domestic market. Consequently, its industrial output must go to foreign markets. This linkage between population size and domestic market translates into an export-driven economy. Canada's small market makes it dependent on world trade, which accounts for Canada's unusually high volume of trade with foreign countries, particularly the United States. Over 70 per cent of Canada's trade is with the United States. No other high-income country comes anywhere close to this level of trade domination.

Much of the explanation for Canada's closely integrated economy with the US is due to the geographic proximity of Central Canada's manufacturing belt to the major market and manufacturing area in the US. A secondary factor is that American industry views Canada's hinterland as a 'natural' source for their energy and raw materials. This view is sometimes described by Americans as a 'continental economy'. A third factor is Canada's stable political environment and its long involvement with the American economy before Confederation (see Chapter 3 for a discussion of reciprocity).

North American Market

The Free Trade Agreement (1989) marked a shift in Canadian economic policy. This agreement and its successor, the North American Free Trade Agreement (1994), call for the integration of the Canadian economy into the North American economy. Since Confederation, the general pattern of trade has been the export of Canadian raw materials for American industry (low-value-added goods) and the import of American-manufactured products for Canadian consumers and industry (high-value-added goods). Since the Free Trade Agreement, the level of trade between the two countries has risen. Now nearly 83 per cent of Canada's exports by value go south of the border, while nearly 68 per cent of Canada's imports come from the United States.

As a result of free trade, Canadian companies now have access to a much larger market, but they must also compete with large firms that have an advantage in economies of scale and an established presence in the American market. Canadian companies have sought to restructure their operations and thereby become sufficiently efficient to compete with other North American firms. Restructuring has placed enormous pressure on companies and their workers. To become more efficient, many businesses have reduced the number of their employees and restructured their companies. Others have simply closed down their operations here and moved to countries where labour costs are lower. The initial impact of industrial restructuring has had mixed results. Certainly there has been an increase in foreign trade, particularly with the United States, which has benefited the resource sector of Canada's economy. The manufacturing sector, however, has had to make major adjustments.[1] These adjustments have demanded a more productive labour force, which has often meant fewer jobs.

Profits and Unemployment

The integrated North American economy is growing, but how is this affecting employment and company profits? When Canada entered an economic recovery in the early 1990s, an increase in company profits and rising demand for labour were anticipated. True to form, many Canadian companies reported large profits. General Motors of Canada Ltd, for example, recorded its greatest profit in 1995. Unlike during the last economic recovery, however, Canada's annual unemployment rate remained extremely high—for a number of years in the 1990s, the

unemployment rate exceeded 10 per cent, and in 1993, the unemployment rate reached 11.2 per cent (Statistics Canada 1998a—see Table 4.14). Why is this?

One reason for the high rate of unemployment is that liberalized trade and globalization have exposed Canadian companies to overseas competitors, while deregulation at home is shaking up once-cozy oligarchies and monopolies. These forces have kept management focused on boosting productivity, which, for most large companies, means cutting jobs, using part-time labour, and contracting work to outside, non-union firms. This strategy of downsizing the workforce makes sense for

Table 4.14

Annual Unemployment Rates by Province, 1986–1997

Province	1986–9	1996	1997
Newfoundland	18.4	25.1	18.8
Prince Edward Island	15.3	13.8	14.9
New Brunswick	12.6	15.5	12.8
Nova Scotia	12.1	13.3	12.2
Québec	11.2	11.8	11.4
British Columbia	10.1	9.6	8.7
Ontario	7.9	9.1	8.5
Manitoba	8.3	7.9	6.6
Alberta	8.6	7.2	6.0
Saskatchewan	7.4	7.2	6.0
Canada	9.5	10.1	9.2

Sources: Scotiabank, 1998:12; Statistics Canada, 1998f:1.

Table 4.15
Profitable Companies Continue to Downsize

Company	Profit 1995 ($M)	Change Percentage 1994	Employment 1995	Lost Jobs
GM Canada	1,391	36	35,000	2,500
CIBC	1,015	14	39,329	1,289
Bank of Montreal	986	20	33,341	1,428
TD Bank	794	16	25,413	354
Shell Canada	523	63	3,920	471
Imperial Oil	514	43	7,800	452
Bell Canada	502	-30	48,333	3,170
INCO (US $)	227	3,281	13,746	1,963
Petro-Canada	196	-25	5,645	564
CP Rail	112	75	22,400	1,500
Maritime Tel	32	-33	3,512	529

Source: Greg Ip, 'Jobs Cut Despite Hefty Profits', *The Globe and Mail*, 6 February 1996:A4. Reprinted with permission from The Globe and Mail.

Vignette 4.4

Unemployment and Restructuring in the G-7 Countries

The number of unemployed people in the leading industrial nations of the world (Canada, France, Germany, Italy, Japan, the United Kingdom, and the United States) reached 23 million in 1996. Such levels of unemployment were not anticipated when the process of economic globalization and regional economic integration of the European and North American economies began. After ten years of economic restructuring, the marketplace has failed to generate enough jobs for the industrial workforce of these countries. Unusually high rates of unemployment in a period of economic growth are common in all industrial nations. But by 1996, when economic restructuring had brought some corporations record profits, and when these same companies announced continued layoffs, public reaction was very negative. For this reason, the French government proposed that the 1996 G-7 meeting study strategies to create jobs for the 23 million jobless in their countries and for international standards for the social protection of workers. The United States did not support this strategy and the French proposal failed.

labour-intensive firms whose very existence is at stake. For instance, by 1997, both textile and furniture manufacturing firms had reduced their labour forces by a third and, by investing in new equipment, increased their productivity. But what about other companies whose profits are rising? These companies are also reducing their labour forces. The issue is not survival but making more money. For example, while General Motors of Canada Ltd laid off 2,500 employees in 1995, it also announced a profit of $1.39 billion (Table 4.15). This company strategy of reducing the size of the labour force while making huge profits is widespread among Canada's largest corporations (Table 4.15). Consequently, the drive for higher productivity and greater profits prevents the unemployment rate from dropping to the lower levels experienced during the last economic boom of the 1970s.

Canadians have not approved of the actions of corporate Canada in this painful restructuring process, but the government has taken no action on their behalf. Instead, Ottawa is hoping that the private sector will generate enough jobs to reduce the unemployment rate to levels that the public will accept but that will nevertheless be well above those of the 1960s. On the other hand, if Ottawa's assumption is incorrect, then unemployment rates will not decline but hover between 8 and 10 per cent—until the next recession when they will rise, at which time the public will likely react.

The problem of high unemployment in an expanding economy is not unique to Canada. Most industrial countries around the world are facing similar levels of unemployment. The United States has somehow managed to keep its unemployment level well below that of Canada and countries in the European Union, while Canada has had the highest rates of unemployment over the past ten years since the Great Depression. Not surprisingly, the highest rates of unemployment are in the less favoured areas of Canada (Table 4.14). Atlantic Canada, for example, has the highest unemployment rates and is more dependent on unemployment insurance payments than any other geographic region. Within Atlantic Canada, by far the highest unemployment rates are in Newfoundland. In 1997, Atlantic Canada had an unemployment rate of nearly 15 per cent, while Newfoundland's rate exceeded 18 per cent, which was more than double the rates in Ontario, Western Canada, and British Columbia. These sharp regional differences have worsened the gap between the have and have-not regions of Canada.

Earnings of Canadians

The Canadian economy is growing, but the average earnings of Canadians—after inflation and taxes—is declining (Little 1997:B1, B5). By 1997, Canadian workers were less well-off in terms of earnings than they were ten years ago. In addition to this dilemma, various levels of government have raised their taxes, leaving less money in the pockets of working Canadians. This decline in earnings has affected lower-income families more so than higher-income families, thereby widening the gap between the rich and the poor. This phenomenon is also taking place in other industrial countries, including the United States.

The earnings of Canadians are not only

Vignette 4.5

Income and Gender

In all major occupational groups, women earn less on average than their male counterparts. Women also dominate the lowest-paying occupational groups—such as clerical and service occupations. In terms of families, there is a large number of female one-parent families in the lower-income group. In 1990, for example, two out of three female-headed one-parent families had a total income of less than $30,000 (Rashid 1994:7). Such incomes fall below the poverty line for families.

---

declining but also vary widely by region. Regional variations in earnings are greatest between Ontario, where they are well above the national average, and Atlantic Canada, where they are well below. In 1990, the average family income in Canada was $51,300. In three geographic regions—Ontario, the Territorial North, and British Columbia—average family incomes were above the Canadian norm (Rashid 1994:57). Western Canada is just below the national average, though the average for Alberta is above the national figure. Québec, at $46,593, is well below the national average. In Atlantic Canada, however, the family income is around $43,000, far below the Canadian average. Within Atlantic Canada, there are significant variations with the lowest average incomes occurring in Newfoundland. In 1990, Newfoundland's average family income was $40,900 compared to $44,001 for Nova Scotia (Rashid 1994:57).

Women in the Labour Force

In the second half of the twentieth century, significant numbers of women began to break away from the traditional homemaker role and entered the workforce. By the 1980s, working wives and mothers became a common and necessary feature of Canadian society. From 1951 to 1996, the percentage of females participating in the wage economy increased steadily. In 1951, for example, the female labour force participation rate was only 24 per cent, but by 1991, it had increased to 60 per cent (McVey and Kalbach 1995:251). With so many husbands and wives working, the term dual-earner family was coined. A measure of the growth of dual-earners is revealed by the fact that, in 1961, less than 20 per cent of all families had both the husband and wife employed; by 1991, this figure had more than tripled to 60 per cent (McVey and Kalbach 1995:253). The dramatic and sudden increase in the participation of women, particularly married women, in the labour force has altered traditional family roles. In a traditional North American nuclear family, the father was regarded as the 'breadwinner' and the mother as the 'homemaker'. One demographic consequence of this social change was a delay in family formation and smaller family units.

Unfortunately, this growing number of women in the workforce has not yet resulted in full pay equity between the sexes. Although women's average earnings have risen, a substantial income gap remains between women and men in the same occupations and with the same education (Gartley 1994:47–8). While income for women does increase as their level of education rises, it remains below incomes obtained by men (Vignette 4.5).

Population Trends and Canada's Faultlines

Having established the predominant demographic and socio-economic characteristics of Canadian society as a whole, we will now examine some key demographic shifts as they relate to the three major faultlines discussed in Chapter 3. The various demographic shifts being experienced in Canada will have significant long-term implications for the residents of the various regions. For example, the population of Atlantic Canada is expected to continue to decline. The number of Aboriginal Canadians is growing rapidly, while the French-Canadian population is growing very slowly. These population changes are affecting Canada's three faultlines. We will now examine how population trends are affecting the Aboriginal population; centralist/decentralist, especially core/periphery, population shifts; and the French/English population balance.

Aboriginal Population

The population figures for Aboriginal peoples have undergone great changes since European contact. Prior to contact, Aboriginals (within the territory that became Canada) may have been as numerous as 500,000. However, less than 100 years ago, the Aboriginal population

Table 4.16
Major Phases for the Aboriginal Population in Canada

Phase	Characteristics
Precontact (1000–1500)	The Aboriginal population in Canada was at least 200,000 and possibly as large as 500,000. This population may have varied in size due to the carrying capacity of the land, which, in a hunting society, is controlled by the availability of game (food). For instance, natural conditions, especially weather, could affect the size and migration routes of animal populations.
Early Contact (1500–1940)	Aboriginal peoples who came into contact with Europeans were exposed to new diseases. Population losses were heavy. Loss of hunting lands also added to their demise. By the end of the nineteenth century, the Aboriginal population was just over 100,000.
Late Contact (1940 to present)	The Aboriginal population now exceeds 1 million. High fertility and low mortality account for this remarkable population rebound. The net result is a population explosion that seems likely to continue into the next century.

had declined to about 100,000 because of the effects of European diseases and the loss of vast hunting grounds which had previously been their means of survival. But during the past five decades, Aboriginal peoples have experienced a population explosion. The total Aboriginal population in Canada now exceeds 1 million. The general population trends from precontact to the present illustrate three major demographic phases (Table 4.16).

When Jacques Cartier sailed into the Baie de Chaleur in 1541, the Indian and Inuit population in Canada may have been as high as 500,000. The exact figure will never be known. What we have are only estimates. By reconstructing the land's capacity to support wildlife and therefore also hunting societies, anthropologist James Mooney (1928:7) estimated that about 220,000 Indians and Inuit lived in Canada at the time of contact. More recently, scholars have revised this figure upwards. Dickason (1992:83) and Denevan (1992:370) estimate that the number of Aboriginal peoples living in Canada was closer to half a million. Whatever the exact figure, initial contact with Europeans resulted in a rapid depopulation of Aboriginal peoples. Factors include loss of hunting grounds and therefore food shortages, increased warfare, and the spread of new diseases from Europe among the Indian tribes. Communicable diseases, such as smallpox, caused great suffering and many deaths among Indian tribes. Epidemics sometimes quickly reduced the size of tribes by half. The result was that by the time of Confederation, Aboriginal peoples numbered about 100,000.

A demographic rebound of the Aboriginal population began in the 1940s and so far, shows little sign of slowing down. In 1996, there were approximately 1.1 million Canadians of Aboriginal descent. The outcome of this increase has not only meant greater numbers but also a greater proportion of Aboriginal peoples within Canadian society. In 1951, Aboriginal peoples comprised less than 2 per cent of Canada's population. By 1996, this proportion had reached 3.9 per cent. What, then, led to this recovery in the Aboriginal population? Many factors contributed to the population's increase; especially important was their relocation to settlements where access to food supplies and medical care was available.

Since the Second World War, the rate of natural increase of Aboriginal peoples has greatly exceeded the national figure. For much of that time, the rate of natural increase was about 3 per cent thanks to high birth rates and low death rates. The death rate for Aboriginal peoples dropped swiftly with the introduction of modern medicine. While communicable diseases, such as tuberculosis, continue to plague Native peoples, their mortality rates have declined below the national average. However, this low rate is somewhat misleading because the Aboriginal population is so much younger than the national average. Access to medical services and food supplies provide a partial explanation for the drop in the mortality rates.

As the Aboriginal death rate declined, the birth rate remained high. The fertility rate for Aboriginal Canadians remains more than double the national rate. While the national fertility rate is below replacement level, the rate for Aboriginal peoples is much higher. In the past decade, the national rate was about thirteen births per 1,000 people while the Aboriginal rate was close to thirty births per 1,000 people. This fertility difference between the national and Aboriginal populations has led to other demographic differences, including a more youthful Aboriginal demographic structure. While only 21 per cent of the Canadian population is under the age of fifteen, the figure for Aboriginal peoples is 38 per cent—almost double. While their rate of natural increase is high, Native Canadians face very high infant mortality and suicide rates, which indicate a social malaise that has affected the general health of that population. Nevertheless, based on the current demographic trends in Aboriginal society, the Aboriginal population is likely to continue increasing. Until Aboriginal parents opt for smaller family units, the gap between the birth and death rates will remain, as will the high rate of natural increase.

The political implications of this high rate of natural increase are profound. Already this rapid population increase has dispelled the myth of the 'vanishing Indian', instead assuring the continued existence of Aboriginal peoples. Aboriginal political power is, in part, linked to population size, so with the percentage of Aboriginal peoples in Canada increasing (perhaps exceeding 4 per cent by the turn of the century), the federal, provincial, and territorial governments will have to heed the political voice of Native leaders more carefully. A rapid population increase can also have negative impacts, such as placing greater demands on scarce

A youthful Aboriginal population indicates that the current population explosion among Aboriginal peoples in Canada will continue. (Corel Photos)

Table 4.17
Aboriginal Population by Geographic Region, 1996

Region	Population	Per cent
Yukon	6,440	
Northwest Territories	39,850	
Territorial North	46,290	4.2
Prince Edward Island	2,400	
Newfoundland	24,590	
New Brunswick	17,095	
Nova Scotia	26,795	
Atlantic Canada	70,880	6.5
British Columbia	184,445	16.6
Saskatchewan	117,350	
Manitoba	138,890	
Alberta	155,650	
Western Canada	411,890	37.4
Québec	142,385	12.9
Ontario	246,070	22.3
Canada	1,101,960	3.7

Note: In censuses prior to 1996, counts of Aboriginal people were derived primarily from responses to a question that asked respondents about their ancestry. The 1996 Census asked both an ancestry and an identity question (see Table 1.6). In 1996, 1,101,960 people reported Aboriginal ancestry, while 799,010 Canadians reported an Aboriginal identity; that is, these Canadians identified themselves as North American Indian, Métis, or Inuit.

Source: Adapted from Statistics Canada, 1998e:15 of 21.

resources. For example, the collective resources of Aboriginal peoples are already facing increased demands from its members for housing. Another implication of population increase has a regional component. Since the distribution of Aboriginal peoples across Canada is uneven, the greatest increases in their numbers will take place in Western Canada, followed by Ontario and British Columbia (Table 4.17). The five provincial governments

in those regions will face considerable pressure from those status Indians seeking a resolution to their land claims and from the non-status Indians and Métis who seek to better their way of life.

Core/Periphery Populations

Canada's population is unevenly distributed across the country's regions. The population remains concentrated in Central Canada,

Table 4.18

Regional Population of Canada by Percentage, 1901–1996

Region	1901	1921	1941	1961	1981	1991	1996
Ontario	40.6	33.4	32.9	34.2	35.4	37.0	37.3
Québec	30.7	26.9	29.0	28.8	26.4	25.3	24.8
British Columbia	3.3	6.0	7.1	8.9	11.3	12.0	12.9
Western Canada	7.8	22.2	21.0	17.5	17.4	16.9	16.6
Atlantic Canada	16.7	11.4	9.8	10.4	9.2	8.5	8.1
Territorial North*	0.9	0.1	0.2	0.2	0.3	0.3	0.3
Canada	100.0	100.0	100.0	100.0	100.0	100.0	100.0
Canada (millions)	5.4	8.8	11.5	18.2	24.3	27.3	28.9

*In 1901, the Yukon population included many associated with the Klondike gold rush. By 1911, most had left. This accounts for the abrupt change in population for the North after 1901. Also, even though the present territories of Alberta, Saskatchewan, and much of Manitoba still belonged to the Northwest Territories in 1901, their populations were assigned to the Prairie region.

Sources: McVey and Kalbach, 1995:46; Statistics Canada, 1997c:Table 1.

though population shifts have occurred in Western Canada and British Columbia. Atlantic Canada, on the other hand, has lost ground. Therefore, the hinterland has increased its share of the national population, but that increase has been geographically uneven—the West has experienced a relative increase in its share of the national population, while Atlantic Canada has seen a decrease (Table 4.18). Ontario has continued to grow rapidly while Quebec's rate of population increase has slowed. The reasons for this distribution in population and population growth can be found in the country's historical migration and settlement patterns, which can be characterized as east to west and rural to urban. Population shifts driven by interprovincial migration are revealed in Table 4.19.

The early history of Canada partly accounts for the powerful population cluster in Central Canada. In 1867, approximately 80 per cent of all Canadians lived in Ontario and Québec, where geographic conditions were most beneficial for settlement. The remainder of the population lived in New Brunswick and Nova Scotia. By 1996, however, these figures had declined to 70 per cent in Ontario and Québec, and 8 per cent in Atlantic Canada (even with the addition of Prince Edward Island and Newfoundland). The principal reasons for this relative decline over the past 130 years are: (1) the large population increase in

Table 4.19
Net Interprovincial Migration, 1981–1996

Geographic Region	1981–6	1986–91	1991–6	Total
Atlantic Canada				
Newfoundland	-16,550	-13,960	-23,240	-53,750
Prince Edward Island	1,530	-855	1,460	2,135
Nova Scotia	6,280	-4,870	-6,450	-5,040
New Brunswick	-1,370	-6,070	-1,965	-9,405
Québec	-63,300	-25,550	-37,450	-126,300
Ontario	99,350	46,955	-47,010	99,295
Western Canada				
Manitoba	-1,550	-35,245	-19,380	-56,175
Saskatchewan	-2,820	-60,350	-19,775	-82,945
Alberta	-27,670	-25,015	3,590	-49,095
British Columbia	9,500	125,880	149,945	285,325
Territorial North				
Yukon	-2,660	780	670	-1,210
Northwest Territories	-755	-1,700	-395	-2,850

Source: Adapted from Statistics Canada, 1998d:7 of 12.

Western Canada and British Columbia; (2) the slow rate of population growth in Québec and Atlantic Canada.

In the late-nineteenth century and early-twentieth century, the lure of 'free' land drew homesteaders from eastern Canada, the United States, and Europe to the Canadian Prairies. At the time of Confederation, only marginal agricultural land remained in Atlantic Canada, Québec, and Ontario. Before the West was opened up for settlement, an urban-to-rural migration was already underway in Ontario and Québec, driven by rural overpopulation and the need to find employment in the new factories. Following the completion of the Cana-dian Pacific Railway, the Canadian Prairies began attracting large numbers of land-hungry migrants. By 1921, Western Canada included 22 per cent of Canada's population, while British Columbia had 6 per cent.

Beginning with the drought of the 1930s, dubbed the Dust Bowl, new migration patterns formed. Families abandoned their homesteads located in the grassland of the three Prairie provinces. Many migrated to urban centres, including those in Ontario and British Columbia, looking for a more prosperous life. Farm labourers also joined this urban migration stream. They had been an invaluable part of the agricultural workforce in the Canadian

Prairies, but mechanization would eventually mean that fewer farmers were required. Since only those farmers with large landholdings could maximize the use of expensive farm implements, rural areas rapidly lost much of their population. By 1991, the rural depopulation of Western Canada had slowed. However, some indication of the extent of the pre-1991 rural-to-urban movement is revealed in Table 4.19. During this period, Saskatchewan suffered the greatest population loss, which is displayed in the out-migration figures for this province in Table 4.19, and which illustrates the limited capacity of the province's cities to absorb its surplus rural population. The net result was a smaller share (down to 16.6 per cent) of the national population for Western Canada (Table 4.18).

After the Second World War, however, a second wave of immigrants went to the West. Some were attracted by the mild West Coast climate, while others sought work, particularly in the resource industries. This population growth varied by province. Manitoba and Saskatchewan did not fare as well. Their populations grew very slowly, mainly because out-migration was so high that it almost equalled the rate of natural increase. British Columbia and Alberta led in population growth. The attraction of western provinces had changed over time from free land in the Prairies to jobs in Alberta and British Columbia. In British Columbia, the forest industry needed many workers, while in Alberta, the petroleum industry attracted many. During this time, British Columbia became the favourite destination for Canadians seeking to relocate (see Table 4.19).

Today, British Columbia is Canada's fastest-growing province thanks to its status as the major destination for migrants from the rest of the country as well as immigrants from Asia. British Columbia has been growing at an astonishing rate of 5.2 per cent per year, followed by Alberta which had the second highest rate at 4.2 per cent. In comparison, Canada has had an annual growth rate of about 2.3. Besides these provinces and the Territorial North, only Ontario grew at a rate greater than the national average (3.4 per cent). By 1996, the magnitude of this population shift revealed that the West accounted for almost 30 per cent of Canada's population, with Western Canada contributing 17 per cent and British Columbia 13 per cent.

All indications suggest that current regional trends will continue until the next census in 2001. Certainly, if Atlantic Canada continues to experience high levels of out-migration, its share of the national population will continue to decline. How much of its population will relocate to Ontario or further west will be determined by job opportunities. Ontario and British Columbia are expected to make the greatest population gains, while Western Canada's increase, driven by Alberta, is expected to be more modest. Together, British Columbia and Western Canada could form over one-third of the national population by 2001, which could result in long-term gains if political representation is re-adjusted to reflect regional populations. As for the trend of urban migration, it is expected to continue, as new Canadians continue to be attracted to cities where relatives and friends live and where society is more diversified.

French/English Balance

French- and English-speaking Canadians represent the traditional duality of Canadian society. This dual relationship has deep historic roots. With the establishment of the Province of Canada in 1841, the relationship flowered into a partnership with English-speaking Canada West and French-speaking Canada East sharing political power. In 1867, the British North America Act declared that both French and English were official languages of this new Dominion. The assignment of language rights to provinces ensured that the Québec government could protect and promote the French language.

Language is the essential element of French heritage and ensures the preservation of French culture. It is from this perspective that francophones interpret the social and political significance of the French/English balance. If the Census of Canada reports that the proportion of French-speaking Canadians is declining, then francophones become concerned. If a Royal Commission (such as the Laurendeau-Dunton, 1963–71) provides evidence that the French language is in 'deadly danger', then francophones become alarmed and want their governments to take action.

Linguistic rights for both French and English as the official languages give them a special place in Canadian society. One implication is that new Canadians whose language is neither French nor English gravitate to one linguistic group or the other. Within Canada, English is the dominant language. English is also the business language of North America and, to a large degree, the world. Immigrants therefore have a compelling reason to learn English. Also, most immigrants settle in English-speaking areas and so are naturally drawn into the anglophone linguistic group. Recent Statistics Canada figures (1996) support this argument. Approximately 83 per cent of Canadians speak English, while only 32 per cent speak French. However, a number of Canadians speak both English and French. The number of bilingual Canadians is increasing. From 1951 to 1996, the percentage of Canadians who could speak both official languages increased from 12 per cent to 17 per cent (Statistics Canada 1997b:11 of 15).

French/English dualism is a fundamental aspect of the geopolitical nature of Canada. As Jacques Bernier, professor of geography at Université Laval, (1991:79) stated: 'Canada's duality is intrinsic, and as long as it is not clearly recognized and dealt with, the issue of Canadian unity will remain.' This duality is a political concept embedded in the historical relationship between the two cultures. The main indicator of the stability of this French/English dualism is language; that is, the stability of this concept depends on a relatively constant number of Canadians speaking each language. But how should we measure duality? Should mother tongue or household language hold the key? Perhaps the population size of the minority language group should be the determining factor? Maybe the number of bilingual Canadians should hold sway?

The French/English balance can be measured by a number of factors relating to language use. In the 1996 census, Canadians were asked three questions about their linguistic characteristics, namely, their mother tongue, home language, and bilingual skills. Mother tongue is the first language learned that is still understood, while home language is the language most often spoken in the home. In that year, 17.1 million Canadians declared that their mother tongue was English, while 19.3 million indicated that the language spoken at home was English (Statistics Canada 1997b:4 and 7 of 15). Furthermore, Canadians who were able to speak English totalled 22.5 million, or 83 per cent. Since most immigrants learn English but have a different mother tongue, the number of Canadians who can speak English is much higher than indicated by mother tongue or home language. In comparison, 6.5 million Canadians stated that their mother tongue was French, but only 6.3 million spoke French at home. Canadians able to speak French totalled 8.5 million, forming 32 per cent of Canada's population.

French-speaking Canadians are distributed in three major geographic areas. The principal one is the province of Québec, where 82.8 per cent of the population speaks French in the home (Statistics Canada 1997b:8 of 15). A second one is in New Brunswick, where Acadians and other French-speakers form 30.5 per cent of the population. A minor cluster exists in the third area, Ontario, where 2.9 per cent speak French at home. The number of French-speakers is increasing in Canada but most of this increase has occurred in Québec (Table 4.20). Québec is the only geographic region where the majority of the population declares French as their mother tongue. It is also the only region where the number of people speaking French at home is greater than those declaring French as their mother tongue. This difference suggests that some immigrants living in Québec are adopting the French language and assimilating into the francophone culture.

The number of French-speaking Canadians is increasing, but at a rate slower than

Table 4.20
Population with French Mother Tongue, 1951–1996

Year	Canada (000)	Québec (000)	Rest of Canada (000)	Canada (per cent)	Québec (per cent)	Rest of Canada (per cent)
1951	4,069	3,347	722	29.0	82.5	7.3
1961	5,123	4,270	854	28.1	81.2	6.6
1971	5,794	4,867	926	26.9	80.7	6.0
1981	6,178	5,254	924	25.7	82.5	5.2
1991	6,562	5,586	976	24.3	82.0	4.8
1996	6,637	5,747	970	23.5	81.5	4.5

Sources: Harrison and Marmen, 1994:Table 2.1; Statistics Canada, 1997b:15 pp.

the national rate of population growth (Table 4.20). Since 1951, the number of Canadians whose mother tongue is French has increased from 4 million to 6.7 million. Within Québec, the increase in francophones is even greater, jumping from 3.3 million in 1951 to 5.8 million in 1996. Even in the predominantly English-speaking areas of Canada, the number of francophones has increased from 722,000 to 970,000. At the same time, however, the proportion of Canadians indicating that French is their mother tongue has dropped from 29 per cent to below 24 per cent (Statistics Canada 1997b:4 of 15). Finally, the number and percentage of bilingual Canadians has increased sharply. In 1981, 3.7 million Canadians spoke both languages, while 4.8 million fell into this category by 1996 (Statistics Canada 1997b:11 of 15).

The French/English duality remains strong, but the demographic position of both linguistic groups is weakening due to the increasing numbers of Canadians whose mother tongue is neither French nor English (Table 4.21). With low fertility levels among anglophones and francophones, neither group can rely on natural increase to maintain their percentage of the Canadian population. The key to maintaining the present level of language duality lies in the language choice made by new Canadians. As revealed in Table 4.21, in 1996, nearly 17 per cent (4.7 million Canadians) stated that their mother tongue was neither English nor French; a 15 per cent increase from 1991. New Canadians hold the key to the balance between French and English because they must choose at least one of the two official languages to function in Canadian society.

Between 1971 and 1996, the anglophone population grew by about 33 per cent compared with only 16 per cent for the francophone population. As a result, the proportion of Canadians who were French-speaking slipped to 23.5 per cent by 1996. Still, the number of French-speaking Canadians within Québec is rising. Two reasons account for this increase in Québec. One is the natural increase of the Québécois and the other is the number of anglophones and the increasing numbers of allophones who are learning to speak French. Until the Québec government passed a series of language laws restricting the use of English, most allophones were choosing to learn English. The reason was simple—it gave them greater economic mobility because they recognized that English is spoken in other parts of Canada and the United States.

Among the Québécois, language laws are very popular because they ensure the place of French in Québec, particularly in Montréal. While such legislation runs against the current of bilingualism in the rest of Canada, Québec governments (both Liberal and Parti Québécois) support the principle of a unilingual province in order to ensure the survival of the French language in Canada and North America. While language laws are not popular with all new immigrants, they are required by law to send their children to French schools. As Québec has an extremely low birth rate, immigration and the requirement that children of immigrants learn French keep the French-speaking population growing.

Québec language laws, particularly the law requiring all signs to be in French, have annoyed many English-speaking citizens of

Table 4.21
Population by Mother Tongue, 1951–1996

Year	English	French	Other
1951	59.1	29.0	11.8
1961	58.5	28.1	13.5
1971	60.2	26.9	13.0
1981	61.4	25.7	13.0
1991	60.4	24.3	15.3
1996	59.9	23.5	16.6

Sources: Harrison and Marmen, 1994:Table 2.1; Statistics Canada, 1997b:1 and 2 of 15.

the province. In fact, it violates the language rights guaranteed by the Canadian Constitution (formerly known as the British North America Act). In 1988, the Supreme Court of Canada ruled that Québec's French-only sign regulation violated the Charter of Rights in the Constitution of Canada. Francophones, ever concerned about the preservation of their culture, reacted strongly, taking to the streets, declaring 'Ne touchez pas à la loi 101.' The Québec Premier, Robert Bourassa, finding himself on the horns of a dilemma, chose to circumvent the Supreme Court's ruling by invoking the 'notwithstanding' clause in the Canadian Constitution. Later, the provincial government amended its sign legislation so that signs could be in English as long as French signs were larger. Nevertheless, ever since the language legislation favouring the French language was passed, English-speaking Quebeckers have felt less comfortable living in that province. In fact, these language laws, combined with two referendums, have caused many anglophone Quebeckers to migrate to

other provinces (see Table 4.19). In 1971, there were 789,000 anglophones in Québec, but twenty-five years later, the number had dropped to 626,000 (Statistics Canada 1997b:2 of 15).

Summary

Since the Second World War, the human face of Canada has changed. Like other modern industrial countries, Canada's society has been transformed from an agrarian to an industrial economy. But this transformation has taken place within a regional context whereby certain more favoured regions have enjoyed greater economic development and experienced faster-growing populations than less favoured regions. This transformation has also been affected by Canada's geographic position within North America. Having the United States, the most powerful nation in the world, as a neighbour and trading partner has resulted in close economic ties and a strong dependency on the American economy.

Major economic changes in Canada stem from the country's economic integration with the United States and, to a lesser degree, from increased involvement in the global economy. Economic restructuring, sometimes painful, took place following the Free Trade Agreement in 1989. It continues under the North American Free Trade Agreement and the General Agreement on Tariffs and Trade, but at a slower pace. Economic restructuring has also affected the Canadian labour force. Three notable labour changes have taken place: more women are employed, more workers are engaged in tertiary occupations, and more people are now unemployed.

Equally important changes stem from Canadian social and political forces, such as the Constitution with its emphasis on bilingual, pluralistic, and civic nationalism. Also, changes in Canada's demography have resulted in the reduction of Canada's rate of natural increase. Canadian politicians have extended immigration opportunities to people from all over the world and increased the annual number of immigrants coming to Canada. In turn, new Canadians, particularly those from non-European and non-Christian countries, have altered the ethnic and religious make-up of Canada. In response, Ottawa initiated a policy of multiculturalism to create harmonious relations among Canada's various ethnic groups and to promote social equality. Such a policy runs counter to the concept of duality between the two founding peoples, the French and the English, and makes francophones, especially in Québec, feel their cultural status is threatened.

Changes to Canada's economy and society have affected Canada's three critical tensions or faultlines: (1) the Aboriginal/non-Aboriginal peoples, (2) the core/periphery populations, and (3) the French/English societies. Key demographic changes are occurring along these divides, reflecting internal forces and revealing population and political shifts. The first faultline is characterized by a much higher rate of natural increase among the Aboriginal population. The second faultline involves changes in the sizes of Canada's regional populations. Since Confederation, population growth has occurred in all regions but at varying rates. Ontario, Québec, and Atlantic Canada's shares of the national population have declined, while those of British Columbia, Western Canada, and the Territorial North have increased. The third faultline has experienced a faster rate of population increase among English-speaking Canadians than French-speaking Canadians; the explanation lies not in fertility rates but in immigration.

The following chapters, 5 through 10, focus on the regional nature of Canada, with each chapter devoted to exploring one of Canada's six geographic regions. While many of the same topics and issues appear in each of the six geographic regions, each is examined from the perspective of the region under discussion. Each regional chapter will illustrate the physical diversity within each geographic region and show how this diversity affects human activities and settlement. Each chapter also has a section on the region's historical development to enrich our sense of the present by providing a link with the past. One recurring theme is the relationship among the six regions. This relationship is cast in the core/periphery model, but it is changing due to continental economic integration (NAFTA) and global trade (GATT).

Notes

1.   As Canada reacts to the new economic realities, industrial restructuring is not without economic pain. Canadian companies that cannot compete in the North American market fail. Some have relocated to Mexico where wage costs are almost ten times lower. In the first few years after NAFTA, job losses in the manufacturing sector amounted to 21 per cent (Jackson 1993:109). This loss was due largely to the closure of branch plants of American parent corporations (Healy 1993:287–94) and to Canadian-owned firms relocating to the United States and Mexico. Canada is not alone in this painful restructuring process. David Harvey (1989: 256–78) refers to global economic restructuring as one element in the worldwide transformation of industrial society from modern to postmodern.

Key Terms

age dependency ratio
The ratio of the economically dependent sector of the population to the productive sector; arbitrarily defined as the ratio of the elderly (those sixty-five years and over) plus the young (those under fifteen years) to the population of working age (those fifteen to sixty-four years).

census metropolitan area
A large urban area with a population of at least 100,000, together with adjacent smaller urban centres and even rural areas that have a high degree of economic and social integration with the larger urban area.

crude birth rate
The number of births per 1,000 people in a given year.

crude death rate
The number of deaths per 1,000 people in a given year.

culture
The sum of attitudes, habits, knowledge, and values shared by members of a society and passed on to their children.

demographic transition theory
The historical shift of birth and death rates from high to low levels in a population. The decline in mortality precedes the decline in fertility, resulting in a rapid population growth during the transition period.

demography
The scientific study of human populations, including their size, composition, distribution, density, growth, and related socio-economic characteristics.

ecumene
The portion of the land that is settled.

ethnic group
The association of people who identify themselves with a particular culture.

fertility rate
The number of births per 1,000 people in a given year; crude birth rate; (not to be confused with the 'general fertility rate' which is the number of live births per 1,000 women who are of childbearing age—fifteen to forty-four years—in a given year).

industrial stage
Each major change in the evolution of the capitalist economic system is called a stage. The industrial stage marks the shift from a predominantly agricultural economy to an industrial one.

mortality rate
The number of deaths per 1,000 people in a given year; crude death rate.

natural increase
The surplus or deficit of births over deaths in a population in a given time period.

net migration
The net effect of immigration and emigration on a country's population in a given time period.

old-age dependency ratio
Similar to the age dependency ratio except the old-age dependency ratio focuses only on those over sixty-four.

population density
The total number of people in a geographic area divided by the land area; population per unit of land area.

population distribution
The dispersal of a population within a geographic area.

population growth
The rate at which a population is increasing or decreasing in a given period due to natural increase and net migration; often expressed as a percentage of the original or base population.

population increase
The total population increase resulting from the interaction of births, deaths, and migration in a population in a given period of time.

population strength
Equates population size with economic and political power.

postindustrial stage
Each major change in the evolution of the capitalist economic system is called a stage. The postindustrial stage marks the shift from an industrial economy based on manufacturing to an economy in which service industries, particularly high-technology industries, become the dominant economic activities.

preindustrial stage
Each major change in the evolution of the capitalist economic system is called a stage. The preindustrial stage identifies an economic system that predates capitalism.

primary activities
Activities engaged in the direct extraction of natural resources such as agriculture, fishing, logging, mining, and trapping.

quaternary activities
Activities engaged in the collection, processing, and manipulation of information.

regional service centres
Urban places where economic functions are provided to residents living within the surrounding area.

resource towns
Urban places where a single function dominates the town's economy; a single-industry town.

secondary activities
Activities that process and transform raw materials into finished goods; the manufacturing sector of an economy.

sex ratio
The ratio of males to females in a given population; usually expressed as the number of males for every 100 females.

tertiary activities
Activities that engage in services such as retailing, wholesaling, education, and financial and professional services.

urban population
Communities that have economic and social functions that differentiate them from rural places; the common practice of defining urban population is by a specified size that assumes the presence of urban economic and social functions. In 1996, Statistics Canada considered all places with a combination of a population of 1,000 or more and a population density of at least 400 per square mile to be an urban area. People living in urban areas make up the urban population. People living outside of urban areas are considered rural residents and, by definition, constitute the rural population.

References

Bibliography

Badets, Jane, and Tina W.L. Chui. 1994. *Canada's Changing Immigrant Population*. Catalogue no. 96–311E. Ottawa: Statistics Canada.

Beaujot, Roderic. 1991. *Population Change in Canada: The Challenges of Policy Adaptation*. Toronto: McClelland and Stewart.

Bernier, Jacques. 1991. 'Social Cohesion and Conflicts in Quebec'. In *A Social Geography of Canada*, edited by Guy M. Robinson, Chapter 4. Toronto: Dundurn Press.

Bissoondath, Neil. 1994. *Selling Illusions: The Cult of Multiculturalism*. Toronto: Penguin Books.

Bourne, Larry S., and David F. Ley, eds. 1993. *The Changing Social Geography of Canadian Cities*. Montréal–Kingston: McGill-Queen's University Press.

Britton, John N.H., ed. 1996. *Canada and the Global Economy: The Geography of Structural and Technological Change*. Montréal–Kingston: McGill-Queen's University Press.

Cameron, Duncan, and Mel Watkins, eds. 1993. *Canada under Free Trade*. Toronto: Lorimer.

Cohen, A. 1990. *A Deal Undone: The Making and Breaking of the Meech Lake Accord*. Vancouver: Douglas & McIntyre.

Denevan, William M. 1992. 'The Pristine Myth: The Landscape of the Americas in 1492'. *Annals of the Association of American Geographers* 82, no. 3:369–85.

Dickason, Olive Patricia. 1992. *Canada's First Nations: A History of Founding Peoples from Earliest Times*. Toronto: Oxford University Press.

Foot, David. 1996. *Boom, Bust and Echo: How to Profit from the Coming Demographic Shift*. Toronto: Macfarlane, Walter & Ross.

Gartley, John. 1994. *Focus on Canada: Earnings of Canadians*. Catalogue no. 96-317E. Ottawa: Statistics Canada.

Harrison, Brian, and Louise Marmen. 1994. *Focus on Canada: Languages in Canada*. Catalogue no. 96-313E. Ottawa: Minister of Industry, Science and Technology.

Harvey, David. 1989. *The Urban Experience*. Baltimore: Johns Hopkins University Press.

Healy, Theresa. 1993. 'Selected Plant Closures and Production Relocations: January 1989–June 1992, Ontario'. In *Canada under Free Trade*, edited by Duncan Cameron and Mel Watkins, Appendix I. Toronto: Lorimer.

Hiebert, Daniel. 1994. 'Canadian Immigration: Policy, Politics, Geography'. *The Canadian Geographer* 38, no. 3:254–8.

Ip, Greg. 1996. 'Jobs Cut Despite Hefty Profits'. *The Globe and Mail* (6 February):A4.

Jackson, Andrew. 1993. 'Manufacturing'. In *Canada under Free Trade*, edited by Duncan Cameron and Mel Watkins, Chapter 5. Toronto: Lorimer.

Kalbach, Madeline A., and Warren E. Kalbach. 1995. 'Ethnic Diversity and Persistence as Factors in Socioeconomic Inequality: A Challenge for the Twenty-first Century'. *Toward XXI Century: Emerging Socio-Demographic Trends and Policy Issues in Canada,* 147–60. Proceedings of the 1995 Symposium Organized by the Federation of Canadian Demographers.

Li, Peter S. 1988. *Ethnic Inequality in a Class Society*. Toronto: Wall and Thompson.

_____, ed. 1990. *Race and Ethnic Relations in Canada*. Toronto: Oxford University Press.

_____. 1996. *The Making of Post-War Canada*. Toronto: Oxford University Press.

Little, Bruce. 1997. 'Are We Better Off under the Liberals?' *The Globe and Mail* (26 April):B1 and B5.

McRoberts, Kenneth. 1997. *Misconceiving Canada: The Struggle for National Unity*. Toronto: Oxford University Press.

McVey, Wayne W. and W.E. Kalbach. 1995. *Canadian Population*. Toronto: Nelson Canada.

Marsh, James H., ed. 1988. *The Canadian Encyclopedia*, 2nd edn. Edmonton: Hurtig Publishers.

Mooney, James. 1928. *The Aboriginal Population of America North of Mexico*. Smithsonian Miscellaneous Collections. Washington: Smithsonian Institution.

Nash, Alan. 1994. 'Some Recent Developments in Canadian Immigration Policy'. *The Canadian Geographer* 38, no. 3:258–61.

Norcliffe, Glen. 1996. 'Foreign Trade in Goods and Services'. In *Canada and the Global Economy*, edited by John N.H. Britton, Chapter 2. Montréal–Kingston: McGill-Queen's University Press.

Randall, Stephen J., and Herman W. Konrad, eds. 1996. *NAFTA in Transition*. Calgary: University of Calgary Press.

Rashid, Abdul. 1994. *Focus on Canada: Family Income in Canada*. Catalogue no. 96-318E. Ottawa: Statistics Canada.

Reid, Scott. 1992. *Canada Remapped: How the Partition of Quebec Will Reshape the Nation*. Vancouver: Arsenal Pulp Press.

_____. 1993. *Lament for a Notion: The Life and Death of Canada's Bilingual Dream*. Vancouver: Arsenal Pulp Press.

Reitz, Jeffrey, and Raymond Breton. 1995. *The Illusion of Differences*. Ottawa: C.D. Howe Institute.

Royal Commission on Aboriginal Peoples. 1996. *Report of the Royal Commission on Aboriginal Peoples*, 5 vols. Ottawa: Minister of Supply and Services.

Scotiabank. 1998. *Global Economic Outlook*. Toronto: Bank of Nova Scotia.

Statistics Canada. 1992a. *Immigration and Citizenship*. Catalogue 93-316. Ottawa: Minister of Supply and Services.

_____. 1992b. *Age, Sex and Marital Status*. Catalogue no. 93-310. Ottawa: Minister of Industry.

_____. 1992c. *Census Division and Subdivisions*. Catalogue 93-303. Ottawa: Minister of Supply and Services.

_____. 1992d. *Census Metropolitan Areas and Census Agglomerations*. Catalogue no. 93-303. Ottawa: Minister of Supply and Services.

_____. 1993a. *Ethnic Origin*. Catalogue no. 93-315. Ottawa: Minister of Supply and Services.

_____. 1993b. *Mobility and Migration*. Catalogue no. 93-322. Ottawa: Minister of Supply and Services.

_____. 1996a. *Canadian Economic Observer: Historical Statistical Supplement 1995/96*. Catalogue no. 11-210. Ottawa: Minister of Supply and Services.

_____. 1996b. *Canada Year Book 1997*. Ottawa: Minister of Industry.

_____. 1997a. The Daily—1996 Census: Immigration and Citizenship, 4 November 1997

[online database], Ottawa. Searched 11 August 1998; <URL:http://www.statcan.ca/Daily/English/>:11 pp.

_____. 1997b. The Daily—1996 Census: Mother Tongue, Home Language and Knowledge of Languages, 2 December 1997 [online database], Ottawa. Searched 14 July 1998; <URL:http://www.statcan.ca/Daily/English/>:15 pp.

_____. 1997c. *A National Overview: Population and Dwelling Counts*, Catalogue no. 93-357-XPB. Ottawa: Minister of Industry.

_____. 1997d. *Mortality—Summary List of Causes, 1995*. Catalogue no. 84-209-XPB. Ottawa: Minister of Industry.

_____. 1998a. Labour Force, Employment and Unemployment [online database], Ottawa. Searched 12 August 1998; <URL:http://www.statcan.ca/english/Pgdb/Economy/Economic/econ10.htm>:1 p.

_____. 1998b. Recent Immigrants by Country of Last Residence [online database], Ottawa. Searched 17 August 1998; <URL:http://www.statcan.ca/english/Pgdb/People/Population/demo08.htm>:1 p.

_____. 1998c. Single and Multiple Ethnic Origin Responses, 1996 Census [online database], Ottawa. Searched 17 August 1998; <URL:http://www.statcan.ca/english/Pgdb/People/Population/demo28a.htm>:2 pp.

_____. 1998d. The Daily—1996 Census: Education, Mobility and Migration, 14 April 1998 [online database], Ottawa. Searched 11 August 1998; <URL:http://www.statcan.ca/Daily/English/>:12 pp.

_____. 1998e. The Daily—1996 Census: Ethnic Origin, Visible Minorities, 17 February 1998 [online database], Ottawa. Searched 16 July 1998; <URL:http://www.statcan.ca/Daily/English/>:15 of 21.

_____. 1998f 1981–1996 Census: Labour Force Activity, 1998 [online database], Ottawa. Searched 2 September 1998; <URL:http://www.statcan.ca/english/census96/mar17/labour/table6/t6p59a.htm>:1 p.

Trewartha, G.T., A.H. Robinson, and E.H. Hammond. 1967. *Elements of Geography*, 5th edn. New York: McGraw-Hill.

United Nations. 1994. *Human Development Report 1994*. New York: Oxford University Press.

Young, R.A. 1989. 'Political Scientists, Economists, and the Canada–US Free Trade Agreement'. *Canadian Public Policy* XV, no. 1:49–56.

Further Reading

Britton, John N.H., ed. 1996. *Canada and the Global Economy: The Geography of Structural and Technological Change*. Montréal–Kingston: McGill-Queen's University Press.

McRoberts, Kenneth. 1997. *Misconceiving Canada: The Struggle for National Unity*. Toronto: Oxford University Press.

McVey, Wayne W., and W.E. Kalbach. 1995. *Canadian Population*. Toronto: Nelson Canada.

Chapter 5

Overview

Ontario is Canada's economic powerhouse, accounting for 40 per cent of Canada's industrial output and 37 per cent of Canada's population. Southern Ontario, which lies in the western sector of the Great Lakes–St Lawrence Lowlands, contains over three-quarters of Ontario's population, is the major sector of Canada's industrial core, and is on its way to becoming a major industrial area in North America. Northern Ontario, on the other hand, with a much more modest population, resembles a declining resource hinterland.

Southern Ontario's automobile industry is examined in this chapter's *Key Topic*. The economic integration of automobile manufacturing in the North American market began under the Auto Pact (1965) and intensified under the Free Trade Agreement (1989) and the subsequent North American Free Trade Agreement (1994). Ontario's trade axis has been reoriented from an east-west alignment to a north-south one.

Objectives

- Describe Ontario's physical geography and historical background.
- Present the basic characteristics of Ontario's population and resources.
- Identify Ontario's dichotomy—an industrial core in southern Ontario and a resource hinterland in northern Ontario.
- Examine the importance of the manufacturing sector, especially the automobile industry, to Ontario's economy.
- Analyse the significance of the Auto Pact and the North American Free Trade Agreement to Ontario.
- Focus on Ontario's changing role within Canada and North America.

Ontario

Introduction

Ontario is often described as the 'Province of Opportunity', but Ontario is a geographic paradox because it has two dissimilar areas, southern Ontario and northern Ontario. Each has very different physical conditions that have formed the basis of two distinct economies. The upbeat description of Ontario as the 'Province of Opportunity' applies to only one part, southern Ontario. Southern Ontario's economy has generated jobs and business opportunities. This area's positive economic environment has drawn people from across Canada and around the world to its cities. The manufacture of automobiles has been a significant part of southern Ontario's economy and constitutes the *Key Topic* in this chapter. While southern Ontario is the industrial and population heartland of the province, northern Ontario is an old resource hinterland, where economic and population growth are stalled. Because so many young people are moving to more prosperous areas in Canada, the population of northern Ontario is losing a valuable segment of its labour force, leaving its cities and towns with disproportionately large numbers of older and very young people. In 1996, northern Ontario had a population of 826,000, while southern Ontario had a population of just over 9.9. million.

Ontario Within Canada

Ontario is the leading economic region in Canada (Figure 5.1). In fact, according to Thomas Courchene, one of Canada's foremost economists, Ontario has evolved into one of North America's leading economic regions and may become the new 'heartland' of North America (Courchene 1998). Four resources—agriculture, forests, minerals, and hydroelectricity—have spurred the province's economic development.

In addition to its economic strength, Ontario is the political linchpin in Confederation. This political role is attributed partly to Ontario's historic role in Confederation and partly to its population size. Almost 38 per cent of Canada's population lives in Ontario, so this province sends more representatives to the House of Commons than any other province.

Canadians living in less prosperous and less powerful regions often see Ontario in a different light, perhaps precisely because of Ontario's economic success and political power. Westerners and Maritimers have accused Ontario of using its political power to

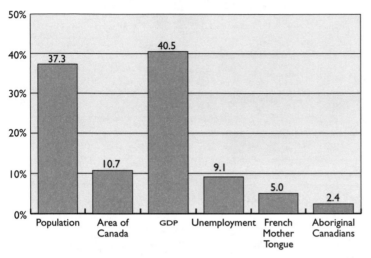

Figure 5.1 Ontario, 1996. Ontario's economic and political strength within Canada is revealed by its share of the nation's GDP and population. Ontario comprises the largest single market in Canada and holds the demographic key to political power in Ottawa, and its economy, by accounting for over 40 per cent of the nation's GDP while having only 37 per cent of its population, exhibits a high level of productivity. Franco-Ontariens and Aboriginal peoples form small minorities.

Sources: Statistics Canada, 1997b:Table 1; 1997e:1 to 3 of 3; 1997f:4 and 5 of 15; 1998a; 1998c:1 and 2 of 10; 1998b:15 to 20 of 21; and Stanford, 1998:185.

maintain Ontario's economic mastery over the rest of the country. Quebeckers have pointed to the striking difference between the two provinces' unemployment rates as proof that Ottawa favours Ontario. Ontarians reject these criticisms, claiming that Ontario has done its share to shoulder the 'burden of Confederation'. For example, Ontario is the major contributor to equalization payments allocated to have-not provinces.

Ontario's Physical Geography

Ontario is larger than most countries (Figure 5.2), stretching out over 1 million km². Extending over Ontario are three of Canada's physiographic regions (Great Lakes–St Lawrence Lowlands, Canadian Shield, and Hudson Bay Lowlands), and three of the country's climatic zones (Arctic, Subarctic, and Great Lakes–St Lawrence) (Figures 2.1 and 2.6). The Great Lakes form the province's southern boundary with the United States. Manitoba lies to its west, Hudson and James bays to its north, while Québec is on its eastern boundary, marked in part by the Ottawa River. This central location within Canada and its close proximity to the industrial heartland of the United States have facilitated Ontario's economic development.

Southern Ontario is located in the southernmost part of Canada. Windsor, for example, is at latitude 42° N, and Toronto is close to 44° N. Southern Ontario lies in the Great Lakes–St Lawrence climatic zone, which has a moderate continental climate. This climate is noted for long, hot, and humid summers followed by short, cold winters. Annual precipitation is about 1000 mm. The greatest amounts of precipitation occur in the lee of the Great Lakes, where winter snowfall is particularly heavy. The Great Lakes modify temperatures and funnel winter storms into this region. During the winter, this region experiences a great variety of weather conditions.

Southern Ontario is the most favoured physical area in Canada. As a result, the vast

Parliament Hill, back view. Because it has the largest population of all the provinces, Ontario has the greatest number of seats—and therefore the greatest degree of representation—in the House of Commons. (*Courtesy R.M. Bone*)

The Niagara Escarpment, located in southern Ontario, was formed by water erosion and glaciation. (*Courtesy Victor Last, Geographical Visual Aids 35664*)

Figure 5.2 Central Canada. Ontario and Québec have the two largest provincial economies and populations in Canada. Toronto and Montréal, the two largest cities in Canada are located in Canada's manufacturing belt, which stretches from Windsor to Québec City.

majority of the province's population, nearly 10 million or 92 per cent, lives in southern Ontario. The area is underlain by slightly tilted sedimentary rocks, which are covered by a thin deposit of glacial till. Except for the Niagara Escarpment, there is little relief topography. This escarpment provides the most spectacular scenery in this subregion. Formed since deglaciation, the Niagara Escarpment is an ero-

sional remnant of the more resistant sedimentary rocks. A mixed forest vegetation flourishes in this temperate continental climate (Figure 2.7). With a long growing season, ample precipitation, and fertile soils, the southern Ontario lowland has the most productive agricultural lands in Canada. The Canadian Shield, however, marks an abrupt end to agricultural land. At this point, northern Ontario begins.

Northern Ontario has a less favoured physical area and only 8 per cent of Ontarians live in this part of the province. Northern Ontario lies in two physiographic regions: the Canadian Shield and Hudson Bay Lowlands. Located in much higher latitudes, its winters are longer and colder than those experienced in southern Ontario. For the most part, northern Ontario lies north of 46° N and extends almost as far north as 57° N. Even along its southern edge, short summers make crop agriculture vulnerable to frost damage. In addition to a difficult climate for agriculture, the rocky Canadian Shield has very little agricultural land. Its rugged, hilly topography is dotted with lakes. Climate, soils, and physiography combine to limit agriculture in northern Ontario. However, this subregion does have a vast forest, superb scenery (which supports tourism), and extensive mineral wealth. As a resource hinterland, northern Ontario's economy is almost entirely based on these three resources, which are found in the Canadian Shield.

Ontario's Historical Geography

Since its beginning as a British colony, Ontario has been the counterweight to Québec. When the American Revolution began in 1775, Ontario was a densely forested wilderness inhabited by a few fur traders and Indians. After losing the American colonies, loyal British subjects returned to England or relocated in one of Britain's colonies. In 1782–3, Loyalists moved northward to Nova Scotia and New Brunswick, while others resettled in the Eastern Townships of Québec. A smaller number, perhaps as many as 10,000 Loyalists, settled

on land along the St Lawrence River upstream from French-speaking *habitants*, colonizing the wilderness area that later became known as Upper Canada. These Loyalists were followed by American, British, and European newcomers, who later flooded into Upper Canada in search of land. In less than a century, this natural landscape was transformed into a British agricultural colony.

Until the War of 1812, many settlers had come from the United States. They, like most other colonists, sought land in the fertile lowlands of the Great Lakes. When hostilities between Britain and the United States escalated, the Americans launched an attack on British North America. The War of 1812 effectively ended the influx of American settlers into Upper Canada. After the war, an increasing number of settlers came from the British Isles, especially Ireland and Scotland. By 1830, Upper Canada's population grew rapidly from natural increase and immigration. Thirty years later, Upper Canada had grown in numbers to nearly 1.4 million. Of these inhabitants, about 33,000 French-speaking Canadians lived along the Detroit River, where their ancestors had put down roots during the French regime in North America.

At first, settlement took place along the coastal edge of Lake Ontario and Lake Erie, where forests were cut down to clear land for farming. Then the colonists moved northward towards the Canadian Shield, which marked the limits of agricultural land. Within thirty years, virtually all the arable land in the Great Lakes Lowlands was cleared of forest and cultivated. By 1830, settlers occupied the best lands. Within a decade, some turned further northward to try their luck in the few pockets

A watercolour, painted in 1804, showing part of York, the capital of Upper Canada and one of the region's growing urban centres. (*Courtesy National Archives of Canada C34334*)

of arable land within the Canadian Shield. By the time of Confederation, there was a land shortage in Upper Canada and settlers began to look westward for land.

While Upper Canada was clearly a rural society during the nineteenth century, a network of urban places began to spring up. With the growing shortage of agricultural land, sons and daughters of farmers moved to these urban centres to start a new and different life. In spite of this rural-to-urban migration, the demographic balance of power clearly remained with the rural community—urban areas, such as villages and towns, contained less than 20 per cent of the population of Upper Canada.

When Upper Canada joined Confederation in 1867, it was renamed Ontario (Figure 3.4). At that time, the geographic extent of Ontario was about 100 000 km²—just a fraction of its present size—but as Canada grew in size, acquiring more territory from Great Britain, both Ontario and Québec benefited by obtaining some of these new lands (Figures 3.5–3.7). Since Confederation, the borders of Ontario have been extended three times. These boundary changes have greatly increased the geographic size of the province. The first expansion occurred in 1874 when Ontario's boundaries were pushed northward to about 51° N and eastward towards the Lake of the Woods. The second expansion took place in 1889. Until then, both Manitoba and Ontario claimed the land around the Lake of the Woods. The Ontario government won this dispute and was able to enlarge the province once more. At the same time, Ontario's northwest boundary was adjusted to the Albany River, which flows into James Bay. As a result of this adjustment, Ontario gained access to James Bay. In 1912, the final boundary modification occurred when the District of Keewatin south of 60° N was assigned to Ontario and Manitoba. As a result, Ontario extended its political boundary to the northwest, stretching from northern Manitoba to the coast of Hudson Bay at the latitude of 56°51' N. Through these boundary adjustments, Ontario reached its present geographic extent of 1 million km².

The Birth of an Industrial Core

The geographic essence of Ontario lies in its industrial base. Manufacturing in southern Ontario began in the nineteenth century and received a boost in 1854, when Britain and the United States signed the Reciprocity Treaty, which allowed trade between Upper Canada and the United States (primarily the states of Michigan, New York, and Pennsylvania). By 1867, Ontario had the largest population of Canada's four founding provinces and a fledgling industrial base. Small manufacturing outfits existed in small villages and often con-

View in King Street [Toronto], Looking East, 1835, by Thomas Young. (Courtesy National Archives of Canada C1669)

sisted of less than five employees (i.e. a village blacksmith or miller). Larger manufacturing activities took place in towns and cities, where sawmill, gristmill, and distillery operations were commonplace. At that time, most manufacturing activities were dependent on waterpower, so most were located near a stream or river.

The Reciprocity Treaty lapsed in 1866, cutting off Upper Canada's access to the American market. Confederation and the National Policy, however, enabled Ontario to secure the Canadian markets. Under the National Policy high tariffs were imposed on imported manufactured goods, which allowed manufacturing in southern Ontario to flourish. (For more about the Reciprocity Treaty and the National Policy, see Chapters 1 and 3.) The consequences of Confederation and protective tariffs for Ontario were threefold: (1) the creation of a Canadian market for Ontario products; (2) an increase in the size of the more successful manufacturing companies; (3) and the growth of the industrial workforce in Ontario. In the rest of the country, however, prices for manufactured goods (made in Ontario) were generally higher than in the adjacent areas of the United States. The regional price variations reflected two factors: Canadian transportation costs and Ontario's more limited economies of scale in comparison to those of US manufacturers.

Confederation therefore provided the impetus that established a national industrial core in southern Ontario as well as Québec, while the rest of the country was relegated to becoming a domestic market for these manufactured goods. Not until the Free Trade Agreement between Canada and the United

States was signed in 1989 did this national core/periphery relationship undergo significant change. (See Chapter 1 for more on the Free Trade Agreement.)

Ontario Today

Ontario remains Canada's economic engine. A measure of Ontario's industrial success is its enormous energy demand. Ontario's powerful economy is fuelled by electricity generated within the province by hydroelectric installations at Niagara Falls and along the Ottawa River, by thermal coal plants using coal from Pennsylvania and West Virginia, and by nuclear-generators in plants at Pickering and Bruce Peninsula. Still Ontario's industry requires more energy. To meet that demand, electricity is imported from Québec and petroleum from Alberta.

Ontario is divided along two quite different economic functions. In 1951, one of Canada's well-known geographers, Donald Putnam (1952:213), considered this division to be based on the physical differences between the Great Lakes Lowlands and the Canadian Shield. These physical differences created distinct economic activities and settlement patterns. Northern Ontario, for example, has the characteristics of a resource hinterland, while southern Ontario is the epitome of an industrial core. As discussed earlier in this chapter, geography goes a long way to explaining this regional variation. With this regional dichotomy in mind, the following economic and population statistics for Ontario apply best to southern Ontario:

- Ontario produces 40 per cent of Canada's GDP.
- Its value of annual output exceeds $300 billion.
- Its average personal income is well above the national average.
- With over 37 per cent of Canada's population, it has the largest population of the six geographic regions.

Ontario's economic success has been facilitated by conditions that promoted trade for the province, with both other provinces and other countries.

Trade

Ontario is well positioned to engage in trade, both domestically and internationally. From northern Ontario, forest products and minerals are shipped to both national and international markets. In Southern Ontario, manufactured goods ranging from aircraft to steel products are produced for both these markets.

The free trade agreements and Ontario's geographic location within North America have allowed the region's business firms to penetrate into the huge US market. For example, Ontario's exports to the United States in the 1980s were about the same value as those to the rest of Canada; now Ontario's exports to the US are two and a half times the value of exports to the rest of Canada (Courchene 1998:276–7). The sheer size of the US market near Ontario has tilted the province's trade and economy in a new direction—a north-south alignment.

The international trade dimension of the Ontario economy is evident in the 1996 export figures. In that year, the leading Canadian exports by value were automobiles and their parts, and the value of motor vehicle exports almost reached $68 billion (Statistics Canada 1997d:1). Since over 90 per cent of these automotive exports came from assembly and parts plants in Ontario, the importance of this industrial sector to Canada and Ontario cannot be overestimated. Furthermore, the automobile industry has a ripple effect on the economy of southern Ontario by creating a demand for steel, rubber, plastics, aluminum, and glass products.

Southern Ontario is, in reality, a northern extension of the North American manufacturing belt. With US automobile plants just across the border in Detroit, it is not surprising that this type of manufacturing spread across the Detroit River into Windsor. Initially this northward diffusion of the automobile industry took the form of branch plants owned by the Big Three (General Motors, Ford, and Chrysler). Until the Auto Pact, Ottawa placed high tariffs on imported automobiles and their parts. In doing so, the federal government encouraged American and other foreign firms to open branch plants that produced similar goods but, because of the size of the Canadian market, were smaller manufacturing plants than those in the United States and various European countries. The net effect was higher prices for Canadian consumers due to higher production costs, the high cost of shipping these goods long distances to regional markets in Canada, and the provincial/federal taxes on automobiles that were higher than those in the United States. Ontario, however, benefited from the expansion of its industrial base.

The Canada–United States Auto Pact eventually integrated Canada's automobile assembly plants into the North American market. The Auto Pact not only altered the nature and purpose of automobile manufacturing in Ontario but also marked a shift in federal policy towards trade relations, especially continentalism (see Chapter 1 for a discussion of continentalism). Today, Ontario has a healthy trade market, but it has been forced to adjust to a number of changes in the world economy to maintain its strong economic position.

The New World Economic Order

During the 1980s, two events propelled Ontario into a new world economic order, forcing a sometimes painful restructuring of its manufacturing sector. **Restructuring** meant that companies had to become more efficient, and in Canada's high-wage economy, that usually involved reducing the size of their labour force.

The first event was the liberalization of world trade. Under the terms of the General Agreement on Tariffs and Trade (GATT), most countries in the world, including Canada, agreed to reduce barriers to international trade. In 1940, for example, the average tariff on manufactured goods had been about 40 per cent. Since 1947, however, a series of tariff reductions were achieved through agreements made by members of GATT. By the eighth agreement in 1990, the average tariff was 5 per cent (Dicken 1992:153). Today the main impediments to trade are not tariffs but non-tariff barriers such as import quotas ('voluntary' export restraints) and health regulations.

The second event to shape the new economic order was the continental **Free Trade Agreement** (FTA) between Canada and the United States. The FTA committed Canada and the United States to an integrated North American economy. In 1994, the FTA was replaced by a similar agreement, the **North American Free Trade Agreement** (NAFTA), which integrated Mexico in the trade agreement. Since tariffs between Canada and the United States are decreasing at a faster rate under NAFTA than under GATT, NAFTA gives Canadian manufacturing firms an opportunity to adjust to the new North American market before being fully exposed to world trade. By creating larger operations, Canadian firms hope to achieve economies of scale that will decrease their per unit costs of manufacturing and therefore allow them to become more competitive.

One result of FTA and NAFTA is that trade between Canada and the United States is increasing, thereby making Canada more dependent on the United States. From this perspective, Canada has hitched its wagon to one horse, which is especially true for southern Ontario (Norcliffe 1996:32). The consequences of this trade dependency are twofold: (1) booms and slumps in the American economy affect Canada more directly than before FTA/NAFTA, and (2) Canada's long-term economic fortunes are even more closely tied to the United States' economy than before FTA/NAFTA.

These trade agreements have raised controversial issues within the country and led to bitter disputes with the country's main trading partner, the US. Leading up to FTA, access to the American market was considered crucial for Canada. Therefore, the Canadian government's objectives were the reduction of protective tariffs and the removal of non-tariff barriers—initiatives that would ensure better access to the American market. Non-tariff barriers included quotas, such as on durum wheat, or special countervailing penalties on certain Canadian export industries in order to protect American producers of the same goods, such as American grain farmers and American lumber companies. Canada insisted that FTA have a trade-dispute settlement mechanism to prevent the use of barriers that contravened the trade agreement. As time has proven, even when this mechanism rules in favour of a Canadian firm, intense lobbying by American firms causes the American government to intervene. Since 1989, there have been several bitter disputes between Canadian and American grain farmers and lumber companies. Each time, the United States government has pressured Ottawa into accepting a 'voluntary quota' on Canadian exports to the United States. For example, in the 1990s, quotas were placed on the amount of durum wheat and lumber that could be shipped to the United States.

Within Canada, FTA committed the country's manufacturing sector to become more competitive. Firms unable to adjust to the highly competitive marketplace in North America would be forced to close. For most firms, adjusting to greater competition meant lowering the costs of production by creating larger manufacturing plants and by increasing the productivity of workers. As part of this economic restructuring, American branch

plants lost their *raison d'être* (MacLachlan 1996:195–214 and Norcliffe 1994:2–17). A few adjusted their operations by specializing in one aspect of the manufacturing process to serve the North American market. In this way, a few branch plants were able to achieve economies of scale, which reduced their costs of production per unit of output. Most companies simply ceased production, thus allowing their parent plant in the United States to supply the Canadian market.

The outcome, while far from clear, is creating a new geography of manufacturing in southern Ontario. As MacLachlan stated:

> The full effect of the agreement [NAFTA] on individual regions and labour markets is still conjectural. It appears that southern Ontario stands the greatest risk of employment losses because of its large secondary manufacturing sector and continued dependence on foreign-owned subsidiaries for employment in key manufacturing industries (MacLachlan 1996:195).

The fear of employment losses continues to haunt Canadian workers. While the total number of employed workers has increased since 1989, unemployment rates have stayed at levels not known in Canada since the Great Depression. The problem is that job growth has not kept pace with the rising number of workers entering the labour market. Furthermore, the increased competition brought on by NAFTA will continue to challenge, sometimes threaten, the province's manufacturers—a sector of the economy that Ontario heavily relies upon—and thereby maintain a level of uncertainty in the labour market.

Ontario's Economy and Industrial Structure

Today, two major developments are affecting the new world order and by extension, Ontario's economy. The first, the globalization of economic activities, involves the integration of world production and trade. Multinational corporations are particularly well suited to engage in global economic activities. For that reason, large international firms have an advantage over smaller firms, which have traditionally been oriented towards domestic markets. In southern Ontario, the trend is towards larger firms. In the auto-parts industry, for example, this trend is well underway, while it is just beginning in the financial world with the proposed mergers of major banks.

The second development is that rates of economic and technological change are creating uncertainty and volatility among firms and their employees. Ontario, for instance, has been experiencing high rates of unemployment even though its economy has been working at a high level (see Chapter 4). Another outcome of this uncertainty is the shift of workers from one sector of the economy to another. Economic geographers divide employment into four sectors: primary, secondary, tertiary, and quaternary. (See Chapter 4 for a discussion of these four sectors of a modern economy.) The allocation of workers into each of these sectors constitutes the industrial structure of an economy. In Ontario, tertiary and quaternary sectors dominate. Those who perform quaternary activities are senior members of a firm or public agency who make key decisions about company operations. These people direct the affairs of businesses in all three sectors of the economy. Unfortunately, the industrial sector data collected by Statistics Canada cannot identify workers in the quaternary sector. Consequently, Table 5.1 has only three industrial sectors—primary, secondary, and tertiary (which includes quaternary).

The tertiary sector has become the dominant form of economic activity in all modern industrial states. Ontario is no exception, although Ontario's economy retains a relatively strong secondary sector because of its manufacturing activities. Employment data in Table 5.1 illustrate both the dominance of the tertiary sector (over 73 per cent) in Ontario's economy and the importance of the secondary sector (almost 24 per cent).[1]

The secondary sector is concentrated in southern Ontario where both heavy industry (steel) and final product industries (automobile and aircraft industries) are found. Final product industries generate a strong demand for manufactured goods. For example, the automobile industry requires steel products. The importance of the automobile industry to Ontario is further explored under the *Key Topic*.

The primary sector has been shrinking. The greatest shift in employment has taken place in this sector, with fewer and fewer jobs available. Most primary activities, such as mining and forestry, take place in northern Ontario. Uncertainty in the workplace has been particularly evident in the primary sector as the skills and job experience gained in primary activities often do not correspond to

Table 5.1
Employment by Industrial Sector in Ontario, 1995

Industrial Sector	Workers (000)	Workers (per cent)
Primary	156	3.0
Secondary	1,236	23.6
Manufacturing	972	18.6
Tertiary	3,839	73.4
Transport	381	7.3
Trade	865	16.5
Finance	353	6.8
Service	1,947	37.2
Public	293	5.6
Total	5,231	100.0

Source: Adapted from Statistics Canada, 1996: Table 40.

the skills and experience needed in the growing service sector.

Southern Ontario

Southern Ontario occupies the southern half of the Great Lakes–St Lawrence Lowlands and accounts for much of Canada's agricultural output by value. Agriculture in southern Ontario represents an important primary industry. The large urban population provides a stable local market for its dairy, livestock, and fresh vegetable products. Although the amount of cropland is relatively small, its farming operations are much more intense than in Western Canada, where extensive grain crops dominate farming operations. The difference between the two agricultural areas is revealed in the size of farms and the types of crops. In 1996, the average size of farms in Ontario was 206 acres (83 ha) compared to an average of 1,152 acres (466 ha) for farms in Saskatchewan (Statistics Canada 1997a: Table 7). Grain farming and cattle ranching dominate in Western Canada, while there is a much more diversified agricultural land use in southern Ontario, where corn, barley, and winter wheat are the important crops along with highly specialized ones such as tobacco and vegetables. Corn and grain often serve as fodder for the hog, dairy, and beef livestock farms that also operate in Ontario.

A farm in Stittsville, southwest of Ottawa. Farming in southern Ontario is characterized by dairy, cattle, and fruit/vegetable production. (*Corel Photos*)

Agricultural land use varies within southern Ontario due to subtle but important physical differences. For example, soils are less fertile and the growing season is shorter east of Toronto than west of Toronto. There are three highly specialized agriculture zones west of Toronto: the Essex–Kent vegetable area, the Norfolk Tobacco Belt, and the Niagara Fruit Belt. All lie in southwestern Ontario where the growing season is extremely long. These zones, located south of 43° N, are the southernmost lands in Canada.

Agricultural changes, however, are affecting southern Ontario farmers. For instance, prior to 1995, feed grain for Ontario's livestock industry was imported from Western Canada. However, with the elimination of the Crow Benefit (a transportation subsidy), western feed grain is now too expensive for Ontario farmers, thus causing a major realignment in the livestock industry. Another change is forecast when Canada and Ontario discontinue marketing boards. Though their purposes vary, Canadian and provincial marketing boards generally control their respective markets by ensuring stable prices and incomes for producers and standard market access. Ontario farmers, especially dairy farmers, are accustomed to selling their products through marketing boards. These boards may be forced to cease their operations because, by controlling agricultural trade, marketing boards may be violating the free trade rules under NAFTA and GATT. Without such protection, Ontario dairy farmers may have difficulty competing with American farm products.

Southern Ontario, with almost half of all manufacturing jobs in Canada, is also the leading financial, manufacturing, and service subregion in Canada (Statistics Canada 1997c:369). Most Canadian-based corporations have located their head offices in Toronto. Toronto's Bay Street is the centre of the Canadian financial world. Other major urban centres also house important sectors of southern Ontario's manufacturing industry. Hamilton is Canada's steel city. Two steel companies, Dofasco and Stelco, are based in Hamilton. Auto-assembly plants are in eight urban centres: Windsor, St Thomas, Ingersoll, Cambridge, Bramalea, Alliston, Oakville, and Oshawa. All these communities have two features: excellent access to major highways, and proximity to major markets in Michigan, Ohio, Pennsylvania, New York, and Québec.

Four factors have led to the successful development of manufacturing in southern Ontario. Using the automobile industry as an example, it becomes apparent how each factor contributed to the growth of manufacturing in Ontario. The first factor is Ontario's proximity to America's manufacturing belt, which led American industries to locate branch plants in Canada and gave Canadian plants ready access to a large American market. This spillover effect first took place in the Windsor–Detroit area in 1904, when the Ford Motor Company established an automobile assembly plant in Windsor. The second factor was trade restriction on foreign manufactured goods (imposed

by the National Policy in 1879). Automobile parts from Ford's plant in Detroit had to be ferried across the Detroit River and assembled in an old carriage factory in Windsor. Access to the markets in the British Empire provided a third advantage. American branch plants in Canada could take advantage of lower tariffs for Canadian-made products in the British Empire. As a result, not only were Canadian-built Ford cars sold in Canada but also in various parts of the British Empire. General Motors followed this same strategy by opening a branch plant in Oshawa. The size of its domestic market provided southern Ontario with its fourth location advantage for the automobile industry. Most cars were sold in southern Ontario, while the remainder were shipped to the other regions of Canada. In this way, transportation costs were minimized.

Under the policy of high tariffs, manufacturing firms in southern Ontario enjoyed a stable economic environment. Since 1989, however, manufacturing firms have had to adjust to competition from American firms. Often these adjustments have resulted in the closure of American branch plants or the consolidation of Canadian firms. Even the most profitable firms in Canada have merged in order to attain a size deemed necessary to compete in the North American and global markets. The purpose is twofold: to compete in the world marketplace and to avoid a takeover by a large foreign company. Mergers, however, often result in the closure of offices to avoid duplication and thus, in the reduction of the number of employees.

Like many towns in southern Ontario, Cambridge shows signs of such economic changes. Created in 1973 by the amalgamation of Galt, Preston, and Hespeler, Cambridge is an industrial centre just west of Toronto. This manufacturing town has suffered from plant closures. Savage Shoes, Inglis, Franklin Manufacturing, InterRoyal Corp., and Dobbie Industries Ltd have shut their doors, laying off hundreds of workers. Savage Shoes simply could not compete with lower-cost shoes from developing countries, where labour costs are a fraction of those in Canada. After FTA, the American owners of the Inglis household appliance plant, the InterRoyal furniture factory, and the Dobbie wool-spinning mill decided to close their branch plants and to produce their entire output from their much larger plants in the United States.

One American branch plant, Rockwell Automation, has managed to survive, but with a smaller labour force. Before FTA, this firm produced electromechanical devices used in automated factories in Canada. To survive in the North American market, the plant had to find a single product for the global market. Now it is the only Rockwell plant in the world that produces medium-voltage products and electrical generators, 70 per cent of which are exported to the United States and other foreign countries. Niche production for the North American market allowed this plant to achieve economies of scale, and now its products can compete for business in all parts of the world (Campbell 1996:D1–2).

Cambridge has also been affected by globalization. In 1985, Cambridge became an important automobile-assembly centre for Toyota Motor Manufacturing Canada Inc. Toyota employs over 2,000 workers, making this plant the city's largest employer. The presence of an automobile-assembly plant has had several benefits for Cambridge: a stable employment base, high-paying jobs, and spinoff effects due to purchases by the company and its employees.

Key Topic: The Automobile Industry

The automobile industry was the key manufacturing activity in North America for most of the twentieth century. Peter Dicken put it best:

> The significance of the industry lay not only in its sheer scale but also in its immense spin-off effects through its linkage with numerous other industries. The motor vehicle industry came to be regarded as a vital ingredient in national economic development strategies (Dicken 1992:268).

The success of the auto industry in southern Ontario today dates back to the 1960s when the Auto Pact was signed.

The Auto Pact

The Auto Pact is an example of a successful single-industry production-sharing agreement between Canada and the United States. Before the Auto Pact, the auto industry in Canada had small, high-cost plants, each a smaller version of the much larger plants in the United States (Vignette 5.1). Even though Canadian wages for auto workers were lower than those for American workers, Canadian automobile prices were considerably higher

Vignette 5.1

Branch Plant Canada

In the United States in the second half of the nineteenth century there was a managerial revolution in the organization of production and distribution. Instead of one factory marketing its products through agents and merchants, corporations emerged that not only had factories but also elaborate networks of warehouses, service facilities, and, in some cases, retail outlets. This new corporate system facilitated the development of a continent-wide market by allowing companies to circumvent tariff barriers by setting up branch plants in the protected market. The Canadian market, especially Ontario, was protected by tariffs in the National Policy of 1879 but was especially attractive to the new American corporations because it was culturally similar, close to the major American manufacturing cities, and could provide a base for the development of British Empire markets. These factors contributed to the emergence of Canada as a popular location for American-owned branch plants.

Source: Anthony Blackbourn and Robert G. Putnam, *The Industrial Geography of Canada* (London: Croom Helm, 1984):139–40. Reprinted by permission of Routledge.

than for the same model in the United States. The Canada–US Auto Pact was signed in 1965. The agreement called for Canada to eliminate the 15 per cent tariff on imported automobiles and US parts for Canadian manufacturers, and for the US to eliminate its corresponding tariff. The agreement included special protection for Canadian plants. Canada was guaranteed a minimum level of automobile production based on production levels for each type of automobile manufactured in Canada in 1964; that is, the ratio of vehicle production to sales in Canada for each class of vehicle had to be at least 75 per cent or the percentage attained in the 1964 model year, whichever was the highest. Furthermore, **value added** in vehicles produced in Canada had to be no less than the dollar amount achieved in 1964.

The Growth of the Auto Industry

The automobile industry drives the Ontario economy. Nearly one in seven Canadian manufacturing jobs depends directly on this industry (Holmes 1996:230). Wages in the assembly plants are relatively high for semi-skilled workers, but somewhat lower for workers in parts firms. Nevertheless, these incomes enable workers to make substantial purchases, thereby stimulating southern Ontario's retail sector. In addition, the automobile industry accounts for nearly one-quarter of Canada's merchandise exports. (Most of Canada's remaining exports are primary products, such as grain and forest products.)

The automobile industry consists of two separate operations: the assembly of automobiles and trucks and the production of their parts. In addition to these two operations are manufacturing firms that supply semi-processed materials. In southern Ontario, there are fabricating firms that produce steel, rubber, plastics, aluminum, and glass parts for automobile assembly and parts plants in Canada and the United States. Then there are the service firms, ranging from the advertisers and designers to the sales and service staff. In short, the auto industry is a final product type of manufacturing, and as such, its added value reaches a maximum.

By being highly efficient and strategically located, automobile-parts firms can operate on a **just-in-time principle**; that is, auto components are produced in small batches and quickly delivered as needed to their customers. In turn, this allows the auto-assembly plants to achieve considerable savings by reducing their inventories, warehousing space, and labour costs. By 1996, Canada's automobile industry was in a very healthy state. Because of the low Canadian dollar, auto and light truck production had increased over the last decade, reaching a record 2.5 million units in 1996. Canada now accounts for 17.5 per cent of North American assembly capacity, though only 7 per cent of automobile sales (Jestin 1996:10).

A similar pattern of growth is emerging in the auto-parts industry. By the late 1980s, this sector was expanding mainly because automobile manufacturers decided to subcontract much of the parts business, a practice called **outsourcing**. Purchasing parts from specialized firms allowed automobile manufacturers to concentrate on assembling automobiles and reducing their costs. Cost savings result from the fiercely competitive bidding process among parts firms and the lower wages that parts firms pay their workers. Their labour force is non-unionized and, as a result, workers receive a much lower pay than unionized workers in the automobile-assembly plants. This wage differential and the savings it provides is the main reason why General Motors, Ford, and Chrysler continue to divert work to auto-parts firms. In Ontario, the Canadian-based Magna International Inc. has grown into the third largest auto-parts company in North America. Encouraged by Canadian workers' improved productivity and a weak Canadian dollar, automobile firms are subcontracting more parts to firms in Canada. The average North American vehicle now contains about $1,500 of Canadian parts, up 27 per cent since the late 1980s (Jestin 1996:10). Employment in auto-parts production reached a new high in 1997, despite ongoing industrial consolidation.

Automobile-assembly plants are concentrated in southern Ontario where transportation links to the major markets of Canada and the United States are readily available and driving distances are short (Figure 5.3). All Canadian-based automobile companies sell most of their cars and trucks in the United States. In 1996, approximately 80 per cent of all vehicles produced in Ontario were sold in the United States. Three companies, General Motors, Ford, and Chrysler, account for over 90 per cent of vehicle production, 41 per cent of the value of parts produced, and 50 per cent of the industry's employment (Holmes 1996: 231). Competition from Japanese and Korean automobile companies with assembly plants located in southern Ontario is fierce. These Asian-based firms have captured 20 per cent

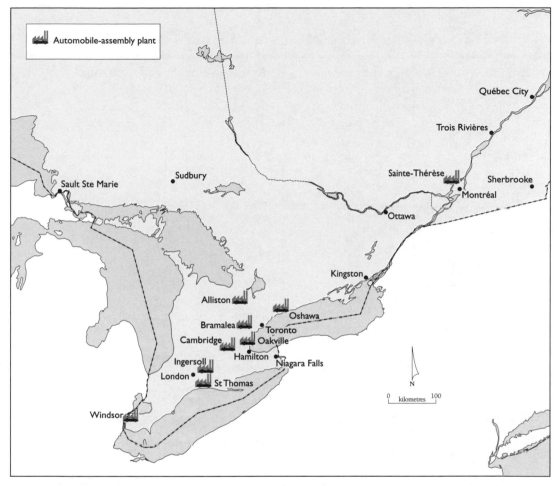

Figure 5.3 Automobile-assembly centres in Central Canada. Eight of Central Canada's automobile-assembly plants are located in southern Ontario. One more is situated in Québec, at Sainte-Thérèse northwest of Montréal.

of Canada's market and an equivalent part of the American market.

While Japanese automobile factories are located in the United States, several advantageous features in Canada have attracted substantial Japanese investment in Ontario. These features include the availability of a highly motivated workforce, the lower value of the Canadian dollar compared to the American dollar, and the existence of a public medical system. In American-based automobile-assembly plants, the cost of providing medical insurance is often built into the wage agreement.

Except for the General Motors plant at Sainte-Thérèse near Montréal and a Volvo assembly plant in Halifax, all auto-assembly plants are located in southern Ontario close to both the Canadian and American markets (Table 5.2). Both the Sainte-Thérèse and the Halifax plants have 'marginal' locations. In

Table 5.2
Automobile-Assembly Plants in Southern Ontario, 1996

Company	Location	Employment	Product
Chrysler Canada	Windsor	2,000	Dodge Ram vans
	Windsor	5,600	Chrysler minivans
Ford Canada	St Thomas	2,700	Crown Victoria
			Grand Marquis
Suzuki Canada	Ingersoll	2,200	Geo Metro
			Geo Tracker
			Suzuki Swift
			Suzuki Sidekick
Toyota Motor	Cambridge	800	Toyota Corolla
Chrysler Canada	Bramalea	3,200	Chrysler Intrepid
			Chrysler Concorde
			Eagle Vision
Honda Canada	Alliston	1,800	Honda Civic
Ford Canada	Oakville	3,600	Ford Windstar
Ford Canada	Oakville	1,000	F-series half-ton truck pickups
GM Canada	Oshawa	3,250	C/K half-ton truck pickups
GM Canada	Oshawa	3,250	Chevrolet Lumina
			Monte Carlo
GM Canada	Oshawa	3,800	Chevrolet Lumina
			Buick Regal

Source: Bruce Little and Greg Keenan, 'Ontario's Economic Future Is the Sum of Its Auto Parts', *The Globe and Mail* (2 March 1996):A1, A10. Reprinted with permission from The Globe and Mail.

fact, in September 1998, Volvo announced that it plans to close its Halifax assembly plant.[2] Auto-parts firms tend to locate their plants in close proximity to automobile-assembly plants. For example, in 1996 Magna International Inc. decided to build a truck frame plant in St Thomas, from which truck frames could be delivered to any of three General Motors truck-assembly plants within one day. One of these assembly plants is located at Oshawa, Ontario, while the other two are located in Pontiac, Michigan, and Fort Wayne, Indiana.

The globalization of the automobile industry has created a highly competitive situation. The consolidation of production in North America is likely to continue. Both assembly and parts production are likely to become more specialized, thus concentrating production in fewer and fewer places. While some automotive units in southern Ontario have closed as part of the consolidation process, major expansion is underway by Honda, Toyota, and the three American automotive firms. The challenge for southern Ontario is to remain an attractive place to invest for automobile manufacturers.

Northern Ontario

Located in two physiographic regions, the Canadian Shield and the Hudson Bay Lowlands, northern Ontario comprises 87 per cent of the geographic area of Ontario yet has less than 10 per cent of Ontario's population. Most live in towns and cities along the two major transportation routes that link Montréal and Toronto with Winnipeg. Both these trans-Canada routes cross the rugged Canadian Shield. The southern route includes the Canadian Pacific Railway and the Trans-Canada Highway. The Trans-Canada Highway connects North Bay with Sudbury, Sault Ste Marie, Nipigon, Thunder Bay, Dryden, and Kenora. The northern route includes the Canadian National Railway and a major highway. This northern highway connects North Bay, Timmins, Kirkland Lake, Cochrane, Kapuskasing, and Nipigon to Thunder Bay.

Like old resource hinterlands elsewhere in Canada, northern Ontario is troubled by a sluggish economy and high unemployment rates. Consequently, its population has two characteristics that are strikingly different from southern Ontario: (1) a slowly increasing but aging population; and (2) a net out-migration, especially of younger members of its population.

Mining, forestry, and tourism are the major economic activities, although many people are employed in the public sector. Like other resource hinterlands, the development of northern Ontario's economy was linked to external markets. Resource exploitation in northern Ontario was possible only in the late-nineteenth century, after the construction of railways that would allow access to external markets. By 1900, resource development formed a narrow zone along the Canadian Pacific Railway (CPR) and the Canadian National Railway (CNR). The same pattern of resource development took place along the Timiskaming and Northern Ontario Railway (also known as the Ontario Northland Railway).[3] The railway extends northward from North Bay to New Liskeard, Kirkland Lake, and

Mining in northern Ontario is a major part of the region's economy. (*Courtesy Victor Last, Geographical Visual Aids 38251*)

Cochrane, which is on the CNR main line. Northern Ontario exports minerals and forest products, including gold, nickel, newsprint, and lumber. Yet all these resource products constitute less than 10 per cent of the value of Ontario's exports.

Unlike in the northern hinterland of Québec, the natural conditions in northern Ontario for hydroelectric developments are not promising. While a number of rivers flow across northern Ontario to James Bay and Hudson Bay, the gentle slope of the land, especially in the Hudson Bay Lowlands, does not make the construction of hydroelectric dams possible. A few established sites for hydroelectric production, such as along the

Ottawa River and at Abitibi Canyon about 100 km north of Cochrane, are important, but they cannot meet the energy demands in Ontario. The much higher elevations in the Canadian Shield area of Québec provide the necessary natural drop to drive the turbines that produce hydroelectric power. Since Québec has a surplus of electrical power, some of it is transmitted to southern Ontario.

Northern Ontario has a strikingly different settlement pattern from that of southern Ontario (Figure 5.2). Long distances separate the four major cities. For example, the distance from North Bay to Sudbury is over 100 km; from Sudbury to Sault Ste Marie is about 300 km; and from Sault Ste Marie to Thunder

Vignette 5.2

Population Distribution in Northwestern Ontario

The large, sprawling Thunder Bay census division of northwestern Ontario stretches from Lake Superior to Hudson Bay and James Bay. This division has less than 160,000 people. As is typical of northern Ontario, most people live in a few large urban centres while the remaining territory is virtually empty. Northwestern Ontario's population geography has three distinct zones:

- *Zone 1*: Most people (71 per cent) reside in Thunder Bay. In 1991, this major population cluster had 114,000 people.
- *Zone 2*: Most of the remainder (27 per cent) live south of the Canadian National Railway line where a number of small towns, reserves, and villages are located.
- *Zone 3*: North of the Canadian National Railway line lies the 'empty' population zone, which is beyond the reach of the national highway system and has only 2 per cent of the region's population.

Bay is about 500 km. The explanation for this oasis-like settlement pattern is due to northern Ontario's physical geography. The rocky terrain of the Canadian Shield discourages continuous settlement; a series of isolated settlements are located at mining sites, pulp plants, and key transportation junctions. Downsizing and even abandonment of resource towns are characteristic of an old hinterland's settlement pattern. Single-industry towns like Cobalt, Kirkland Lake, and Porcupine rose and fell as the cycle of resource exploitation ran its course.

Beyond these towns and cities lies a sparsely populated area where fewer than 10,000 people live. Cree and Ojibwa Indians live on isolated reserves or small Native settlements in this vast region of shrub forest and muskeg. Only a few gold mines, trapping cabins, and fishing lodges dot the landscape. For the Indian residents, fishing and hunting are an important source of food, while trapping and guiding provide cash income. Since this cash income usually does not meet their basic needs, transfer payments supplement their income.

Mining Industry

The Canadian Shield provides ideal geological conditions for the formation of hard rock minerals such as gold, nickel, and uranium. In the 1990s, the annual value of Ontario's mineral production was about $4 billion. Most of that production took place in northern Ontario.

Mining has always played an important role in the economy of northern Ontario. In 1883, the copper-nickel ores of the Sudbury area were discovered during the building of the Canadian Pacific Railway. In 1903, a rich silver deposit near Cobalt was detected during the construction of the Temiskaming and Northern Ontario Railway. Rich gold deposits were uncovered near Timmins in 1909 and at Kirkland Lake in 1911. New mineral finds keep the mining economy of Ontario flourishing. During the 1960s, a large lead-zinc deposit was discovered in Timmins. A decade later, rich gold deposits were found at Marathon (the Hemlo gold mine) and at Pickle Lake.

Promising deposits are currently being evaluated for commercial production, but unlike in the past, new towns are unlikely to spring up to house the mining workforce (unless the mineral deposit is as large as the nickel deposit at Sudbury).

Minerals are a non-renewable resource that is depleted over time. Therefore, mining communities can have a short life span. A mine closure can occur without warning, causing the sudden demise of a single-industry centre. Over the years, a number of ore bodies have been exhausted or production has ceased because new, lower-cost mines have been discovered. These two circumstances have forced some mines to close, resulting in a great outflow of miners and their families. For instance, Elliot Lake's uranium mine was closed, not because its ore was exhausted but because open-pit uranium mines in northern Saskatchewan could produce uranium oxide at a much lower price. Ontario Hydro, the principal buyer of Elliot Lake uranium, decided to purchase its supplies from the uranium mines in northern Saskatchewan. Other mines that have closed include an iron mine at Steep Rock and the silver mines at Cobalt. While former mining towns may remain, they undergo a drastic downsizing. With a stock of low-priced houses and an attractive wilderness setting, Elliot Lake made a modest recovery as a retirement centre. However, its population has declined steadily and is now only about 13,000 (see Table 5.4).

Forest Industry

Across northern Ontario, the boreal forest is the largest natural vegetation zone in the province. Forests are a renewable resource. Sustainable logging changes the forest, but does not destroy it. In this way the forest is renewed, allowing generation after generation to continue to harvest timber. Communities based on such timber harvesting practices are assured of a stable economy and a long existence, but clear-cut logging has threatened the boreal forest in northern Ontario. Since the 1950s, the volume of timber harvesting has more than tripled. This level of logging has put so much pressure on the boreal forest that original species are disappearing. As a result, the boreal forest is shifting from coniferous to broadleaf species. Spruce, pine, and fir have been replaced by poplar and birch, which are both pioneer species, well adapted to regenerate in clear-cut portions of the boreal forest. The long-term consequences of this species shift are unknown, but such a shift clearly illustrates how forest companies have replaced forest fire as the main agent of change.

Northern Ontario is a major logging area in Canada. In fact, as a renewable resource, forestry is the most durable economic activity in northern Ontario, where many communities depend on the boreal forest. By volume of wood cut, Ontario ranks just behind British Columbia and Québec. Mills in northern Ontario produce pulp and paper, lumber, fence posts, and plywood. The pulp and paper industry in northern Ontario accounts for about 25 per cent of the national production.

Ontario is a leading exporter of newsprint and pulpwood to the United States. In fact, most pulp and paper firms operating in northern Ontario are American-owned companies. Road access is an extremely important factor

The boreal forest stretches across most of northern Ontario, providing for a stable forest industry and spectacular scenery. (*Courtesy Victor Last, Geographical Visual Aids 44254*)

in the forest industry. Trucks must bring the pulpwood to the mill and then the final product must be shipped to market. Pulp and paper mills are located in communities that have access to transportation networks—Thunder Bay, Sault Ste Marie, Sturgeon Falls, Kenora, Fort Frances, Dryden, Marathon, Iroquois Falls, Kapuskasing, and Terrace Bay.

Technology in the forest industry has advanced over the years and greatly increased the productivity of workers in the mills and in the woods. When loggers first arrived in northern Ontario, they had only axes and saws. A logger with such equipment might cut

five to ten trees a day. By the middle of the twentieth century, the technology had advanced. Chain-saws then enabled a logger to cut fifty to seventy trees a day. Mechanical harvesters, which can harvest hundreds of trees per day, have now replaced the chain-saw.

Transportation of timber to mills has also changed. In the past, logging often took place in the winter months. Horses pulled the logs to frozen rivers where they were stored until the river melted in the spring. Tractors later took over the task of hauling the logs to the river. In the spring, the logs were floated to the sawmills, which were often located at the mouths of the rivers. Today, most timber is used in the pulp and paper industry. These huge plants require a steady supply of pulpwood. This demand and the technological changes that have taken place have altered the nature of logging in four ways: (1) logging has become a year-round affair; (2) most logs are hauled to the mills by trucks; (3) trees are harvested before they reach maturity; and (4) mechanical tree harvesters are employed and usually clear-cut a wooded area.

Today, the forestry industry faces several challenges. One is maintaining a balance between logging and the regeneration of the forest. Efforts to hand-plant seedlings are helping to speed up the process of reforestation, but there are concerns that these efforts are insufficient to assure the same level of logging twenty years from now. The Ontario government, like most other provincial governments, grants forest leases to logging companies. The practice of granting forest companies long-term timber leases is based on the assumption that it is in the companies' self-interest to manage the forest well. It remains to be seen whether this assumption and corresponding leasing policy will ensure the protection and regeneration of Ontario's forests.

A second problem facing the forest industry is the age of pulp and paper plants. Many mills in Ontario were built before the Second World War and continue to use old technology, which results in much higher discharges of toxic wastes into the environment. The consequences of such practices can be life-threatening. The worst cases of massive toxic discharges into streams and rivers occurred in the 1960s. At that time, a chemical plant that pre-pared the bleaching solution for the pulp plant at Dryden was releasing effluents containing mercury into the English-Wabigoon River. Eventually, the mercury entered the food chain. In 1970, mercury was discovered in the fish near the Grassy Narrows Indian Reserve, which is about 500 km downstream from the pulp mill. Levels of methyl mercury in the aquatic food chain were ten to fifty times higher than those in the surrounding waterways (Shkilnyk 1985:189). These levels were similar to those found in the fish of Minamata Bay, Japan. Over 100 residents of this Japanese village died from mercury poisoning in the 1960s, and over 1,000 people suffered irreversible neurological damage. Because they depend on fish and game, the Ojibwa at the Grassy Narrows Reserve ate fish on a daily basis and many complained of mercury-related illnesses. While, unlike the Minamata incident, no one died, the economic and social impact on the Ojibwa was nevertheless an industrial tragedy of immense proportions. Since then, forestry mills have updated their operations to ensure a cleaner and safer natural environment.

Ontario's Urban Geography

Ontario is the most highly urbanized province in Canada. Almost 9 million people—approximately 82 per cent of the province's total population—live in towns and cities. As well, many of Canada's largest cities are located in Ontario. For example, ten of the twenty-five largest cities in Canada are in Ontario (Table 5.3). With the exceptions of Sudbury and Thunder Bay, these census metropolitan areas (CMAs) are located in southern Ontario.

The pattern of urban growth from 1981 to 1996 varied considerably for these CMAs (Table 5.3). The fastest-growing cities in Ontario were Oshawa, Ottawa, Toronto, and Kitchener. These four centres had growth rates well above the provincial average of 29 per cent for CMAs. London and Hamilton fell just below the provincial average. The remaining four centres (Sudbury, Thunder Bay, Windsor, and St Catharines–Niagara) had much lower growth rates. Sudbury and Thunder Bay are both loc-

ated in the resource hinterland of the Canadian Shield. Their small population increase over the last fifteen years is indicative of the declining state of the resource economy.

Ontario's two subregions—southern Ontario and northern Ontario—each have a different urban geography. Southern Ontario has most of the province's urban population. The four largest cities in Ontario (Toronto, Ottawa, Hamilton, and London) are located in southern Ontario (Figure 5.4). These four cities form the core of the three urban clusters in southern Ontario, namely, the Golden

Table 5.3

Census Metropolitan Areas in Ontario, 1981–1996

Census Metropolitan Area	Population 1981	Population 1996	Change (per cent)
Thunder Bay	121,379	125,562	3.4
Sudbury	149,923	160,488	7.0
Oshawa	186,446	268,773	44.2
Windsor	250,855	278,685	11.1
Kitchener	287,801	382,940	33.1
St Catharines–Niagara	342,429	372,406	8.8
London	326,817	398,616	22.0
Hamilton	542,095	624,360	15.2
Ottawa*	558,379	763,426	36.7
Toronto	3,130,392	4,263,759	36.2
Total	5,896,516	7,639,015	29.6

*The population of the Ottawa–Hull CMA was 743,821 in 1981 and 1,010,498 in 1996.

Sources: Statistics Canada, 1982:Table 1; 1997b:Table 2.

Horseshoe, southwestern Ontario, and the Ottawa Valley. Each of southern Ontario's three major urban clusters has a population of at least 1 million. As mentioned earlier, northern Ontario has fewer and smaller cities, each separated by long distances. It has two major cities (Sudbury and Thunder Bay) and no single urban cluster. Sudbury, the largest

centre, has only 160,000 residents, while Thunder Bay's population is 125,000.

The Golden Horseshoe

Over 5 million Canadians live, work, and play in Canada's largest population cluster, the Golden Horseshoe. Located at the western end of Lake Ontario, this area has more than a

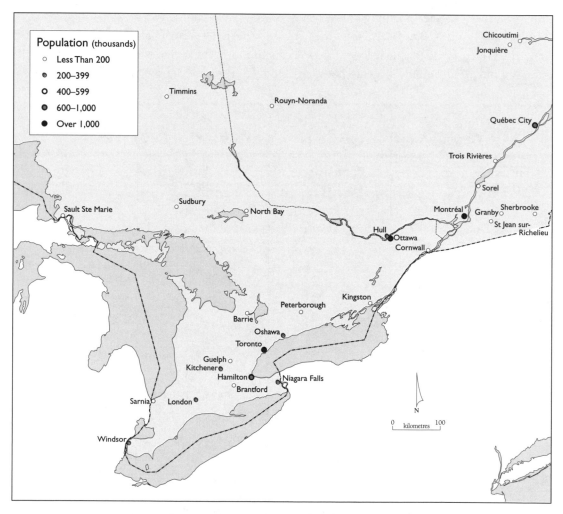

Figure 5.4 Major urban centres in Central Canada. Most large urban centres in Central Canada are located in the Great Lakes–St Lawrence Lowlands, especially in southern Ontario.

dozen towns and cities. Called the Golden Horseshoe because of its prosperous past and its horseshoe-like shape, this urban concentration is the most densely populated area of Canada. Even more significant, most manufacturing and financial activities discussed earlier take place within this highly developed area along the edge of Lake Ontario.

The Golden Horseshoe begins near the Canada–United States border at St Catharines. A major American city, Buffalo, is just across the Niagara River. The Golden Horseshoe extends northwest to Hamilton and then northeast to Toronto. This major population cluster follows the shore of Lake Ontario and ends at Oshawa. Toronto, the largest city in Canada, is its urban anchor, while Hamilton, with its steel plants, is the focus of heavy industry (Vignette 5.3), and Oshawa is Canada's leading automobile-manufacturing city. Toronto is

the dominant city and its place within the Golden Horseshoe is critical.

Toronto

With a population exceeding 4 million, Toronto is Canada's largest and most ethnically diverse city. It is the seat of government for Ontario, the province's major cultural and entertainment centre, and the financial capital for Canadian business. Toronto houses the main offices of national and international banks and investment firms, as well as the Toronto Stock Exchange. Because Toronto is also a hub for the entertainment industry, which ranges from live theatre to professional sports, it plays an essential role in shaping Canadian culture.

Like other major cities in North America, Toronto is trying to cope with a rapidly expanding population, much of which has

Vignette 5.3

Hamilton: Steel City

Hamilton, the second largest city in Ontario, is situated at the west end of Lake Ontario on Burlington Bay. Hamilton is Canada's largest steel producer and ranks high in industrial production.

Originally, Hamilton was a textile town. Streams flowing down the Niagara escarpment provided power for the mills, and the city's location at the head of navigation on Lake Ontario gave it access to other towns and cities. In the early 1850s, the Great Western Railway was extended to Hamilton from Toronto and Montréal, encouraging more industrial developments. Steel production flour-

ished with the railway boom on the Canadian Prairies in the early 1900s. Hamilton's steel industry languished during the Great Depression but during both the First and Second World Wars, military demand for steel again created boom conditions.

After the Second World War, Hamilton's steel industry shifted its production to the rapidly expanding consumer market, especially for appliances and automobiles. Hamilton's reliance on the steel industry increased when, in the 1960s, the last textile mills closed. Two of Canada's largest steel firms (Stelco and Dofasco) are located in Hamilton.

Toronto's financial district includes the head offices of some of Canada's major banks, as well as the Toronto Stock Exchange. (*Corel Photos*)

moved into adjoining urban areas. These two processes—rapid population growth and geographic expansion of the urban population—have created a number of problems:

- *Traffic congestion*: With so many people living outside of downtown Toronto but working in the city centre, massive numbers of them commute daily, creating traffic jams during the morning and afternoon rush hours.
- *Time lost travelling to and from work*: Travelling time to and from work often takes more than one hour and sometimes as much as two hours.

- *Lower land values and rents in neighbouring cities*: The spread of Toronto's population and businesses into neighbouring areas is partly due to lower land values and rents. Land values and rents are highest in downtown Toronto. Cities surrounding Toronto have much lower land values and can therefore attract businesses by offering lower office rents. This leads to further geographic expansion.
- *An overlap of city and municipal governments*: Prior to 1998, seven city governments functioned in the Toronto area.[4] In 1997, the Ontario government passed legislation to create a single city govern-

ment for Toronto. While many opposed this approach to mega-urban government, the objective is to reduce administrative overlap and thereby create a more efficient urban government. Whether or not this single administration approach will succeed in serving such a large population remains to be seen.

Toronto is evolving. Land use within the city centre continues to change. Manufacturing firms have left the city in recent years. Deindustrialization—the relocation of manufacturing and service functions from the 'old' city to outlying areas—weakened Toronto's grip on manufacturing. Norcliffe (1996) has demonstrated that a substantial transfer of manufacturing activities took place in Toronto over a twenty-year period. In 1971, the city of Toronto accounted for 36 per cent of the manufacturing jobs, while 42 per cent were in the inner suburbs and 22 per cent in the outer suburbs. Twenty years later, Toronto suffered a sharp decline to 16 per cent of the manufacturing jobs, while the inner suburbs dropped to 38 per cent. The greatest growth both relatively and absolutely occurred in the outer suburbs. By 1991, the outer suburbs had 46 per cent of the manufacturing jobs (Norcliffe 1996:266-7).[5]

Ottawa Valley

With a population of just over 1 million, Ottawa–Hull is a major population cluster. It is also the fourth largest metropolitan area in Canada. Ottawa is relatively close to both Montréal and Toronto.

On the Ontario side, Greater Ottawa is the third largest urban cluster in Ontario. The federal government, located in the national capital, is the major employer in the region. The federal government requires a wide variety of goods and services in its daily operations. This demand provides an opportunity for many small and medium-size firms in the Ottawa Valley. By locating its departments and agencies in both Ottawa and Hull, the federal government has ensured that Ottawa's economic orbit extends to a number of small towns on both the Ontario and Québec sides of the Ottawa River. Most people live on the Ontario side, especially in the city of Ottawa. Other urban centres on the Ontario side include Nepean, Gloucester, Kanata, and the municipalities of Rockcliffe and Vanier. On the Québec side of the Ottawa River, Hull is the major urban centre, while Gatineau and Aylmer are smaller towns. Around Ottawa, in its periphery, are a number of smaller centres, including Pembroke, Cornwall, Brockville, and Kingston. However, with the exception of Pembroke, these cities are more closely tied to economic and social activities in Toronto and Montréal.

As the capital of Canada, Ottawa is the focus of national and international affairs and has subsequently developed into a national administrative centre with a large federal civil service. In its early days, the forest industry played a strong economic role, but it was eventually overshadowed by the activities of the federal government. Still, the geographic location of Ottawa–Hull makes it an ideal place for a pulp and paper mill: pulpwood can easily be harvested from the interior forest lands

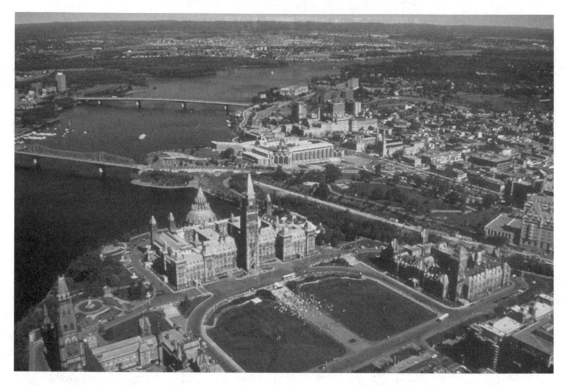

Aerial view of downtown Ottawa with Parliament Hill, and behind it, the National Gallery of Canada. The bridges that span the Ottawa River between Ontario and Québec attest to the close links between Ottawa and Hull. (*Corel Photos*)

and then transported along the Ottawa and Gatineau rivers to the Eddy Pulp and Paper Mill in Hull. An added advantage of this site is the availability of low-cost electrical power from the hydroelectric installation on the Gatineau River near the Chaudière Falls.

In recent years, Ottawa has become an industrial leader in high technology. Described in the press as the 'Silicon Valley of the North', the key to Ottawa's success has been the emergence of high-technology firms specializing in telecommunications, computers, software and computer services, and electronics (Britton 1996:266). In 1997, Northern Telecommunications (Nortel) announced plans to significantly

expand its operations in the Ottawa area. Over a four-year period, Nortel plans to invest $250 million and hire nearly 5,000 new employees (McCarthy 1997:B4). Such a commitment demonstrates the attractiveness of Ottawa as a location for the high-tech industry.

Southwestern Ontario

Southwestern Ontario is the third major urban cluster. About 1 million people live in this area. Though this region does not have a major metropolitan city, it is close to Toronto, Buffalo, and Detroit. The main urban centres within southwestern Ontario are Cambridge, Kitchener, London, and Waterloo. Windsor

The Nortel complex is located in the growing high-technology area in Kanata, near Ottawa. (*Courtesy R.M. Bone*)

and Sarnia, while part of southwestern Ontario, are on the edge of this urban cluster.

London is the unofficial capital of southwestern Ontario. It provides administrative, commercial, and cultural services for the larger region. London is also the headquarters of several insurance companies, including London Life Insurance Company. Among its economic activities, London is noted for its manufacturing, including the production of armoured personnel carriers and diesel locomotives by General Motors. London, therefore, has a sound industrial foundation based on insurance, manufacturing, and high-tech industries. Such industries pay relatively high wages to their employees. Consumer spending by these employees supports a strong retail sector.

Automobile-assembly plants are located in Cambridge, Ingersoll, St Thomas, Alliston, and Windsor (Table 5.2). Auto-parts plants also operate in southwestern Ontario. There are also a number of high-tech firms, particularly in the Cambridge, Kitchener, and Waterloo areas. High-tech industries in this region, which is commonly known as Canada's 'Technology Triangle', employ innovative technology that is often developed in research institutes or universities. Such technology is then used to create commercial products. Bathelt and Hecht provide the following description of this technology zone:

All [high-tech firms] are surprisingly strongly tied, in terms of input and output linkages, to economic activities within a 100-km radius, which includes the Toronto area. Skilled labour and local residence/education are the strongest location factors. The relatively favourable industrial climate in the Cambridge Technological Triangle region should enable key technology industries to flourish in the area (Bathelt and Hecht 1990:225).

Cities of Northern Ontario

Timmins, located in the mineral-rich Canadian Shield, is situated in the centre of a gold belt, while Sudbury is in a world-famous nickel belt, where the mining and smelting of nickel and copper ores are core functions. Both cities began as single-industry towns, but over time, have become important regional centres. Both cities have developed service industries, which have diversified their economies and made their communities more stable.

Sault Ste Marie is a steel town located on the Great Lakes between Lake Superior and Lake Huron. The town's major firm is the Algoma steel mill, which is located at some distance from its major industrial customers, thus putting it at a disadvantage. However, Algoma makes use of local railway and seaway connections to secure raw materials, such as iron ore, and to market its steel products. Major markets are in the United States, particularly along the Great Lakes.

Thunder Bay, situated at the western end of Lake Superior, is a major transshipment point. Bulky products—particularly grain, iron, and coal—are shipped to Thunder Bay for loading onto lake vessels. In recent years,

however, this function has diminished. In the past, for example, iron from mines at Atikokan passed through Thunder Bay on its way to steel mills located along the shores of the Great Lakes. This mining operation is now defunct. Grain from the Canadian Prairies is still the major product handled at Thunder Bay. The Canadian Wheat Board sends Prairie grain by rail to Thunder Bay and then by grain carrier to eastern ports for export to foreign countries. Until 1996, Ottawa heavily subsidized grain shipments under the Crow Benefit. With the elimination of that subsidy, the advantage of sending grain by rail to Thunder Bay has diminished, leaving its grain transportation role in doubt. (For more on this subject, see Chapter 8.)

Northern Ontario's population is growing very slowly. For example, from 1981 to 1996, population growth in the major cities of northern Ontario amounted to only 4.7 per cent (Table 5.4). In comparison, cities in southern Ontario have increased at a much higher rate—about 29 per cent. More specifically, the major cities and towns in northern Ontario—with the exceptions of North Bay and Sudbury—show few signs of growth. In fact, from 1981 to 1996, Kirkland Lake, Elliot Lake, and Kapuskasing suffered significant population losses. Another four (Timmins, Sault Ste Marie, Kenora, and Thunder Bay) had very low rates of growth.

Cities and towns in northern Ontario are struggling to diversify their economies. One option being pursued is tourism. Many private firms, such as motels, summer camps, and fishing and hunting lodges, have sprung up (especially in smaller communities) to meet

Table 5.4
Urban Centres in Northern Ontario, 1981–1996

Centre	Population 1981	Population 1996	Change (per cent)
Sudbury	149,923	160,488	7.0
Thunder Bay	121,379	125,562	3.4
Sault Ste Marie	82,697	83,619	1.1
North Bay	51,268	64,785	26.4
Timmins	46,114	47,499	0.3
Elliot Lake	16,723	13,588	(18.8)
Kirkland Lake	12,219	9,905	(18.9)
Kapuskasing	12,014	10,036	(16.5)
Kenora	9,817	10,063	2.5
Total	502,154	525,545	4.7

Sources: Statistics Canada, 1997b:Table 1; 1982:Table 1.

the demand from the growing number of tourists coming to experience the wilderness of northern Ontario.

The urgency to diversify is driven by the continuing reduction in the size of the primary labour force. For example, the workforce at the nickel mines in Sudbury has fallen by half over the last thirty years. Sudbury has reacted by aggressively seeking to expand its service industry. The city has met with some success, partly because of its strategic location at the junction of three highways—the Trans-Canada Highway, Highway 69 (which leads to Toronto), and Highway 144 (which connects Sudbury with Timmins)—and partly because of its efforts to have federal and provincial

agencies (Federal National Tax Data Centre and Provincial Department of Mines) and institutions (Laurentian University and the Science North Museum) locate in Sudbury.

More isolated towns, such as Kapuskasing, remain very dependent on a single industry and are therefore vulnerable to fluctuations in world demand for resource products. Another problem that single-industry towns face is that fewer and fewer workers are required to operate the mills and mines. Unless resource towns diversify, their populations will dwindle. Kapuskasing provides a case-study.

Like many single-industry towns in northern Ontario, Kapuskasing came into existence in 1920 as a railway station on the National

Transcontinental Railway. As it was located in the boreal forest, it soon became involved in the forest industry. In the early 1920s, the Spruce Falls Power and Paper Company was established at Kapuskasing to supply the *New York Times* with newsprint. In 1990, the US owners of the mill decided to close the plant at Kapuskasing. As this mill was the main employer, 1,200 workers would lose their jobs, and the community would suffer a fatal loss. In 1991, the mill management, employees, and other residents of Kapuskasing bought 59 per cent of the company, while Tembec Inc., an experienced pulp and paper manufacturer, bought the remainder. Spruce Falls employed 1,400 workers in the mill and surrounding woodlands, but to return to a profitable operation, its workforce had to be reduced from 1,400 to 670 over three years.

Ontario's Future

As Ontario enters the twenty-first century, it remains the most powerful geographic region in Canada. Ontario has the largest population and economy of the six geographic regions. It is especially dominant in the financial and manufacturing sector of Canada's economy, but its resource industries face troubling times as the quality and quantity of their resources diminish. Such a decline is natural in the mining industry as the higher-grade ore is mined, but the renewable resources should not undergo such a decline. In the future, southern Ontario's economic prospects seem bright, but those in northern Ontario are less so. Great uncertainty lies ahead for northern Ontario.

Some uncertainty is related to Ontario's changing place in the Canadian economy. As foreign goods become available in Canada's hinterland, Ontario's grip on its domestic market is loosening. At the same time, Ontario's industry is penetrating the North American market. Much of this change is epitomized by the automobile industry. Before 1965, Canadians purchased their cars from manufacturers in southern Ontario. Today, they obtain their cars from assembly plants in different regions of North America. While this trade shift reflects the emergence of a north-south pull, for the moment the east-west alignment remains in place.

Like other regions of Canada, Ontario faces the challenge of remaining competitive with other countries. Dangers lurk in the global economy. In spite of its strong economy, Ontario's levels of unemployment remain stubbornly high. Fierce competition from other countries could force more firms to relocate to low-wage countries, such as Mexico, where labour costs are but a fraction of those in Ontario, or to reduce their labour costs by some other means. The steel industry, for example, must continually invest in new technologies in order to stay competitive with off-shore, low-wage steel producers. By substituting capital for labour, these manufacturing firms become more efficient and require fewer employees.

On a more optimistic note, Ontario's economy has become more competitive within the North American market. Ontario's manufacturing firms may be poised to capture a larger share of that market, as well as gain a foothold in the growing global marketplace. Much depends on the performance of southern Ontario's economy. Northern Ontario, on the other hand, may experience little or no economic growth. Much depends on world prices for its primary products and on competition from other resource regions and countries. As mentioned earlier, the discovery of new ore bodies can quickly and significantly affect communities in northern Ontario, like the uranium deposits in northern Saskatchewan that forced the closure of the mines at Elliot Lake. The recent discovery of Voisey Bay's nickel/copper/cobalt deposit in Labrador may threaten production at Sudbury. As Ontario adjusts to new circumstances, its resource hinterland and industrial core will continue to define this province. All things considered, however, Ontario's leading economic position within Canada depends heavily on the continued success of its industrial core in southern Ontario.

Summary

Ontario is the most powerful province in Canada. Its subregions of southern Ontario and northern Ontario have very different physical and human geographies. Southern Ontario is not only the industrial and population heartland of the province and nation but also an increasingly important sector of the North American economy. Southern Ontario's population is increasing rapidly. Canadians and immigrants are attracted to southern Ontario's robust economy. In sharp contrast, northern Ontario is an old resource hinterland whose economy is stalled by greater competition in the resource industry. Its population is growing very slowly. As young people move to more prosperous areas of Canada, especially to the major cities in southern Ontario, the population of northern Ontario is aging.

Ontario remains the dominant of the six geographic regions in Canada, both economically, and by population. Ontario is experiencing growth in its financial and manufacturing activities, but its resource industries face troubling times. With the elimination of tariffs, Ontario will face continued challenges from manufacturing firms and resource companies in other parts of North America and the world. Sustainable development of its resources will also become a crucial challenge.

With the emergence of a powerful north-south trade alignment, Ontario sees its economic future within North America, thereby diminishing Ontario's traditional role as the economic heartland of Canada. Ontario's new economic role may have both economic and political consequences for the rest of Canada.

Notes

1. As a general rule, wages in the primary and secondary sectors are often higher than the national average, while wages in certain tertiary jobs are below the national average. In fact, many jobs in the service sector are associated with low hourly wages and part-time employment. These are often disparagingly referred to as 'hamburger-flipping' jobs.

2. AB Volvo opened its Halifax automobile-assembly plant in 1963 to take advantage of the Canada–

US Auto Pact, which allowed the Swedish firm to import cars and parts, duty-free, to Canada. The Halifax plant is being closed for three reasons: (1) Volvo has an excess manufacturing capacity around the world, so its most marginal plants will be closed; (2) the Halifax plant is known as a 'kit plant', meaning that it is dependent on parts manufactured in Sweden and shipped to Halifax; and (3) the Halifax plant is a small operation that lacks the flexibility needed to assemble new models (Keenan 1998:B1, B4).

3. At the beginning of the twentieth century, the Ontario government sought to develop its northeastern territories, especially its agricultural lands in the Clay Belt, its forest stands, and its mineral deposits in the Canadian Shield. Between 1903 and 1909, the government-financed Temiskaming and Northern Ontario Railway was built from the Canadian Pacific Railway at North Bay northward to the small town of Cochrane on the National Transcontinental Railway (now the CNR). Over the next decade, branch lines were extended to Cobalt, Timmins, and Iroquois Falls. Overall, the Ontario government was pleased with the railway's impact on resource development. Plans were made to build the railway farther north—into the Hudson Bay Lowlands. By 1932, the Temiskaming and Northern Ontario Railway stretched from Cochrane to Moosonee at the southern tip of James Bay, but further developments did not occur in the resource-scarce Hudson Bay Lowlands.

4. Toronto has tried various forms of regional government. About forty years ago, the Ontario government established a form of regional government by combining the City of Toronto with the surrounding centres of Etobicoke, North York, Scarborough, York, and East York into the municipality of Metropolitan Toronto (Metro Toronto). These six jurisdictions became one large urban area for regional planning purposes, but each retained its city government. As Metro Toronto continued to grow, it spread into adjacent jurisdictions. In 1988, the Ontario government created the Greater Toronto Area (GTA). Often called the Metro Toronto Region, it consists of Metro Toronto and the regional municipalities of Halton, Peel, York, and Durham. In 1997, the provincial government passed a bill to change the municipal government structure to ensure more efficient, cost-effective services. The legislation came into effect in 1998, amalgamating six former municipalities (Etobicoke, York, Toronto, East York, North York, and Scarborough) and the regional municipality into the new city of Toronto.

5. Norcliffe defines inner suburbs as the (now former) municipalities of East York, Etobicoke, North York, Scarborough, and York. The outer suburbs lie beyond the boundary of the Greater Toronto Area (GTA).

Key Terms

Free Trade Agreement
 Trade agreement between Canada and the United States enacted in 1989.
just-in-time principle
 System of manufacturing in which parts are

delivered from suppliers at the time required by the manufacturer.
North American Free Trade Agreement
 Trade agreement between Canada, the United States, and Mexico enacted in 1994.

outsourcing
  Arrangement by a manufacturing firm to obtain its parts from other firms.
restructuring
  Economic adjustments made necessary by fierce competition; companies that are driven

to reduce costs often resort to reducing the number of workers at their plants.
value added
  A difference between a firm's sales revenue and the cost of its materials.

References

Bibliography

Bathelt, Harald, and Alfred Hecht. 1990. 'Key Technology Industries in the Waterloo Region: Canada's Technology Triangle (CTT)'. *The Canadian Geographer* 34, no. 3:225–34.

Blackbourn, Anthony, and Robert G. Putnam. 1984. *The Industrial Geography of Canada*. London: Croom Helm.

Bluestone, B., and B. Harrison. 1982. *The De-industrialization of America*. New York: Basic Books.

Campbell, Murray. 1996. 'Galt 25 Years Later'. *The Globe and Mail* (13 April):D1–2.

Canada. 1994. *1993 Canadian Minerals Yearbook: Review and Outlook*. Ottawa: Ministry of Natural Resources.

Ceh, S.L. Brian. 1996. 'Temporal and Spatial Variability of Canadian Inventive Companies and Their Inventions: An Issue of Ownership'. *The Canadian Geographer* 40, no. 4:319–37.

Dear, M.J., J.J. Drake, and L.G. Reeds, eds. 1987. *Steel City: A Geography of Hamilton and Region*. Toronto: University of Toronto Press.

Dicken, Peter. 1992. *Global Shift: The Internationalization of Economic Activity*, 2nd edn. New York: Guilford Press.

Economist, The. 1996. *Country Profiles: Canada*. London: The Economist Intelligence Unit.

Francis, R. Douglas, Richard Jones, and Donald B. Smith. 1996. *Destinies: Canadian History Since Confederation*, 3rd edn. Toronto: Harcourt Brace.

Gad, Gunter. 1991. 'Toronto's Financial District'. *The Canadian Geographer* 35, no. 2:203–7.

Gertler, Meric S. 1998. 'Negotiated Path or "Business as Usual"?: Ontario's Transition to a Continental Production Regime'. Paper presented at the Annual Meeting of the Association of American Geographers, Boston, 17 March.

Gilbert, Anne, and Johan Marshall. 1995. 'Local Changes in Linguistic Balance in the Bilingual Zone: Francophones de l'Ontario et anglophones du Québec'. *The Canadian Geographer* 39, no. 3:194–218.

Holmes, John. 1996. 'Restructuring in a Continental Production System'. In *Canada and the Global Economy: The Geography of Structural and Technological Change*, edited by John N.H. Britton, Chapter 13. Montréal–Kingston: McGill-Queen's University Press.

Jestin, Warren. 1996. *Global Economic Outlook*. Toronto: Scotiabank.

Keenan, Greg. 1998. 'AB Volvo to Close Halifax Plant'. *The Globe and Mail* (10 September):B1, B4.

Little, Bruce, and Greg Keenan. 1996. 'Ontario's

Economic Future Is the Sum of Its Auto Parts'. *The Globe and Mail* (2 March):A1, A5.

McCarthy, Shawn. 1997. 'Ottawa Trades Bureaucrats for Bytes'. *The Globe and Mail* (30 August):B1, B4.

MacDonald, N.B. 1980. 'The Future of the Canadian Automotive Industry in the Context of the North American Industry'. Working Paper no. 2. Science Council of Canada. Ottawa: Minister of Supply and Services.

McKnight, Tom L. 1992. *Regional Geography of the United States and Canada*. Englewood Cliffs: Prentice-Hall.

MacLachlan, Ian. 1996. 'Organizational Restructuring of U.S.-Based Manufacturing Subsidiaries and Plant Closure'. In *Canada and the Global Economy: The Geography of Structural and Technological Change*, edited by John N.H. Britton, Chapter 11. Montréal–Kingston: McGill-Queen's University Press.

Marsh, James H., ed. 1988. *The Canadian Encyclopedia*, 2nd edn. Edmonton: Hurtig Publishers.

Matthew, Malcolm R. 1993. 'The Suburbanization of Toronto Offices'. *The Canadian Geographer* 37, no. 4:293–306.

Norcliffe, Glen. 1994. 'Regional Labour Market Adjustments in a Period of Structural Transformation: An Assessment of the Canadian Case'. *The Canadian Geographer* 38, no. 1:2–17.

_____. 1996. 'Mapping Deindustrialization: Brian Kipping's Landscape of Toronto'. *The Canadian Geographer* 40, no. 3:266–73.

Paterson, J.H. 1994. *North America: A Geography of the United States and Canada*. New York: Oxford University Press.

Putnam, D.F., ed. 1952. *Canadian Regions: A Geography of Canada*. Toronto: Dent & Sons.

Relph, Edward. 1991. 'Suburban Downtowns of the Greater Toronto Area'. *The Canadian Geographer* 35, no. 4:421–5.

Shkilnyk, A.M. 1985. *A Poison Stronger Than Love: The Destruction of an Ojibwa Community*. New Haven: Yale University Press.

Stanford, Quentin H., ed. 1998. *Canadian Oxford World Atlas*, 4th edn. Toronto: Oxford University Press.

Statistics Canada. 1982. *Census Metropolitan Areas and Census Agglomerations*, Catalogue no. 95-903. Ottawa: Minister of Supply and Services.

_____. 1992. *Census Metropolitan Areas and City Agglomerations*. Catalogue no. 93-303. Ottawa: Minister of Supply and Services.

_____. 1996. *Labour Force Annual Averages 1995*. Catalogue 71-220-XPB. Ottawa: Statistics Canada.

_____. 1997a. *Historical Overview of Canadian Agriculture*. Catalogue no. 93-358-XPB. Ottawa: Industry Canada.

_____. 1997b. *A National Overview: Population and Dwelling Counts*. Catalogue no. 93-357-XPB. Ottawa: Industry Canada.

_____. 1997c. *Canada Year Book*. Catalogue no. 11-402-XPE/1997. Ottawa: Minister of Industry.

_____. 1997d. Canada, Value of Domestic Exports [online database], Ottawa. Searched 3 September 1998;<URL:http.nrcan.gc/nms/efab/mmsd/trade/tbalance>:1 p.

_____. 1997e. 1996 Census: Nation Tables—Population by Mother Tongue, Showing Age

Groups, for Canada, Provinces and Territories, 1996 Census—20% Sample Data, 2 December 1997 [online database], Ottawa. Searched 15 July 1998; <URL:http://www.statcan.ca/english/census96/>:3 pp.

_____. 1997f. The Daily—1996 Census: Mother Tongue, Home Language and Knowledge of Languages, 2 December 1997 [online database], Ottawa. Searched 14 July 1998; <URL: http://www.statcan.ca/Daily/English/>:15 pp.

_____. 1998a. *Canadian Economic Observer*. Catalogue no. 11-010-XPB. Ottawa: Statistics Canada.

_____. 1998b. The Daily—1996 Census: Ethnic Origin, Visible Minorities, 17 February 1998 [online database], Ottawa. Searched 16 July 1998; <URL:http://www.statcan.ca/Daily/English/>:21 pp.

_____. 1998c. The Daily—1996 Census: Aboriginal Data, 13 January 1998 [online database], Ottawa. Searched 14 July 1998; <URL:http://www.statcan.ca/Daily/English/>:10 pp.

Further Reading

Britton, John N.H. 1996. 'High Tech Canada'. In *Canada and the Global Economy: The Geography of Structural and Technological Change*, edited by John N.H. Britton, Chapter 14. Montréal–Kingston: McGill-Queen's University Press.

Courchene, Thomas J. 1998. *From Heartland to North American Region State: The Social, Fiscal and Federal Evolution of Ontario*. Monograph series on public policy, Centre for Public Management. Toronto: Faculty of Management, University of Toronto.

Holmes, J. 1983. 'Industrial Reorganization, Capital Restructuring and Locational Change: An Analysis of the Canadian Automobile Industry in the 1960's'. *Economic Geography* 59:251–71.

Chapter 6

Overview

Québec, like Ontario, has both a northern resource hinterland and a southern industrial core. Québec, as part of Canada's industrial heartland, is seeking a place in the North American economy but its population and economy are growing slowly. Its outstanding feature is its Québécois culture. As the home of most Canadians of French origin, Québec represents a distinct cultural and linguistic region of Canada. Québec is also home to a nationalist, political, separatist movement.

Québec's society and economy were modernized during the Quiet Revolution. Hydro-Québec played a prominent role in that economic and social transformation, harnessing the water resources of the Canadian Shield and strengthening the core/periphery relationship between northern and southern Québec. Hydro-Québec's James Bay Project resulted in the first modern land-claim agreement, the James Bay and Northern Québec Agreement.

Objectives

- Describe Québec's physical geography and historical roots.
- Present the basic elements of Québec's population and economy within the region's physiography.
- Examine Québec's cultural and economic development within the context of the core/periphery model.
- Discuss Québec's position within Canada, its francophone culture, and the resurgence of the separatist movement.
- Outline Québec's economic role in Canada and North America.
- Examine the effect of the James Bay Project on the Cree and Inuit of northern Québec.
- Focus on Hydro-Québec's strategy of developing low-cost electrical power in northern Québec to stimulate industrial activities in southern Québec.

Québec

Introduction

By virtue of its history and geography, Québec occupies a special place in Canada. First, and foremost, Québec is the homeland of the Québécois. Most Québécois also have a strong attachment to Canada, but their first loyalty is often to Québec. This sense of dual loyalty is unique in Canada. Second, Québécois culture is largely derived from the historical experience of living in North America for nearly 400 years and of being Canadians for 130 years. Third, Québec is recognized throughout Canada and the United States as a cultural hearth of French language, customs, and heritage. Many French-speaking Canadians in the rest of Canada and Franco-Americans come to visit Québec to renew their sense of ethnicity. In sum, Québécois history and geography give Québec its *raison d'être* and its vision of Canada as a nation of two founding peoples. Canada's cultural duality has led to tensions between French- and English-speaking Canadians, tensions that have often strained the bonds of Confederation. (See Chapters 3 and 4 for a broader discussion of the French/English faultline.)

The vast majority of people living in Québec speak French. In 1996, Québec's population was 7.1 million and over 82 per cent of Quebeckers declared French as their mother tongue (Figure 6.1). These Quebeckers are

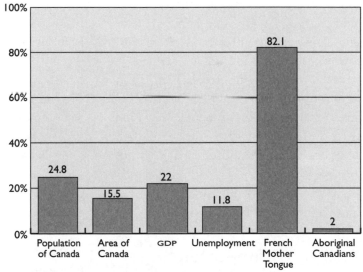

Figure 6.1 Québec, 1996. Québec's economic strength, geographic and population size, and cultural importance in Canada is indicated by its share of the country's GDP, area of Canada, population, and French by mother tongue. Aboriginal Canadians form a small minority in this region.

Sources: Statistics Canada, 1997a: Table 1; 1998a: Table 39; 1997b: 1 to 3 of 3; 1997c: 4 and 5 of 15; 1998b: 1 and 2 of 10; 1998c: 15 to 20 of 21; and Stanford 1998:185.

known as the Québécois. The remaining Que-beckers include Aboriginals and those whose mother tongue is English (anglophones) or a foreign language (allophones). English-speaking Quebeckers are concentrated in Montréal, Estrie (the Eastern Townships), and Outremont (the Ottawa Valley). Allophones are concentrated mainly in urban centres, like Montréal, where immigrants tend to settle. In northern Québec the majority of the population are Cree and Inuit. Social tensions often occur along linguistic lines between French- and English-speaking Quebeckers and are most obvious in Montréal.

Québec is a large province with three of Canada's physiographic regions extending over the province (Figure 2.1). Each has a different settlement and land-use pattern. The St Lawrence Lowlands region is the agricultural, industrial, and population core of Québec. The Canadian Shield extends from the St Lawrence Lowlands to Hudson Bay and is a resource hinterland. The Appalachian Uplands region, which borders on New Brunswick and three American states—Vermont, New Hampshire, and Maine—lies south of the St Lawrence Lowlands and extends eastward to the Gaspé Peninsula. It is divided into two areas: an agricultural/industrial area in Estrie and a resource hinterland in Gaspésie.

Québec Within Canada

Québec's economic strength within Canada is considerable. Ranking second among the ten provinces, Québec accounted for 22 per cent of all the goods and services produced in this country in 1996 (Figure 6.1). Because of its diverse climatic conditions and three physiographic regions, Québec harvests and produces a variety of products, including agricultural products, manufactured goods, and a variety of raw materials. Like Ontario, Québec is divided into two regions—the resource hinterland of northern Québec, and the industrial core of southern Québec. Unlike Ontario, Québec has an energy surplus, thanks to the enormous amounts of hydroelectricity produced in northern Québec. Québec also has an exceptionally strong service sector that caters to the population and the growing tourist trade. Both Montréal and Québec City are popular tourist destinations.

For over a century, high tariffs protected Québec's manufacturing firms from external competition, thereby allowing manufacturing to take root and flourish. Then as now, Montréal is the eastern anchor of Canada's manufacturing belt. In the new world of free trade, the province of Québec and its largest city, Montréal, face a new challenge—adjusting to new trading arrangements with the United States and other countries. The necessary restructuring has already taken its toll, forcing some labour-intensive firms to cease operations and causing others to substitute machines for workers. The textile industry, for example, has survived the initial shock of trade liberalization, but has had to eliminate much of its labour force. The net result of such restructuring is an extremely high unemployment rate in Québec that is one-third higher than that of Ontario (Figures 5.1 and 6.1).

However, not all the economic news is negative. The liberalization of trade has had positive impacts on Québec's economy. Forcing manufacturing firms to become more efficient and more conscious of quality control, has allowed some to gain a larger foothold in the American market. High-technology companies, particularly those in aerospace, metal refining, and pharmaceuticals, have benefited from this more competitive environment.

Resource companies remain dependent on external markets, especially the American market. A number of companies in the forestry and mining sectors of the resource economy are not only US-owned but gear their production to the American market. The iron mines in northern Québec, for example, were developed with capital from US steel mills through their subsidiary, the Iron Ore Company of Canada, in order to secure a stable supply of iron ore for US steel mills in Ohio and Pennsylvania.

Resource industries face strong competition from other countries, and are vulnerable to decreasing demand and prices in times of global economic slow-downs. After the 1970s, for example, iron-producing companies in Québec faced competition from two sources: plastic products that provided a less expensive alternative to steel products, and iron produced from new mines in Australia and Brazil. All were vying for a share of the lucrative American market. Such fierce competition pressures resource companies to reduce costs, a reduction that is often achieved by substituting capital for labour.

As Québec approaches the twenty-first century, its many assets—a proven workforce, a diversified economy, and an abundance of hydroelectric power—promise a bright future. To achieve this future, the province must re-solve several issues that affect its economic and social well-being, including double-digit unemployment rates and net out-migration. The causes of high unemployment are complex but related to the reduction in the number of workers in labour-intensive manufacturing operations and resource industries. In turn, out-migration is largely related to the inability of Québec's economy to generate sufficient jobs and also to the unstable political climate caused by the threat of separatism. For instance, from 1971 to 1996, the period in which separatism became a real political force and a separatist party was elected to the province's legislature, the net out-migration from Québec to the rest of Canada reached its highest point in modern Québec history. During the same time, the relocation of many large corporations from Montréal to Toronto had a negative impact on Montréal and on the province. The challenge facing Québec and the rest of Canada is to find ways to reduce tensions between French- and English-speaking Quebeckers. Unless English speaking Quebeckers feel comfortable within the new Québec society, they are unlikely to remain in Québec let alone invest in its economy. The challenge for Québec in the coming years is to create a more robust economy and address the issues surrounding French/English tensions.

Québec's Physical Geography

Québec, the largest province in Canada, has a wide variety of natural conditions. Its climate ranges from the mild continental climate in the St Lawrence Valley to the cold arctic climate in the most northerly reaches of Québec

(Figure 2.6). Three of Canada's physiographic regions extend over the province's territory—the Canadian Shield, the Appalachian Uplands, and the St Lawrence Lowlands—and each has a different resource base and settlement pattern.

The heartland of Québec lies in the St Lawrence Lowlands. New France was established within this physiographic region and spread into the interior of North America. It remains the cultural core of Québec. The St Lawrence Lowlands offers several natural advantages that give the region its central role in modern Québec. Arable land is one such advantage. By far the best agricultural land is in a small area between Montréal and Québec City. As well, most industrial plants are located in this same area. As the economic core of Québec, the densely populated and highly urbanized St Lawrence Lowlands encompasses Montréal and Québec City. Within this physiographic region, the St Lawrence River provides a key waterway into the heart of Canada. This great river, which figured so prominently in Québec's past, continues to play an essential role in Québec's economy. The creation of the St Lawrence Seaway, however, has eroded some of Montréal's function as a transshipment point because ocean-going ships can now reach Toronto and other ports on the Great Lakes (Vignette 6.1).

The Appalachian Uplands physiographic region is a northern extension of the Appalachian Mountains. Most arable land is in Estrie, where the Loyalists settled in the late-seventeenth century and were largely replaced by French-speaking farmers by the end of the nineteenth century. In the Gaspé Peninsula, there are a number of small communities along the coast. Many people combine farming, fishing, and logging with part-time employment in the villages and towns. Because of the area's spectacular scenery, tourism has become an important source of income during the summer in the Gaspé Peninsula. Mining and forestry are other economic activities in this physiographic region. With the exception of the Lake Champlain gap in the Appalachian Uplands, access to the populous parts of New England is blocked by these rugged uplands. The Lake Champlain gap has therefore become a very important north-south transportation link between Montréal and New York City.

As the largest physiographic region in Québec, the Canadian Shield occupies over three-quarters of the province's territory (Figure 2.1). The Canadian Shield is noted for its forest products and hydroelectric production—it has most of the hydroelectric sites in Canada (Vignette 6.2). Near the major cities of Montréal and Québec City, the southern part of this rolling, rugged, forested upland with its numerous lakes has become a popular recreation site for urbanites and tourists. Further north, single-industry towns based on mining and forestry dot the landscape. Beyond the commercial forest zone lies James Bay, where the Cree coexist with a massive hydroelectric project. In northern Québec, there are a dozen Inuit settlements along the Arctic Coast. Both the Cree and the Inuit continue to hunt and trap, though the majority of Aboriginal peoples now participate in the wage economy.

Vignette 6.1

The St Lawrence River

The St Lawrence River provides a natural waterway into the interior of North America. Cities along its shores, particularly Québec City and Montréal, have benefitted greatly from the waterway's role as a major trading route. Montréal's favourable location on the St Lawrence gave it an economic advantage and fuelled the city's growth.

The St Lawrence has two major obstacles: the Lachine Rapids near Montréal, and Niagara Falls, which prevented ships from passing from Lake Ontario to Lake Erie. The building of the Lachine Canal (1825) and the Welland Canal (1829) allowed ships to reach the Great Lakes. Over time, these canals and locks were improved to accommodate the increasing size of ships and barges.

However, after the Second World War, there were vast improvements in ocean transportation. This was the era of the 'supertanker'. Even the average size of freighters had increased substantially. These larger ships required much greater depth of water than their predecessors in order to float. Consequently, more and more ocean-going ships were unable to enter the St Lawrence River past Montréal. The solution to this transportation bottleneck was the St Lawrence Seaway, which, when it was inaugurated in 1959, would allow ocean vessels to travel from the Atlantic Ocean to the Great Lakes. Consequently, Montréal lost the advantage it had for transshipment, as other cities became more accessible trading points.

The St Lawrence River, the route of early explorers, remains Canada's most important waterway. (*Courtesy Dirk de Boer*)

Vignette 6.2

Water-power Resources

A region's potential for hydroelectric development depends on climate (significant precipitation), topography (sloping landforms), and access to markets. The development of hydroelectric power in Canada varies considerably from province to province. Québec, particularly its northern regions, has by far the greatest potential for hydroelectric development.

Almost all power sites in Canada that are close to markets were developed by the 1960s. Since then, several large-scale hydroelectric projects have been developed in more remote areas, namely, in northern parts of Québec, British Columbia, and Manitoba. Canada's principal hydroelectric generating stations and their installed capacity are: LG–2 on the La Grande Rivière, Québec (5328 MW), Churchill Falls on the Churchill River, Labrador (5225 MW), LG–4 on the La Grande Rivière, Québec (2650 MW), and Gordon M. Shrum on the Peace River, BC (2416 MW).

The main advantages of hydroelectric developments over alternate sources of electricity are the long life of the facility, low operating costs, and zero air pollution. The drawbacks include the initial high capital investment, the long construction period, and the environmental effects of flooding. Another problem is the effect of flooding river valleys on the Aboriginal peoples that use them for hunting and fishing. The impact of flooding prime wildlife habitat is detrimental to the economy and culture of Aboriginal peoples, not to mention the serious consequences of lost habitat for the wildlife.

Québec's Historical Geography

Québec's history is complex. It began in 1608, when Champlain founded a fur-trading post at the site of Québec City and the French colony was established. France's historic impact on modern Québec is still evident in the people, who are French-speaking and Catholic, and in the St Lawrence Lowlands, where the settlement and land-use patterns are distinct from those in the rest of Canada. Village after village was built around a Catholic church and the surrounding farmland was arranged in long lots based on the French seigneurial system.

After the conquest of New France in 1760, Britain exerted an enormous influence on Québec and its people. The British parliamentary system, for example, is firmly ingrained in Québec's political life, and British criminal law remains an important component of Québec's justice system, even after the Québec Act of 1774 recognized the cultural uniqueness of this British colony and thereby ensured the survival of Québec's French heritage within the British Empire.

After the Act of Union in 1841, English-speaking and French-speaking Canadians were forced to work together within a single Assembly. Leaders of these two cultural groups, represented as Upper Canada (Ontario) and Lower Canada (Québec) had to find political solutions to tensions between

French- and English-speaking citizens of the Province of Canada.

After Confederation, Canadian political leaders faced the same challenge of finding a compromise that would be acceptable to the two linguistic groups. But there was an important difference, namely, that considerable powers were assigned to provincial governments under the British North America Act. Under this act, the division of powers between the federal government and the provinces ensured that provincial governments had control over cultural affairs, such as education and language, which allowed Québec to maintain its french heritage.

Québec's historical geography can be divided into three periods: New France, after the British Conquest, and Confederation.

New France

Between the seventeenth and eighteenth centuries, France had control over vast areas of North America. Its core, however, was the St Lawrence Valley, from where New France developed a vast fur-trading empire. The wealth from the fur trade was enormous and provided the reason for France's interest in the New World. Almost every French settler wanted to participate in the fur trade, which left only a few to clear the forest and till the land. For example, several canoes full of beaver pelts could make a man extremely wealthy compared with the meagre returns obtained from the back-breaking toil of clearing land and breaking the soil. Frenchmen who were *coureurs du bois* (trappers) often lived with the Indians. A few were extremely successful and returned to France to enjoy

their good fortune. Others remained in the fur trade or settled in New France.

Geography played a part in France's success both as a fur-trading empire and as an agricultural colony. The St Lawrence River provided a route to the interior, which gave the French explorers and fur traders an advantage over their English rivals, who had to contend with crossing the Appalachian Mountains. With further exploration of the interior of North America, the French were able to expand their territory and secure more and better fur-trading routes. For example, by portaging from the Great Lakes to the Ohio River and then the Mississippi, the famous French explorer, René-Robert Cavelier, sieur de la Salle, reached the mouth of the Mississippi in 1682. Early in the eighteenth century, French fur traders, led by Pierre Gaultier de Varennes, sieur de la Vérendrye, established a series of fur-trading posts in Manitoba. These trading posts, especially Fort Bourbon on the Saskatchewan River at Cedar Lake (just east of the present-day community of The Pas), made it convenient for Indians to trade their furs with French traders rather than travel farther to the British fur-trading posts along the Hudson Bay coast.

New France also established a successful agricultural society. Once the land was cleared, the fertile soils in the St Lawrence Lowlands provided a solid basis for a feudal agricultural settlement known as the seigneurial system. Farming took hold in New France, particularly following the efforts of Jean Talon, the greatest administrator of New France. By the early-eighteenth century, the French had turned to farming, leaving the lure of the wilderness to a relatively small number of more daring souls. By then, New France had existed for over 100 years, and many of its people had been born and raised in the New World. They were no longer 'French' but *Canadiens*.

By the middle of the eighteenth century, almost all the lands in the St Lawrence Lowlands were under cultivation. Farmlands stretched in a continuous belt from Québec City to Montréal. Like the feudal agricultural system in France, peasant farmers (**habitants**) worked the land, paying their lords (**seigneurs**) both in kind and in labour. Through the efforts of the *habitants* and their seigneurs, New France became a successful agricultural colony, but one whose system of land ownership and rural life were quite different compared with those of the British colonies along the Atlantic seaboard, (a difference which would eventually cause Britain to split Québec into Upper and Lower Canada in 1791 [see Chapter 3]).

The Seigneurial System

When the first intendant of New France, Jean Talon, arrived in New France in 1665, he encountered a population of only 3,000 inhabitants, most of whom were men engaged in the fur trade. Talon had been instructed by Louis XIV to create an agricultural society resembling that of rural France in the seventeenth century. Talon undertook three measures to achieve this goal. First, he recruited peasants from France. Second, Talon sent for young women—orphaned girls and daughters of poor families in France—to provide wives for the men of the colony. Third, he imposed the French feudal system of land ownership, known as the seigneurial system. In the seigneurial system, huge tracts of land were granted to those favoured by the king, namely, the nobility, religious institutions of the Roman Catholic Church, military officers, and high-ranking government officials. The seigneur was obliged to swear allegiance to the king and to have his tenants cultivate the lands on his estate. The tenants owed certain obligations to their seigneur: paying yearly dues (*cens et rentes*) to their seigneur, working the seigneur's land (*corvée*), and paying rent for using the seigneur's grinding mill and bake ovens (*droit de banalité*).

The goal of the seigneurial system was to create a feudal society in New France that resembled that of the mother country. To that end, the seigneurial system was controlled by the king of France, who awarded land to his followers, who were frequently nobles, bishops, and officers. In the days of New France, the seigneurial system offered several advantages: (1) it encouraged the establishment and functioning of an agricultural society; (2) it provided access to the seigneur's mill where the *habitants* could grind their grain; and (3) the mill served as a centre of defence for the *habitants* when under attack.

By 1760, there were approximately 200 seigneuries. Seigneuries, which were usually 1 by 3 leagues (5 by 15 km) in size, were generally divided into river lots (*rangs*). These long, rectangular lots were well adapted to the St Lawrence Valley for several reasons, the most important of which was that each *habitant* had access to a river, either a tributary of the St Lawrence River or the river itself. At that time,

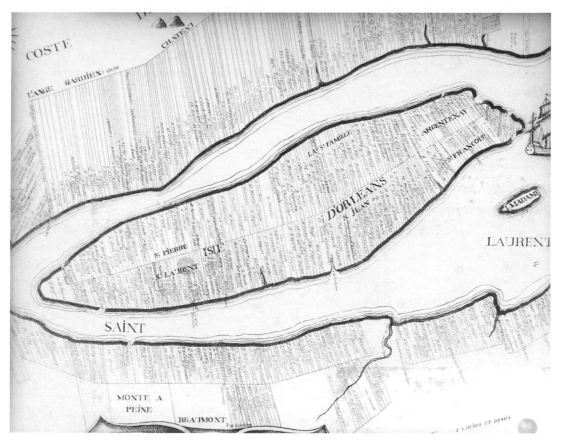

Historic map of the Île d'Orléans and the north and south shores of the St Lawrence. Note the *rangs*, the long, narrow lots cultivated by *habitants* in New France. (*Courtesy National Archives of Canada NMC0048248*)

most people and goods were transported along the river system in New France. For that reason, river access was vital for each *habitant* family.

After the British conquest of New France, the seigneurial system was retained more for political reasons—to gain the support of the principal source of power (the seigneurs and the Roman Catholic Church)—than for economic reasons. By the early-nineteenth century, however, agriculture in Lower Canada had become a commercial venture, making

the seigneurial system an anachronism. The seigneurial system was abolished in 1854 by the Legislature of the Province of Canada.

After the British Conquest

Following the defeat of the French Army in 1760, the British ruled Québec for over 100 years. The British governor was installed at Québec City along with a regiment of British troops, while the fur trade continued to flourish and the agricultural economy went unchanged. Most French Canadians were peasant farmers. After the British conquest, their life on the land remained much the same. Their social and economic lives revolved around their parish church and a landholding system centred on the seigneuries.

By the middle of the nineteenth century, however, French Canadians were migrating out of the St Lawrence Lowlands due to a land shortage. Birth rates were so high in this rural society that there was not enough land left for

This Roman Catholic church at L'Anse-Saint-Jean on the Saguenay River is one among many that dot the province's landscape, illustrating the church's role in the historic development of Québec. (*Courtesy R.M. Bone*)

the children of the *habitants*. French Canadians migrated in three directions: to the Appalachian Uplands, where they either purchased farms from English-speaking farmers or took possession of abandoned land; to the Canadian Shield, where they tried to exist on extremely marginal agricultural land; or to New England's industrial towns, where most were employed in textile factories. Then, in the late-nineteenth century, a small number of French Canadians settled in the West. Through all of these economic and political changes, the vast majority of French Canadians maintained their language and Catholic religion. They turned to the Roman Catholic Church for both spiritual and political leadership. The Church, in turn, encouraged the *habitants* to stay on the land, far from the secularizing influences in towns and cities, where the English Protestants lived.

During this same period, however, political unrest was growing in both Upper and Lower Canada. Each colonial government was headed by a governor who was appointed in England and had absolute power. The governor administered the colony by appointing leading members of the community. This cozy arrangement not only concentrated power in the hands of a few but also led to blatant abuse by powerful élites. In Upper Canada, the political élites were known as the Family Compact, while in Lower Canada they were the Château Clique. In 1837, rebellions took place in each colony (Vignette 6.3). Both rebellions, which called for political reforms, were crushed by the British Army. Britain, however, was determined to remedy the political situation in its two colonies. In an attempt to identify the main sources of discontent in Upper and Lower Canada, the British government sent Lord Durham, a politician, diplomat, and colonial administrator, to North America. Durham recognized that the political solution lay in an elected government, where power was dispersed among elected representatives rather than concentrated among an appointed élite. Durham also observed the French/English faultline, which he described as 'two nations warring in the bosom of a single state'. In Durham's report he recommended responsible (elected) government and the union of English-speaking people in Upper Canada with the French-speaking settlers of Lower Canada.[1] Lord Durham believed that assimilation of the French was desirable and possible, claiming that the French Canadians were 'a people with no literature and no history' (Mills 1988:637). He recommended that English be the sole language of the new Province of Canada, and that a massive immigration of English-speaking settlers be launched in order to create an English majority in Lower Canada. In response to Durham's recommendations, the Act of Union was

---

Vignette 6.3

The Rebellions of 1837

In Lower Canada, Louis-Joseph Papineau, a lawyer, seigneur, and politician, was the leader of the French-speaking majority in the Assembly of Lower Canada. In 1834, Papineau issued a list of grievances known as 'The Ninety-Two Resolutions'. At this time, the economy was depressed and tensions between the French-Canadian majority and the British minority were growing. Papineau sought to shift political power from the British authorities to the elected Assembly of Lower Canada. He planned to use his majority in the Assembly to pass legislation, including tax bills. The British government rejected 'The Ninety-Two Resolutions'—it was just a matter of time before an armed uprising broke out. When it did, the British reacted with force. Even with the strong support of rural areas, Papineau and his *Patriotes* were soundly defeated. Nearly 300 rebels were killed in six battles. Papineau fled to the United States. The British then hung twelve captured rebels and exiled another fifty-eight to Australia. A rebellion also took place in Upper Canada at the same time and it too was crushed. The British government sought to remedy the unrest in both of its colonies. It began this process with a fact-finding mission headed by Lord Durham. The result and Britain's solution was the Act of Union in 1841.

passed by the British Parliament in 1841, uniting the two colonies into the Province of Canada and thus creating a single elected assembly.

Lower Canada was now known as Canada East, and in spite of the political changes, the French Canadians were not assimilated. Under the new form of British administration, the task of maintaining their French culture and language was not easy, but it was achieved because of several factors, the most important being their strong desire to remain Catholic and French-speaking. A second factor was the institutional support provided by the Roman Catholic Church. By providing spiritual guidance and schooling in French, the clergy played an essential role in cultural preservation. Geography and demography were other factors essential to the survival of French Canada in the nineteenth century. The overwhelming number and concentration of French-speaking people in Canada East was a critical geographic reality. A high birth rate and high rate of natural increase for the *Canadiens* ensured an expanding population of French Catholics. Other factors were the rural nature of the French-speaking population, which isolated them from English-speaking residents of the major cities, and the emergence of an intellectual élite whose writings preserved the history and literature of French Canada. One of the most popular novels in Canada East was *Jean Rivard*. Written by Antoine Gérin-Lajoie, *Jean Rivard* is the story of a young French Canadian who is advised by his *curé* on the advantages of becoming a farmer rather than a lawyer. The novel promotes the virtue of living in a harsh and re-mote land, which is superior (in God's eyes) to living in an urban centre with its many worldly temptations.

Confederation

Confederation, achieved in 1867, sought to unite two cultures—English and French—within a British parliamentary system. For Québec, Confederation provided a political framework offering three benefits: an economic union with Ontario, Nova Scotia, and New Brunswick; a political environment where Roman Catholicism and, to a lesser degree, the French language were guaranteed protection by Ottawa; and provincial control over education and language. George-Étienne Cartier, one of the Fathers of Confederation and a French-Canadian leader, viewed these provincial powers as a way for Québec to shape its own destiny within Confederation. Cartier may have identified a fourth benefit—since Québec and Ontario often had mutual economic interests, they could, by working together, influence federal policies and thereby shape the future of Canada.

Confederation also led to the expansion of the geographic size of Québec (Figures 3.4 and 3.6). Since Confederation, Québec's geographic size has increased greatly. It is now 1.5 million km². As Canada acquired more territory from the British government, Ottawa assigned to Québec parts of Rupert's Land lying north of the St Lawrence drainage basin. In 1898, the Québec government received the first block of Rupert's Land. The second was obtained in 1912.[2] Some of this land, however, was claimed by another British colony, Newfoundland. In its argument, Newfoundland demanded all of Québec's territory that drained into the Atlantic Ocean. Though the Imperial Privy Council of Britain awarded this land, known as Labrador, to Newfoundland in 1927 (Figure 3.7), to this day, the Québec government does not recognize the decision.

Prior to the Second World War, Québec continued to project an image of a rural, inward-looking, church-dominated society. This image masked social changes that provided the foundation for the transformation of Québec into a modern industrial society. This historical period of transformation is known as the Quiet Revolution. While many social changes began long before the post-Second World War era, the Quiet Revolution was officially triggered in 1960 with the election of the Liberal government of Jean Lesage. His government initiated major political innovations that accelerated the process of social and economic change within Québec society. By doing so, the provincial government replaced the Catholic Church as the leader and protector of French culture and language in Québec.

The Quiet Revolution instilled a sense of pride and accomplishment among Quebeckers. (See Chapter 3 for more on the Quiet Revolution.) By 1960, Québec was ready for change and the Lesage government was the instrument of that change. The main reforms of the Lesage government were a policy of state intervention in the Québec economy through Crown corporations and the expansion of the provincial civil service. The government's principal achievements were:

- the nationalization of private electrical companies under Hydro-Québec

- the modernization and secularization of the education system, making it accessible to all
- the investment of Québec Pension Plan funds in Québec firms, thereby stimulating the francophone business sector
- and the establishment of Maisons du Québec (quasi-embassies) in Paris, London, and New York, thus signalling to Ottawa that the Québec government wanted to represent Québec interests to the rest of the world.

With these accomplishments behind them, Quebeckers felt confident about their future. Lesage's slogan, 'Maîtres chez nous' (Masters in our own house), became a reality. For federalists in Québec, these achievements proved that a strong Québec could function within Canada, but for separatists, they were not enough. The rise of separatist views in Québec during this time signalled that some Québécois felt only an independent Québec could adequately represent French-Canadian interests. For them the slogan became 'Le Québec aux Québécois' (Québec for the Québécois).

Québec Today

Québec is now a modern industrial society operating within the North American and global economies. It has both a stable resource industry and a solid manufacturing industry. However, economic changes have, and will continue to challenge the province, forcing industries to adapt to an evolving market.

As a modern industrial society, Québec exhibits the customary economic and social characteristics that other such societies do. For instance, most Quebeckers, like their counterparts in the rest of Canada, live in urban settings. Québec also has an extremely low rate of natural increase. Attracting immigrants to the province ensures a higher rate of population growth, but some worry that immigrants could change the linguistic balance between French- and English-speaking Quebeckers. In fact, before the language laws were introduced, almost all non-French-speaking immigrants chose to speak English and send their children to English schools. The language laws now require immigrants to send their children to French schools, thereby ensuring the growth of the French-speaking population.

Today, language continues to shape Québec's economy, politics, and its relationship with Ottawa and the other provinces. It is the issue around which most other issues are framed—in other words, most political, social, and economic issues in Québec are often analysed from the perspective of language. The French/English faultline is therefore a crucial component of Québec and must be well understood if it is to be adequately addressed.

The French/English Faultline in Québec

The two main components of the French/English faultline in Québec are: relations between the French-speaking Québécois and English-speaking Quebeckers, and the state of the Canadian federation and Québec's place in it. Each of these components interacts with, and affects the other. As Québécois society continues to evolve within a Canadian political context, the question of Québec's place within the country remains a contentious political issue that deeply divides Québec society along linguistic lines.

Québécois tend to favour a new arrangement with the rest of Canada, while the English-speaking minority wish to retain the status quo. The division, however, is more complex than it appears. Politically, there are two camps—those who support federalism and those who back separatism. Generally, federalists include both some Québécois and the English-speaking minority. They would like to remain in Canada, but many would like more political powers for their province. Separatists, which include many Québécois and a small number of the English-speaking minority, seek a new political arrangement with Canada. Like the federalists, they too do not hold a single political view. Within separatism, moderates seek a partnership with Canada or, at least, some form of economic union, while hardliners insist on independence. This ambivalence stems from the dual attachment to both Canada and Québec discussed previously.

Though linguistic tensions date back to the early days of Upper and Lower Canada, linguistic political divisions within Québec have really solidified in the past few decades. A number of events contributed to this division. After the Quiet Revolution, attitudes between French- and English-speaking Quebeckers began to harden. The Quiet Revolution had addressed the economic-based linguistic division of a French-speaking underclass (the majority) being led by an English-speaking based business class (a minority). A power shift occurred in the province—more and

Figure 6.2 Less sovereignty, more assocation. In the weeks before the 1995 referendum in Québec, the low birth rate among the francophone population became an issue, causing the separatist leader, Lucien Bouchard, to call on Québécois women to have more children. This cartoonist, in a play on words, made use of the separatist's slogan 'sovereignty association' with Canada.

Source: The Globe and Mail, 17 April 1997:A16. Reprinted with permission from The Globe and Mail.

more French-speaking Québécois were running businesses and attaining positions of power. The English-speaking minority no longer had the economic advantage and also had to adapt to new nationalist policies like the language laws.

French/English relations were then made more complex by events outside of Québec, which included a growing sense of nationalism among English-speaking Canadians and, like the Québécois, a new sense of pride. One symbol of that Canadian nationalism was the patriation of the Canadian Constitution from Britain. (To **patriate** is to give a nation direct control over government decision making.)

In 1982, the patriation of the Canadian Constitution was the culmination of a frenzied period of constitutional debate, set in motion by the defeat of the Québec referendum on sovereignty association in 1980. Patriation of the Constitution achieved a number of things. By giving Ottawa control over constitutional amendments, the Canadian Parliament eliminated the last vestiges of Canada's colonial status. The Canadian Constitution also provided a Canadian formula for future amendments to the Constitution, extended English/French minority language rights in education, and gave Canada a constitutional Charter of Rights and Freedoms. However, the act had one fatal flaw. It was signed over the bitter objections of the Québec government. Since the Canadian government had campaigned for the rejection of independence in the 1980 Québec referendum by promising renewed federalism to the Québécois, many people in Québec felt betrayed. For separatists, this was the final straw. In their view, Ottawa and the other provinces should not have patriated the Constitution without the Québec government's consent.

The combination of events surrounding patriation recharged the separatist movement in Québec. It also led Québec federalists to question the relationship between Ottawa and Québec. A long-time federalist, Léon Dion, concluded that 'English Canada will only yield—and even this is not assured—if there is a knife at its throat' (Simpson 1993:312). Dion was stating an old idea in a new context, namely, that Ottawa would not negotiate a new relationship for Québec within Canada unless Québec was prepared to break up the Canadian federation. In Dion's metaphor, the threat of separation was the knife and the method was a referendum. The goal was a new partnership with Canada. Other Quebeckers, like Jacques Parizeau, had a different strategy, one that would actually lead to an independent Québec.

Because patriation had caused such controversy in Quebec, succeeding federal governments made efforts to find a compromise that would have the Québec government sign the Canadian Constitution. Unfortunately, the two ensuing rounds of negotiations and their respective agreements—the Meech Lake Accord (1987) and the Charlottetown Accord (1992)—ended in failure. However, these attempts and the negotiations, parliamentary committees, and other measures that were pursued during this period demonstrated three realities of Confederation:

- the difficulty in reaching a political compromise when conflicting interests exist between French- and English-speaking Canadians (Vignette 6.4)
- Ottawa's commitment in seeking a compromise
- and, the support of the majority of Canadians, both English- and French-speaking, to support efforts to find such an elusive compromise.

However, the failures of the constitutional negotiations of the late 1980s and early 1990s had further exacerbated the political 'insult' caused by patriation, leading to a resurgence of separatist feelings in Québec. By 1995, a separatism government, led by Jacques Parizeau, was again appealing to its electorate through a referendum to give it a mandate to pursue sovereignty for Québec.

The 1995 Referendum

The 1995 referendum was a political watershed for Canada. The question was: 'Do you agree that Quebec should become sovereign, after having made a formal offer to Canada for a new Economic and Political Partnership within the scope of the Bill respecting the future of Québec and of the agreement signed

Vignette 6.4

The Meech Lake Accord

The Meech Lake Accord rekindled a broad debate about the fundamentals of Canadian state and society and about the nature of Canadian federalism—relations between national and provincial political communities, between federal and provincial governments, and between citizen and state.

In May 1986, the Québec government revealed five conditions that must be met if Québec were to accept the 1982 Constitution Act:

- constitutional recognition of Québec as a distinct society
- a provincial role in immigration
- a provincial role in Supreme Court appointments
- limitations on the federal power to spend in areas of provincial jurisdiction
- an assured veto for Québec in any future constitutional amendments (Simeon 1988:59)

These became the five main components of the Meech Lake Accord as well as the focus of controversy among the media and interest groups. Opposition to the doomed accord was strong among Aboriginal peoples, women, and centralists, who feared distinct society would give Québec special status.

on June 12, 1995?' While one can debate the meaning of the referendum question, all would agree that Quebeckers came very close to separating from Canada—the *Non* side won by a very slim majority. If there had been a *Oui* majority, Premier Parizeau's strategy would have been to immediately declare Québec an independent nation.

The deep language divide within Québec society is shown clearly in the results of the referendum. In 1995, the political map of Québec showed strong support for the *Oui* side in rural Québec where the francophone population dominates. In fact, support for the *Oui* side was strongest among Québec francophones, with estimates placing it as high as 60 per cent, and weakest among anglophones and allophones. Only four areas of Québec—Montréal, the Ottawa Valley, Estrie, and north-

ern Québec—supported the federalist side, but these were the areas with the majority of the voters. In the first three areas, there is a high proportion of anglophones and allophones, and in northern Québec, Cree and Inuit peoples are the vast majority. The Cree have been particularly outspoken on the issue of sovereignty, arguing that they have the right to determine their political future (just like Québec), whether it means remaining part of Canada or joining an independent Québec. The issue of separation is a complex political matter not defined in the Constitution of Canada. However, on 20 August 1998, the Supreme Court of Canada ruled that under both Canadian and international law, Québec (or any other province) cannot declare its independence unilaterally, thus bringing to an end the threat of separation by decree. Never-

Figure 6.3 Referendum fatigue. Following the 1995 referendum, Quebeckers and other Canadians grew weary of the political wrangling over Québec's place in Confederation. Economic issues had become more important to the average citizen. The issue of Québec's separation from Canada was placed on the political back burner.

Source: The Globe and Mail, 31 October 1995:A24. Reprinted with permission from The Globe and Mail.

theless, the Supreme Court accepted the concept that separation was possible through negotiation *after* a 'clear majority' voted for separation.

The French/English faultline is a prominent feature of Québec politics and society, a feature that will continue to dominate linguistic relations within the province, as well as federal-provincial relations in the country. In short, constitutional events since 1980, combined with already existing linguistic tensions in Québec, have led to a national crisis that must be resolved in order for both

Québec and the rest of Canada to function normally.

Québec's Economy

Like Ontario, Québec has a strong and diversified economy. Also like Ontario, Québec's economic activities can be divided in two by geographic location: a manufacturing and agricultural core operates in southern Québec, and a resource-based periphery exists in northern Québec. The similarities between Ontario and Québec can be attributed to their similar physical geography and close proximity in terms of geographic location. Both provinces share two physiographic regions—the fertile Great Lakes–St Lawrence Lowlands and the Canadian Shield (Figure 2.1). Consequently, both provinces have a similar range of resources available for development. There are differences, however. The climatic conditions in Ontario's part of the Great Lakes–St Lawrence Lowlands provide for a longer growing season, thus giving Ontario an advantage. On the other hand, Québec's part of the Canadian Shield gives that province ideal physical conditions, such as high elevations and abundant precipitation, for the production of hydroelectricity. These conditions are not present in Ontario's part of the Canadian Shield.

Québec and Ontario both contain parts of the Canadian manufacturing belt, which extends from Windsor to Québec City. Proximity to Canada's greatest trading partner, the United States, is extremely important for manufacturing. Windsor and other centres in southern Ontario have had the advantage of proximity to the American automobile-manufacturing centre of Detroit. Montréal, on the other hand, is farther from major American manufacturing cities. Nonetheless many firms in Canada's manufacturing belt supply products to other manufacturing companies in this belt, indicating a high degree of economic integration. An example of such integration is the automobile industry. Many firms in the Greater Montréal area produce automobile parts and then ship them to automobile-assembly plants in Québec, Ontario, and the United States. General Motors of Canada has an assembly plant in Sainte-Thérèse that makes the Pontiac Firebird and Chevrolet Ca-

maro sports cars. As well, there are twenty-three auto-parts firms in Quebec. Overall, the automotive industry in Quebec employs nearly 15,000 workers.

Québec-based firms have specialized in certain industrial sectors, including aerospace, biotechnology, clothing design and manufacturing, furniture manufacturing, metal refining, printing, textiles, and transport equipment. Its aerospace firms have been particularly successful in recent years. For example, the aerospace company, CAE Inc., employs about 4,000 highly trained workers in the Montréal area. Approximately one-third are engaged in research and engineering; that is, designing and testing new products like a robot that strips paint from aircraft bodies. Its main product, however, is a flight simulator that is manufactured at its plant in Saint-Laurent near Montréal. Unlike most manufacturing firms, most of CAE's sales do not go to the United States but to customers in Europe, the Middle East, and Asia.

Industrial Structure

Québec employment figures by economic sector reveal an industrial structure that is strikingly similar to Ontario's (Table 6.1). The employment figures indicate that nearly three-quarters of Québec's labour force is engaged in tertiary industries, 23 per cent in the secondary or processing industries, and 3.5 per cent in the primary sector. This general pattern of employment holds true across the country, though there is more variation in the latter two categories. The tertiary sector, however, tends to remain more constant. For example, the Québec figure of 73.5 per cent for tertiary employment is almost identical to the national figure of 73.4 per cent (Table 6.2). What makes Québec's tertiary sector stand out from the national figures is the size of its public sector, which reflects the strength of the provincial civil service.

Table 6.1
Employment by Industrial Sector in Québec and Ontario, 1995

Industrial Sector	Québec (per cent)	Ontario (per cent)	Difference (per cent)
Primary	3.5	3.0	0.5
Secondary	23.0	23.6	(0.6)
Tertiary	73.5	73.4	0.1
Total	100.0	100.0	0.0

Source: Adapted from Statistics Canada, 1996: Table 40.

Table 6.2
Employment by Industrial Sector in Québec, 1995

Industrial Sector	Québec (000)	Québec (per cent)	Canada (per cent)	Difference (per cent)
Primary	112	3.5	5.4	(1.9)
Secondary	737	23.0	21.2	1.8
Manufacturing	593	18.5	15.3	3.2
Tertiary	2,355	73.5	73.4	0.1
Transport	240	7.5	7.5	0.0
Trade	555	17.3	17.3	0.0
Finance	186	5.8	5.8	0.0
Service	1,171	36.5	37.1	(0.6)
Public	203	6.3	5.9	0.4
Total	3,204	100.0	100.0	0.0

Source: Adapted from Statistics Canada, 1996: Table 40.

Québec's primary and secondary sectors make up only 3.5 per cent and 23 per cent of the employed labour force respectively. Again, the key point is the relative importance of secondary employment, particularly employment in manufacturing firms. These employment figures confirm the critical importance of manufacturing in the Québec economy.

Southern Québec

Southern Québec is the economic, social, and political core of Québec, while northern Québec is a sparsely populated resource hinterland. The relationship between the two regions is a provincial version of the core/periphery model.

Southern Québec occupies the more favoured lands of Québec. They consist of two physiographic regions: the Appalachian Uplands and the St Lawrence Lowlands. Bordered on the north by the Canadian Shield and on the south by the United States, southern Québec is only a small part of the territory of Québec, but it contains over 90 per cent of Québec's population and is the most productive area of the province. The human and physical characteristics of each physiographic region in southern Québec are presented below.

Appalachian Uplands

The Appalachian Uplands region lies west of the Atlantic Ocean, and north of the United States, along the southern and eastern edges of the St Lawrence Valley. Because of this geographic situation, the Appalachian Uplands region was settled at different times and in dif-

ferent ways. In the sixteenth century the waters off the Gaspé Peninsula attracted fishers. Even today, fishing in the Atlantic Ocean provides a way of life, though many supplement their income with farming and wage employment in the small coastal settlements. The main settlements are scattered along the Gaspé Coast and the south shore of the St Lawrence River. The largest urban centres are along the south shore. Rimouski, with a population over 30,000, is by far the largest city in the region. Rivière-du-Loup and Matane are medium-size towns that have populations roughly half the size of Rimouski. Towns are much smaller along the Gaspé Coast, as the area has a very limited resource base. Percé, now an important tourist town, is the largest of the small centres along the Gaspé Coast with a population of about 4,000. This French-speaking area has place names like New Richmond, New Carlisle, and Chandler, the legacy of early English settlers, some of whom had relocated along the Gaspé Coast after the American Revolution. Over the last half century, the lack of jobs caused many to migrate out of this area, which seriously reduced the size of English-speaking communities. In 1996, English-speaking residents constituted less than 5 per cent of the region's population.

Estrie (the Eastern Townships) is a pocket of communities in the rolling land of the Appalachian Uplands located east of Montréal. Estrie was also settled after the American Revolution by British Loyalists. The British organized the surveyed land into rectangular townships rather than the long lots of the St Lawrence Valley. Much of the land was ill-suited for cultivation. Within several generations, the more marginal lands were abandoned when English-speaking owners left to look for jobs in Montréal and Boston or to try homesteading on the American frontier. Because of a land shortage in the St Lawrence Valley, French Canadians began to move into the Eastern Townships. French-Canadian migrants, often sons of farmers living in the St Lawrence Lowlands, either bought or took over abandoned farms from the original English-speaking settlers.

The physical geography of Estrie is much more conducive to economic development and agricultural settlement than are the Gaspé coast and south shore of the St Lawrence. From the 1870s to the 1970s, mining at Thetford Mines and Asbestos was a key sector of the regional economy. Since the 1970s, asbestos mining has fallen on hard times because the use of this

mineral has proven hazardous to human life. Still, mining of this deposit continues. Agriculture and forestry have also proven to be durable in this area. Dairy farming takes place in the broad valleys, and logging in the forested uplands. Overall, Estrie is the most prosperous area of the Appalachian Uplands. Sherbrooke exemplifies the relative well-being of the region. Sherbrooke has grown over the years, and is the largest urban centre in the Appalachian Uplands. With a population of nearly 150,000 in 1996, Sherbrooke has become an important regional centre (Table 6.3). From 1981 to 1996, its population increased by 17.7 per cent, making Sherbrooke the second-fastest-growing city in Québec. This demographic increase over the years is reflected in its economic growth. Proximity to Montréal has often worked in Sherbrooke's favour,

Table 6.3
Census Metropolitan Areas in Québec, 1981–1996

Census Metropolitan Area	Population 1981	Population 1996	Change (per cent)
Montréal	2,828,286	3,326,510	17.6
Québec City	583,820	671,889	15.1
Hull*	185,442	247,072	33.2
Chicoutimi–Jonquière	158,229	160,454	0.1
Sherbrooke	125,183	147,384	17.7
Trois-Rivières	125,343	139,956	11.7
Total	4,006,303	4,693,265	17.1

*The population of the Ottawa–Hull CMA was 743,821 in 1981 and 1,010,498 in 1996.

Sources: Statistics Canada, 1982; 1997a: Table 1.

allowing it to engage in the textile industry in the nineteenth century and now in the high-technology industries.

St Lawrence Lowlands

Most of Québec's agricultural and industrial production is in the St Lawrence Lowlands. In fact, this area functions as the province's core. It contains Québec's largest market and is close to transportation networks that facilitate trade.

The St Lawrence Lowlands is the most geographically favourable region for agricultural activities. Livestock farms specializing in beef cattle, hogs, or sheep are common, while some farmers concentrate on poultry and egg production. Livestock farmers grow forage crops for winter feed. During the summer, the cattle graze on pastures. Specialized crops, particularly vegetables and fruit, are also popular and are sold mostly within the major urban centres of the province.

Though farmers in this region engage in a variety of agricultural activities, dairy and vegetable farming predominate. In fact, under the Canadian industrial milk marketing board, dairy farmers in Québec supply 48 per cent of the Canadian market. Moreover the processing of agricultural products provides an added benefit for the Québec economy. For instance, large butter and cheese firms rely on milk products from the dairy farms. Often the processed food product is designed for the provincial market. For example, because of

Québec has Canada's largest dairy products industry. (*Corel Photos*)

the market for unpasteurized cheese products in Québec, a few cheese firms began producing *fromage au lait cru*, cheese made with unpasteurized milk. By the early 1990s, unpasteurized cheese from these Québec firms was competing with European imports, including popular brands from France.

Dairy farmers have fared relatively well under Québec's fluid milk marketing board and a national marketing system that allocates producers a share of the Canadian market for milk. Under NAFTA and GATT agreements, however, Canada may be forced to dismantle its marketing boards in the near future. Without a regulated dairy industry, dairy farmers in Québec would face stiff competition from American dairy imports. Furthermore, the agricultural economy has little room for growth. So in order for Québec's economy to grow, the province will have to rely on its manufacturing sector.

Manufacturing is concentrated in the Montréal area, but extends eastward to Estrie and northeastward to Québec City. As part of the Canadian manufacturing axis, this industrial zone serves as the engine driving both the provincial and the national economies. Most of Québec's manufacturing sector is in Montréal, where various industrial activities are pursued. The textile, knitting, leather, and clothing sector was established here many years ago and remains a part of this industrial area. Now struggling to compete with foreign imports, the textile industry is undergoing

Montréal harbour and downtown at night. Montréal is Québec's manufacturing core. (*Corel Photos*)

massive restructuring. Other labour-intensive manufacturing firms face the same challenge and, unfortunately, the solution lies in reducing the size of their labour force. This trend partly accounts for the high unemployment figures in Québec. In spite of these difficulties, Montréal remains the focus of Canada's clothing and textile industry.

Automobile assembly and parts production make up another important sector. There is an automobile assembly plant at Sainte-Thérèse just north of Montréal, and a bus-assembly operation near Saint-Eustache, a suburb of Montréal. Since the 1980s, Québec's aerospace industry has also become highly competitive on the world market. Canadair, a company now owned by Bombardier, produces the Challenger, an executive jet. The success of this firm is due largely to its ability to sell its product in the global market.

Those firms that are under pressure to be more competitive in world trade have undergone significant economic restructuring, which has had an adverse impact on employment. During this period of adjustment, economic growth in this region of Québec has slowed. One measure of this malaise has been persistent high unemployment rates. The fundamental task for both the provincial and federal governments in the next years will be to support the revitalization of Québec's manufacturing sector. This revitalization strategy should promote key industrial sectors (including aerospace, pharmaceutical, and metal-processing industries), and encourage industrial development in southern Québec by offering firms low-cost electrical power.

Key Topic: Hydro-Québec

Since the Quiet Revolution, successive Québec governments have played an active role in shaping the province's industrial economy. This approach, initiated by the Lesage government, has pursued two goals—one economic, the other political. Its economic objective has been to stimulate economic growth in the province. Its political goal is to increase Québec's ownership and control of its economy within the francophone business community. The most direct way to achieve these two goals was to create public (Crown) corporations. The best example of the success of this industrial development strategy is Hydro-Québec. Now the prize Crown corporation in Québec, Hydro-Québec has played a fundamental role in achieving these two goals, and, in so doing, has helped transform Québec's economy into a solid modern industrial economy and build a strong French-speaking business community. Today Hydro-Québec is Canada's largest electric utility.

Created in 1944, Hydro-Québec was a minor force in the Québec economy until the Lesage government came into power in 1962. At that time, the Québec government announced its intention to purchase the private electricity companies in Québec and place them under the umbrella of Hydro-Québec. The Crown corporation soon extended its activities to cover the whole province by purchasing the shares of nearly all remaining privately owned electrical utilities and taking over their debts. With a virtual monopoly to generate and distribute electricity in the province, Hydro-Québec undertook the task of expanding Québec's hydroelectric capacity. In the 1960s, Hydro-Québec built dams on the Manicouagan River Complex, creating the huge Manic-Outardes hydroelectric complex. The design and construction of this massive project were carried out by Québec firms and workers, thereby creating highly specialized engineering firms capable of designing and building huge hydroelectric projects.

Hydro-Québec harnessed the vast water resources of the Canadian Shield and transported this electrical energy from the dams on the Manicouagan River to Québec City and Montréal. In order to transmit the complex's annual production of about 30 billion kWh over a distance of nearly 700 km, Hydro-Québec became the first utility in the world to transmit electricity at very high voltages (735 kv), which reduces the amount of electrical energy lost in the transmission process. This same technology now enables Hydro-Québec to transmit electrical power from Churchill Falls[3] and James Bay to markets in southern Québec and New England.

Following the success of the Manicouagan River complex, Hydro-Québec embarked on an even larger project—the James Bay Project (Figure 6.4). As it turned out, the James Bay Project, which includes La Grande, Great Whale, and Nottoway river basins, was the crowning success of Hydro-Québec, but also its *bête noire*. This was the largest hydroelectric project in the world and transmitted power over long distances never attempted before. Québec could now sell massive amounts of power to the United States and, in so doing, recoup much of the construction costs. The

natural environment, however, suffered quite a bit when it was altered to divert and change the seasonal flow of rivers to create huge reservoirs. The project negatively affected the fish stocks and the prime wildlife habitat along the river valleys and lakes. The major social impact was the harmful effect on the Cree's hunting and trapping lifestyle. Although Hydro-Québec made efforts to minimize the environmental damage and compensate the Cree, Inuit, and Naskapi, the Cree strongly opposed further hydroelectric developments in northern Québec. (See the section on 'Hydro Power' later in this chapter for further discussion on these topics.)

Regional Linkages

Hydroelectric power became an industry that united the economies of both northern and southern Québec. With vast amounts of potential electrical power in northern Québec, the provincial government saw an opportunity to attract energy-hungry industries into southern Québec by offering them special low electricity rates. For example, in 1957 Reynolds Aluminum built a smelter at Baie-Comeau because Hydro-Québec provided it with low-cost power. In return, the company added to the value of production in the province and, more important, it provided jobs. In 1996, Reynolds Aluminum employed approximately 3,000 workers.

The construction of the James Bay Project brought the potential for much more surplus power, which could be used to lure other industrial companies into southern Québec. The first step in this core/periphery strategy had been the construction of the first phase of the

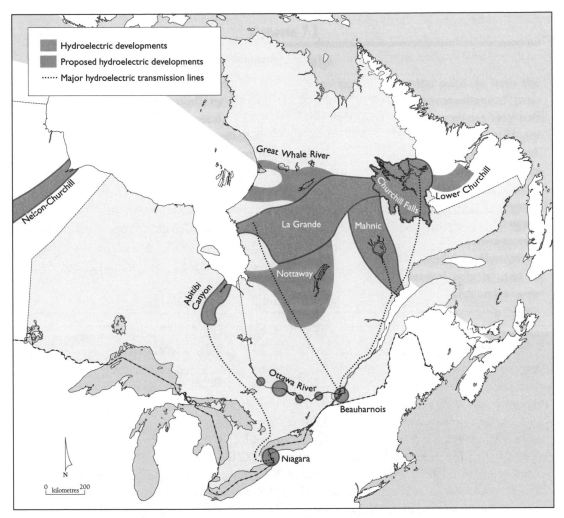

Figure 6.4 Hydroelectric power in Central Canada. Québec's dominant role in the production of hydroelectric power is due to two natural factors: (1) the heavy annual precipitation in northern Québec, and (2) the high elevation of the Canadian Shield in Québec.

James Bay Project in the 1970s. The second step took place in the 1980s. The provincial government took a more aggressive stance by offering risk-sharing contracts and electric-power rebates to industrial firms requiring large amounts of power in their processing operations, such as metallurgical companies. As a result, industrial firms either expanded their operations or new firms located along the St Lawrence River. Thirteen companies, including Norsk Hydro, took up the government's offer.[4] Norsk Hydro requires vast quantities of electricity in its magnesium-smelting operation. Since the mid-1990s, the demand for magnesium die castings for automobile engines has been growing at an annual rate of 15 per cent, causing Norsk Hydro to expand its smelter on the south shore of the St Lawrence River near Trois-Rivières, which is midway between Montréal and Québec City (Vignette 6.5).

Hydro-Québec is able to provide industrial firms with low-cost energy for three reasons: (1) northern Québec can produce vast quantities of low-cost power; (2) Hydro-Québec has a long-term contract to buy power from the Churchill Falls Project in Labrador at 1969 prices; (3) Hydro-Québec has control over its price structure and can set extremely low power rates for its industrial customers. In fact, these rates were so low that American

firms complained that these rates were a subsidy and therefore gave Canadian-based firms an unfair advantage in the US market.

Northern Québec

Northern Québec lies beyond the ecumene of the St Lawrence Valley. This sector of the Canadian Shield, which comprises over 90 per cent of the land mass of Québec, is sparsely populated and fragmented because there are few and highly localized opportunities for development. As Québec's resource hinterland, it is best suited for forestry, mining, and hydro-

---

Vignette 6.5

Norsk Hydro Canada Inc.

In 1988, Norsk Hydro signed a twenty-five-year contract with Hydro-Québec for electrical power at special rates. Norsk Hydro buys hydro power at rates that fluctuate with the world price of magnesium. For the first three years, the company's magnesium smelter at Bécancour was eligible for electrical power rebates of up to 50 per cent of consumption. The purpose was to help defray the start-up costs of the smelter during its first three years of operation.

The Norwegian state-owned company spent $600 million to build its magnesium smelter at Bécancour. Production began in 1990. By 1991, Norsk Hydro had captured 24 per cent of the American market. Its magnesium metal products are used by automobile parts firms. An American firm, Magnesium Corp. of America, complained to the US Com-

merce Department that Norsk Hydro used unfair trade practices by selling its products on the US market below the cost of production. This allegation was based on the fact that Norsk Hydro obtained its electrical power at rates below the cost of producing that power. In 1992, the US Commerce Department placed a 32.7 per cent anti-dumping duty on Norsk Hydro's magnesium exports to the United States. This duty was later withdrawn after Hydro-Québec agreed to remove its electric rebates, but a countervailing duty of 4.48 per cent remains. In spite of the duty, Norsk Hydro made plans in 1997 to double its total capacity at its Bécancour smelter from 43,000 tonnes of magnesium products to 86,000 tonnes by 2000. At that time, employment at the smelter is anticipated to increase from 360 workers to 430.

Source: Konrad Yakabuski, 'Norsk to Expand Quebec Plant', The Globe and Mail, 12 June 1997:B6. Reprinted with permission from The Globe and Mail.

electric power. Tourism has also prospered in this region. Attempts at agricultural settlement have had marginal success (Vignette 6.6). The characteristics of these sectors of northern Québec's economy are discussed below.

Agriculture

The settlement of the Clay Belt demonstrates the difficulty of farming in the northern area of Québec. The Clay Belt occupies an enormous area of northwestern Québec and northeastern Ontario. The Canadian Shield acquired this relatively thick layer of sand, silt, and clay sediments near the end of the last ice age about 10,000 years ago. The rivers flowing to Hudson and James bays were blocked by the remnants of the Laurentide ice sheet. Gradually, a huge glacial lake called Lake Barlow-Ojibway was formed and lasted for several thousand years over an area of about 200 000 km². Lake sediments, especially minute particles of clay, sand, and silt, were deposited. When the glacial lake drained, much of the land was covered by peat, making it unsuitable for agricultural land. In fact, less than 5 per cent of the Clay Belt contains arable land, only a portion of which has been cultivated. As a result, farmland in the Clay Belt is in scattered pockets, separated by large areas of forested country.

Efforts to settle land in the Clay Belt began in the late-nineteenth century (Vignette 6.6), but received a boost during the Great Depression of the 1930s. At this time, both the Québec and Ontario governments encouraged a 'back to the land' movement in the Clay Belt. In Québec, the Catholic Church took a lead role by organizing settlements in Catholic parishes. Farms were small in size and had

---

Vignette 6.6

Agricultural Settlement of the Clay Belt in Québec

By the late-nineteenth century, the growing rural population in the St Lawrence Valley began to move into the Clay Belt of northern Québec in search of arable land. French-Canadian migrants, often sons of farmers in the St Lawrence Lowlands, tried to clear small pockets of land in the Canadian Shield. By the 1880s, settlers had reached the Little Clay Belt around Lake Temiskaming. Agricultural settlement began following the construction of the Grand Trunk Railway (now the Canadian National Railway) from Québec City to the Little Clay Belt and then across the Canadian Shield to Winnipeg.

The Québec government and the Roman Catholic Church encouraged settlers to take up 'free' lands in both the Little Clay Belt around Lake Temiskaming and the Great Clay Belt around Lake Abitibi. Part of their concern was the thousands of French Canadians who were migrating to New England instead and soon becoming assimilated into American society. The Québec Government and the Roman Catholic Church interpreted this migration as a 'loss' to French Canada. For that reason, they made special efforts to attract French Canadians to the last remaining agricultural lands in Québec—the so-called Clay Belt—where the french language and heritage could be maintained.

Only a small portion of the Clay Belt, located in northern Ontario and northern Québec, is suitable for farming. (*Courtesy Victor Last, Geographical Visual Aids 17646*)

little arable land. As they were far from urban markets, these settlers were destined for a hard life with few prospects for developing a commercial farming operation.

In Québec, the resettlement program had two goals: to combat unemployment in the cities of southern Québec, and to ensure that these settlers followed a French-Catholic lifestyle. The provincial government and the Catholic Church gave little attention to the long-term prospect of commercial farming in the Clay Belt; rather, the emphasis was on defusing the unemployment problem in Québec and 'saving souls'. At its peak, agricultural colonization resulted in fewer than 15,000 farms. Most of these farms were small with little commercial production.

Since the Second World War, agriculture in the Clay Belt has undergone three changes: (1) the emergence of a local market for dairy products in the nearby urban centres; (2) the establishment of a dairy marketing board, thereby providing price stability for farmers; and (3) the consolidation of most farms into fewer but larger farms. The move to land consolidation is revealed by the reduction in the number of farms in the Clay Belt from 15,000 in 1936 to just under 2,000 in 1996. Land consolidation was driven by economic pressures. For instance, a commercial dairy farm needs sufficient land for summer grazing and for the production of fodder crops for winter feed. As well, a commercial family farm must have at least 100 milking cows to achieve a minimum level of economies of scale and modern milking machinery. Though the changes undergone in the Clay Belt have improved the prospects for commercial farmers, the potential for agriculture in northern Québec remains limited.

Forest Industry

The vast forest found in Québec represents a major natural resource. In total, Québec has 22 per cent of Canada's productive forest lands. These lands exists in all three physiographic regions of Québec but the vast majority of commercial forest land is in northern Québec in the Canadian Shield. In terms of productive forest, Québec ranks first among Canada's geographic regions, but it ranks second behind British Columbia in terms of total volume of wood cut. Virtually all productive forest lands are located south of 52° N. In total, Québec accounts for nearly 760,000 km² of forest lands. The boreal forest is concentrated in the Canadian Shield, while mixed and hardwood forests are found in the St Lawrence Lowlands and the Appalachian Uplands. Most timber is cut in the boreal forest, especially in the Lac Saint-Jean and Saguenay regions and along the North Shore. Pulp and paper mills are mainly located further south, along the St Lawrence, Ottawa, and Saguenay rivers.

The forest industry contributes significantly to the province's economy both in value of production and employment. In 1994, the direct value of forest production was $15.6 billion while the added value was $6.8 billion (Québec 1998a). Over 80 per cent of Québec's forest products are exported, mainly to the United States. Forest exports make up 21 per cent of Québec's total export, providing another measure of the significance of the forest industry to the Québec economy. The forest industry generates many jobs with logging accounting for 9,000, wood processing for almost 33,000, and pulp and paper for 34,000 (Québec 1998a and 1998b).

In recent years, the value of forest output has risen sharply because more wood is now processed into higher-value products. In fact wood processing and paper manufacturing represent the most important manufacturing sector in the province by value of production. While the value of forest products has increased, the introduction of advanced woodcutting machinery and computerized mills has dampened the demand for forest workers. This relationship between more technological efficiency and a drop in the demand for labour is particularly troublesome for workers in northern Québec, where few employment alternatives exist.

Mining Industry

Mining has always been important in northern Québec. While the value of mineral production is only about 8 per cent of Canada's output, northern Québec communities often rely on mining for their existence. Though the mining industry experienced a boom in the 1960s, events in the marketplace during the 1980s severely affected this sector. The mining industry in Québec will face further challenges now with trade liberalization and increased competition.

Commercial mineral deposits in the hard rock of northern Québec often consist of gold, iron, and copper. There are two main mining areas within this area of the Canadian Shield. In the west, gold and copper are mined by Noranda and Val-d'Or, the centres for much of this production. In the northeast, iron is mined. Deposits of iron were first reported in 1895 by A.P. Low of the Geological Survey of Canada, the first geologist to investigate the region's mineral potential. However, these iron ore deposits had no commercial value because more accessible mines could supply the needs of the iron and steel companies.

All that changed in the 1940s. American steel producers could no longer count on domestic supplies of iron ore, so they sought more reliable sources, including those in northern Québec and Labrador. Through a process of market integration, initiated by US steel companies, Québec's resource hinterland became dependent on a particular group of steel companies in the industrial heartland of the United States for its economic well-being. Two mining companies, Québec Cartier Mining Company and the Iron Ore Company of Canada, developed iron mines in isolated areas of northern Québec and Labrador. By 1947, plans were laid for an open-pit mine in northern Québec near the border with Labrador. The Iron Ore Company built a town (Schefferville) for miners and their families; transmitted power from Churchill Falls to operate the mine and the town; and built a railway (the Québec North Shore and Labrador Railway) to deliver the iron ore to the port at Sept-Îles, from where the ore was eventually transported to supply US steel mills in Ohio and Pennsylvania.

The demand for iron ore rose in the 1960s, resulting in the establishment of three more mining towns at Wabush, Labrador City, and Fermont. At the same time, Québec Cartier built a similar iron-mining operation by constructing the Cartier Railway from Port-Cartier on the St Lawrence River to the resource town of Gagnon. But, by the 1980s, world steel production had surpassed the demand, causing a severe slump in the demand for iron ore. To add to this economic problem, US steel plants were now less efficient than the new steel mills in Brazil, Canada, Korea, and Japan, and lower-cost iron mines had opened in Australia and Brazil. As lower-priced steel from these countries undercut the price of US steel, American steel companies had to reduce their output, close plants, and sell their shares in the two mining companies. The repercussions for the iron ore mining workers in Québec and Labrador were severe. Production from these northern mines fell by half in the early 1980s and hundreds of workers were laid off. In 1983, the mine at Schefferville was closed.

Hydro Power: The James Bay Project

The James Bay Project is a massive project that calls for the production of hydroelectricity from all the rivers that flow into James Bay from Québec territory. This hydroelectric project was announced in 1971 by Premier Robert Bourassa. The James Bay Project is divided into three separate river basins (La Grande, Great Whale, and the Nottaway-Broadback-Rupert basins). The project involves about twenty rivers and affects an area one-fifth the size of Québec. Construction of the first phase of the James Bay Project, La Grande Project, began in 1972, was completed in 1985, and cost $16 billion in 1985 dollars (Bone 1992:172). La Grande Project involves diverting waters from three other rivers (Eastmain, Opinaca, and Caniapiscau) into La Grande Rivière. Electrical energy generated from the three powerplants in La Grande Basin is more than 10 000 MW each year. Plans for the second phase of the project (James Bay II) were announced in 1985. These plans called for the construction of additional dams in La Grande Basin, a generating station (LG1) at the mouth of La Grande Rivière, and hydroelectric development of the rivers in the Great Whale Basin.

The first phase of the James Bay Project, however, raised considerable controversy. It evoked an unprecedented response from Aboriginal peoples and environmental organizations. For them, the James Bay Project unleashed social and environmental problems that remain unresolved. For example, the project resulted in a high content of mercury in the reservoirs, which, though unexpected, has had serious implications for the Cree, who consume fish on a regular basis. Hydro-Québec maintains that the environmental impacts have been mitigated to an acceptable level through modifications to the design of the project. Remaining environmental impacts, such as high mercury content in the waters of the reservoirs, will, they argue, diminish over time.

Continued opposition by the Cree will hamper future developments. From the very beginning, the Cree opposed the James Bay Project because of its effect on their hunting grounds. The Cree, joined by the Inuit of Arctic Québec, forced a land-claims settlement

known as the James Bay and Northern Québec Agreement.

The James Bay and Northern Québec Agreement

In 1971, 6,000 Cree in northern Québec lived as eight bands scattered across 375 000 km² of rivers and forest. They were under the administration of the federal Department of Indian Affairs and Northern Development. The James Bay Project threatened to flood their lands. This threat united the eight bands. When construction began in 1972, the Cree asked the Inuit to join them in taking legal action to halt the construction until the Cree and Inuit land claims were addressed. This action forced the Québec government and the Aboriginal claimants to the bargaining table. The result was the **James Bay and Northern Québec Agreement**. Under this agreement, both the federal and Québec governments became responsible for providing the 'treaty' benefits. As the first modern land-claim agreement in Canada, this 1975 agreement provided land, cash, and the power to administer cultural matters (education, health, and social services) to Aboriginal peoples. In exchange, the Cree and Inuit surrendered their Aboriginal claims to northern Québec and agreed to allow the construction of La Grande Project to proceed.

In combination, these events—the negotiations, the agreement, and the construction project—have forever altered the lives of the Cree and Inuit. Both groups, now living in settlements, are more involved in the modern industrial society than ever before. Many are employed in businesses run by Cree and Inuit organizations, while others work in construction activities in the growing Cree and Inuit settlements and for Hydro-Québec. Still others are involved in the administration of their cultural affairs through the Cree Regional Authority and the Kativik Regional Government. In comparison with other Aboriginal peoples in Québec, the economic situation of the Cree and Inuit is much improved and certainly much better than that of those Indians who do not have the benefit of such an agreement (Simard et al. 1996).

Nevertheless, the Cree have expressed dissatisfaction with the environmental results of past phases of the project and with the James Bay and Northern Québec Agreement, or at least the fulfilment of the two governments' obligations under the agreement. The

Cree are therefore firmly opposed to the next phase of the James Bay Project (the Great Whale River Project), while the Inuit are more open to negotiations. (The Great Whale River Project affects the Cree, not the Inuit.) For the Cree, the Great Whale River Project will mean more loss of hunting land.

Tourism

In Canada and in the rest of the world tourism is becoming an extremely important economic sector. Québec is no exception. In fact, Québec combines its natural beauty, historic past, and francophone culture to draw more and more tourists each year. For example, from 1993 to 1997, the number of tourists visiting the province increased every year, and tourist dollars spent in Québec over that five-year period

As Montréal's cottage country, the Laurentides provides an accessible recreation retreat to many urban dwellers. (*Courtesy R.M. Bone*)

increased by 50 per cent (Québec 1998c). Most tourists in Québec come from other regions of Canada and from the United States, France, the United Kingdom, Germany, and Japan. Tourism in Québec accounts for over 21,000 businesses that employ more than 250,000 workers (Québec 1998d).

The centres of Montréal and Québec City are major attractions for tourists seeking an urban vacation with a francophone atmosphere, while the Laurentides attract visitors looking for a summer or winter playground in the forests and lakes of the Canadian Shield. In addition to its natural beauty, the Laurentides has several advantages as an all-season recreational area. The Laurentides is but a short distance from Montréal and relatively close to major American cities in New England, such as Boston and New York. The climate in this area is ideal for the tourist business—the summers are hot, and heavy snowfall provides ideal conditions for winter sports. Tourism is the major business in the Laurentides, and its towns and villages cater to the tourists who flock there to enjoy water sports in the summer, skiing and snowboarding in the winter, and the numerous restaurants and shops all year long.

A ski hill in the Canadian Shield, approximately 20 km north of Québec City. (*Courtesy R.M. Bone*)

Québec's Urban Geography

Over 80 per cent of Québec's population lives in urban centres (Figures 5.2 and 5.4). Two of the largest metropolitan cities in Canada, Montréal and Québec City, are located in the province. Other major population centres are Hull, Chicoutimi–Jonquière, Sherbrooke, and Trois-Rivières. In total, these six urban clusters have a population of nearly 4.7 million.

As in Ontario and most other regions of Canada, migration played a major role in Québec's urbanization. Push and pull factors (described in Chapter 4) have attracted rural Quebeckers to cities. The key push factors in rural Québec were limited job opportunities, a shrinking labour force in the primary sector, and an increasing number of young people entering the workforce. Even though the birth rate dropped sharply in the 1980s, higher birth rates in earlier decades resulted in a flood of young people entering the workforce in the 1970s and 1980s. The main pull factor was the strong demand for workers in urban centres, which led young rural men and women to seek their fortunes in the cities. The net result was a considerable rural-to-urban migration.

From 1981 to 1996, the population of the major cities, led by Hull, Sherbrooke, and Montréal, grew by over 17 per cent (Table 6.3). Hull's population growth was largely due to the employment and business opportunities generated by the federal government's activities. The slowest-growing city in Québec was Chicoutimi–Jonquière. In fact, from 1991 to 1996, this city actually suffered a small population loss. Chicoutimi–Jonquière is located in the resource hinterland of the Lac Saint-Jean region. The two main economic activities are the aluminum plants and the forest mills. As in other resource hinterlands, employment prospects in these two industries are not bright.

Montréal

Montréal, the metropolis of the province, is the industrial, commercial, and cultural focus of Québec. For a century and a half, Montréal was the largest and most prosperous city in the Dominion. Within Canada today, this historic city has slipped behind Toronto in size and importance. Within Québec, however, Montréal remains the dominant city. In 1996, Montréal had a population of 3.3 million, making it the largest population concentration in Québec.

Like other great cities, Montréal is surrounded by other cities and towns. Such centres are often called satellite towns because they fall within the trade orbit of the larger city. Trois-Rivières, a major urban centre located over 100 km east of Montréal, is a satellite city. Smaller centres that are within 50 km of Montréal, like Sainte-Hyacinthe, St-Jérôme, and Salaberry-de-Valleyfield, are even more closely tied to that metropolis. Laval and Longueuil serve as bedroom communities for Montréal. In most cases, urbanites have fled to the suburbs, attracted by lower-cost housing and the amenities of suburban life. These suburbanites often work in Montréal and live within commuting distance. Each weekday morning and late afternoon, these commuters generate huge volumes of slow traffic and crowded buses in and around Montréal.

Given its strategic location and economic size, Montréal serves as the transportation hub of Québec, making this city the regional core of the province and the rest of Québec the periphery. At the national scale, Montréal is part of the national core because it is part of Canada's manufacturing belt.

Following the Free Trade Agreement, Montréal's manufacturing sector had to respond to strong foreign competition. Labour-intensive manufacturing firms experienced great difficulty in competing with foreign firms that had substantially lower labour costs. Montréal's manufacturing firms made two major changes: labour-intensive plants substituted machinery for workers to increase productivity, and high-technology firms expanded. By the late 1990s, Montréal's economy has become much more specialized in aerospace, telecommunications, and other areas of industrial research and development. However, the city suffers a persistent high unemployment rate that affects its economic stability.

Historic Context

Geography and history lie behind Montréal's success over the years. Montréal's strategic location on the St Lawrence River was a key factor in its initial growth as a transportation route and as a source of energy, though today most power comes from tributaries flowing into the St Lawrence. In the 1820s, the Lachine Canal near Montréal provided the energy to drive the grist mills. Early in this century, power was transmitted to Montréal first from the hydroelectric installation on the St Maurice River at Shawinigan and then from the Beauharnois power site on the St Lawrence River.

In the early years of New France, Québec City was not only the administrative capital but also the centre of the fur trade and the principal port of New France. French ships brought supplies to Québec City and loaded beaver pelts for the return voyage home. By the beginning of the eighteenth century, however, Montréal took over both roles as the centre for the fur trade and the principal port

on the St Lawrence River. After the British conquest of New France, Montréal's economy remained tied to the fur trade, which was now in the hands of British merchants, who established the North West Company. This Montréal-based firm prospered until 1821, when it joined forces with the Hudson's Bay Company. At that time, the St Lawrence fur-trading route into the fur-rich northwest region of Canada was abandoned in favour of the route to Hudson Bay.

By 1820, Montréal's population had surpassed that of Québec City, which it supplanted as the principal transshipment point between Europe and Canada. Wealth accumulated from the fur trade provided the capital necessary for new industrial ventures in manufacturing, resource development, and transportation. By the end of the nineteenth century, Montréal was the premier city in Canada. With the largest population of any city in Canada, Montréal was Canada's transportation hub and financial centre. For example, the construction of railway lines, particularly the Grand Trunk Railway, confirmed Montréal's role as a transportation centre and facilitated the development of railway repair depots and the manufacturing of engines and railcars.

At the time of Confederation, Montréal was the largest and most prosperous city in Canada. With the introduction of high protective tariffs in 1878, Montréal's industrial base increased, thus propelling the city's economic and demographic growth. As Montréal's transportation and manufacturing industries expanded, its role as a financial and cultural centre solidified. At the time of Canada's centenary in 1967, Montréal was the premier city in Canada and the undisputed headquarters of Canada's textile and financial enterprises. In the years that followed, Montréal's demographic and economic growth continued, but at a slower pace than that of Toronto.

Montréal and Toronto

After the 1970s, Toronto quickly replaced Montréal as the premier city in Canada. It was now the largest city and the financial capital of the country. The reasons for this shift in metropolitan power are due mainly to the stronger economic growth in southern Ontario and, to a lesser degree, to the economic and demographic fallout from the threat of Québec separating from Canada. The three main economic factors that led to the shift in metropolitan status were:

* Montréal's economy was much more dependent on labour-intensive manufacturing (which was in decline) like the textile industry.
* Montréal's industrialization had begun much earlier than Toronto's and, for that reason, its manufacturing firms tended to be older and less efficient than those in Toronto.
* Toronto benefited from Ontario's automobile industry, which, because of the Auto Pact, has a secure manufacturing sector. (See discussion of the Auto Pact in Chapters 1 and 5.)

The main political factor affecting Montréal was the rise of the separatist movement in the 1970s. That political change, coupled with the 1980 Québec referendum, resulted in a sense of economic uncertainty that led many anglophones to move to Toronto and other Canadian cities. Between 1975 and 1990, for example, the net interprovincial migration reveals that Québec lost nearly 280,000 people.

For those reasons, Toronto's growth rate exceeded that of Montréal. Even in the ten-year period from 1951 to 1961, when both cities grew at phenomenal rates—Montréal at 4.3 per cent per year and Toronto at 5.1 per cent—Toronto's rate was higher. From 1971 to 1981, the rate of population increase slowed in both cities, with Montréal's close to zero. During that time, Montréal's annual growth rate was only 0.3 per cent, while Toronto's was 1.9 per cent. This gap continued in the early 1990s with the annual rates for Toronto and Montréal corresponding to 1.9 per cent and 1.3 per cent respectively.

Québec City

Québec City has the next largest urban concentration in the province, totalling nearly 700,000 people. Close to the Laurentides, Québec City has a magnificent physical setting on high banks just above the St Lawrence River. It is the only walled city in North America and features buildings over 300 years old. In 1985, Québec City was selected as a World Heritage Site by UNESCO.

The economic base of Québec City revolves around three functions. First, it is a government town. As the seat of government for the province, Québec City employs a large number of civil servants. Second, Québec City has become a world-class tourist centre. The Old-World charm of Québec City draws tourists from around the world, while special

The Québec Winter Carnival attracts a number of tourists from around the world annually.
(*Corel Photos*)

events such as its Winter Carnival are very popular. Third, Québec City is only minutes away from excellent seasonal recreation areas—from skiing in the winter to water sports in the summer. The economic base of Québec City remains heavily dependent on its political and cultural roles, but it also has a number of other economic functions: it is a port and rail centre, as well as a centre for resource processing, metal fabricating, and manufacturing.

Québec's Future

Québec is the heartland of francophones in Canada. As well, it is an essential part of Canada, contributing to the economic, social, and cultural well-being of the nation. As Québec is one of the principal industrial regions in Canada, the economic challenge it faces now and in the next century is to secure a position in the North American marketplace.

Over the past several decades, Québec's economy has grown at a slightly slower rate than those of Alberta, British Columbia, and Ontario. According to the Conference Board of Canada, the Québec economy is 'une économie en perte de vitesse' (an economy losing steam) (Bussière 1995:21). Much of this contrast in growth is due to more rapid economic growth in the resource sectors of the two western provinces and in the manufacturing and service sectors in Ontario. As in northern Ontario, northern Québec's forestry and mining activities have little room to expand and, because workers are increasingly replaced by machinery, population growth is extremely low. Although hydroelectric projects boost provincial energy output, they employ few people after their construction. The prospects for the Canadian Shield and the Appalachian Uplands are dim. Their resource-based economies are likely to generate sluggish economic growth, little population increase, and a small but steady out-migration of young people to places where employment prospects are brighter. The economy of the St Lawrence Lowlands remains stable, though the manufacturing sector is undergoing massive restructuring. Like the neighbouring manufacturing area in Ontario, many American branch plants in Québec have closed their operations. Beyond that, the economic structure of manufacturing in Québec must change even further to adjust to the new economic circumstances of a North American marketplace. On the bright side, several high-tech industries, namely the aerospace and pharmaceutical industries, are expanding and have found a niche in the global market.

Beyond the main economic questions of restructuring, unemployment levels, and debt reduction looms a larger question—Québec's place within Canada. The economic and social stakes are extremely high for all Canadians. For instance, just the threat of Québec separation led many corporations to move from Québec to Ontario and other Canadian provinces during the 1980s. This corporate shift from Montréal to Toronto was particularly hurtful to

Montréal's financial role. At the same time, a revolution was occurring in the business community of Montréal—French was replacing English as the business language, and French-speaking business people had taken control of the business community in Montréal. During this period, French/English relations hardened.

In the past, efforts to find a political solution to Quebec's place in Canada, like Meech Lake, have almost succeeded in recognizing Québec's special place within the country. Such efforts must continue. Indeed, this very process has shaped Canadian society and its identity.

Within Confederation, Québec possesses considerable political, economic, and cultural powers. Since the Quiet Revolution, the Québec government has become a strong force in shaping its economy and society. Through Crown corporations and agencies, such as Hydro-Québec, the provincial government has become a major player in the economic affairs of Québec. Language laws have solidified the place of French in Québec. Control over immigration has ensured the steady influx of French-speaking immigrants into Québec. The French language and Québécois culture are flourishing. The Québécois regard these developments with pride. Still, independence is an attractive option for those who are strongly dissatisfied with Québec's position within Canada.

Bridging the gap between the French and English visions of Canada is not a simple task. The first step is to find a political compromise that recognizes the French Canadians' overwhelming desire to survive as a French collectivity within North America. A view of this nationalism was expressed by Beauregard (1980:7): 'Le Québec veut, quant à lui, faire reconnaître son particularisme et son destin culturel' (Québec wants, for itself, recognition of its uniqueness and its cultural destiny).

Summary

Québec began as New France. It prospered after the British conquest and struggled to maintain a French presence in British North America, and later in Canada. This ongoing struggle has become the principal source of tension within Québec and Canadian society. Québec, with its French language and society, remains a francophone bastion in Canada and North America. Within Québec, however, anglophone, allophone, and Aboriginal minorities exist. Tensions between these minorities and the francophone majority often spread to the national French/English faultline.

Geography and history have marked Québec with a special sense of place. The St Lawrence Lowlands played a key role in Québec's past and now, as the industrial heartland of Québec, the St Lawrence Lowlands remains the province's dominant economic region. More recently, the emergence of the Canadian Shield as the primary source of energy for the cities and factories of southern Québec strengthened the economic relationship between the two geographic parts of the province and, with an Aboriginal majority in northern Québec, exposed a different economy and culture. Like Ontario, Québec is part of the industrial heartland of Canada and an important industrial area of North America, and as such, Québec faces an economic challenge: how to define its economic place within the North American and global markets.

As the francophone minority within Canada, Québec often expresses dissatisfaction with its political place within Confederation. While most Québécois have a strong attachment to Canada, their first loyalty is often to Québec. Separatism, ever a threat to Confederation, remains an unresolved issue. For that reason, the French/English faultline continues to be the most contentious political issue in Canada.

Notes

1. By responsible government, Lord Durham meant a 'political system in which the Executive is directly and immediately responsible to the Legislature, in which the ministers are members of the Legislature, chosen from the part which includes the majority of the elected representatives of the people' (Lucas 1912:1, 138).

2. In 1912, Québec gained northern territories inhabited by the Inuit and Cree. Ottawa ceded these lands to Québec with the understanding that the Québec government would be responsible for settling land claims with the Aboriginal peoples in these territories. The Cree in northern Québec, in response to separatists' claim to territorial independence, have declared that they have the right to secede from Québec. They argue that if Québec has the right to secede from Canada, then the Cree have the right to secede from Québec. From a geopolitical perspective, the partitioning of Canada

or Québec makes sense only to those supporting ethnic nationalism.

3.   The case of Churchill Falls is an interesting one. The divide between the Atlantic Ocean and Hudson Bay marks the Labrador–Québec boundary. For historical reasons, the Québec government does not formally recognize this boundary, but it does treat the area as part of Newfoundland. Newfoundland owns the large Churchill Falls hydroelectric project, but because of the high cost of transmitting the electrical power from Churchill Falls across the Strait of Belle Isle to Newfound-land, virtually all this power is purchased by Hydro-Québec. Hydro-Québec then transmits it across Québec to markets in the St Lawrence Lowlands and the United States.

4.   The thirteen companies with risk-sharing contracts are: Norsk Hydro Canada Inc., Aluminerie Alouette Inc., Quebec-Cartier Mining Co., Cafco Industries Ltd, Timminco Ltd, QIT-Fer et Titane Inc., PPB Canada Inc., Reynolds Metals Co., Argonal, Hydrogenal, SKW Canada Inc., ABI Inc., and Aluminerie Lauralco Inc.

Key Terms

habitants
> French peasants who settled the land in New France under a form of feudal agriculture known as the seigneurial system. After the British conquest, the seigneurial system continued until the mid-nineteenth century, marking a significant difference between Upper and Lower Canada.

James Bay and Northern Québec Agreement
> The 1971 announcement of the James Bay Project triggered a series of events that quickly led to a negotiated settlement and, in 1975, an agreement. In this modern treaty, Aboriginal title was surrendered by the Inuit and Cree of northern Québec in exchange for specific rights, including self-government and benefits (cash and financial support for the hunting economy). Signatories to the agreement included Québec, the Société d'énergie de la Baie James, the Société de développement de la Baie James, the Commission hydro-électrique de Québec, the Grand Council of the Crees of Québec, the James Bay Cree, the Northern Québec Inuit Association, the Inuit of Québec, the Inuit of Port Burwell, and the Government of Canada.

patriate
> To bring legislation, especially a Constitution, under the authority of the autonomous country to which it applies.

seigneurs
> Members of the French aristocracy who were awarded land in New France by the French king. A seigneur was an estate owner who had peasants (*habitants*) to work his land.

References

Bibliography

Bone, Robert M. 1992. *The Geography of the Canadian North*. Toronto: Oxford University Press.

Bussière, Luc. 1995. 'Une économie en perte de vitesse'. *Provincial Outlook: Economic Forecast* 10, no. 1:21–9.

Canada. 1993. *Canada Year Book 1994*. Ottawa: Minister of Industry, Science and Technology.

Chevalier, Jacques. 1993. 'Toronto–Ottawa–Montréal: Concentrations majeures Canadiennes de l'innovation par la recherche-developpement'. *Le géographe Canadien* 37, no.3:242–57.

Coffey, W.J., and Réjean Drolet. 1994. 'La décentralisation des services supérieurs dans la région métropolitaine de Montréal, 1981–1989'. *Le géographe Canadien* 38, no. 3:215–29.

De Benedetti, George J., and Maurice Beaudin. 1996. 'Linguistic Minority Communities' Contribution to Economic Well-being: Two Case Studies'. *Canadian Journal of Regional Science* XIX, no. 2:175–92.

Kaplan, David H. 1994. 'Two Nations in Search of a State: Canada's Ambivalent Spatial Identities'. *Annals of the Association of American Geographers* 84, no. 4:585–606.

Lucas, C.P. 1912. *Lord Durham's Report of the Affairs of British North America*, vol. 1. Oxford: Clarendon Press.

Marsh, James H., ed. 1988. *The Canadian Encyclopedia*, 2nd edn. Edmonton: Hurtig Publishers.

Mills, David. 1988. 'Durham Report'. In *The Canadian Encyclopedia*, edited by James H. Marsh, 637–8. Edmonton: Hurtig Publishers Ltd.

Nemni, Max. 1994. 'The Case Against Quebec Nationalism'. *The American Review of Canadian Studies* 24, no. 2:171–96.

Québec, 1998a. Les Forêts, Le Québec forestier/ Portrait statistique: Le secteur forestier dans l'économie, [online], Gouvernement du Québec, Ministère des Ressources naturelles. Searched 3 February 1999; <URL:http://www. mrn.gouv.qc.ca/3/30/301/économie.asp>:3 pp.

Québec, 1998b. Les Forêts, Le Québec forestier/ Portrait statistique: Key Figures: Québec's Forest Industry, [online], Gouvernement du Québec, Ministère des Ressources naturelles. Searched 3 February 1999; <URL:http://www. mrn.gouv.qc.ca/3/30/301/intro.asp>:8 pp.

Québec, 1998c. Pour donner au monde le goût du Québec: Résumé de la politique de développement touristique, [online], Gouvernement du Québec, Ministère du Tourisme. Searched 3 February 1999; <URL:http://www.tourisme.gouv. qc.ca/francais/mto/publications/poldevtour. html>:4 pp.

Québec 1998d. Le tourisme au Québec: enjeux et orientations, [online], Gouvernement du Québec, Ministère du Tourisme. Searched 3 February 1999; <URL:http://www.tourisme.gouv. qc.ca/francais/mto/enjeux.html>:2 pp.

Rose, D., and M. Villemaire. 1997. 'Reshuffling Paperworkers: Technological Change and Experiences of Reorganization at a Quebec Newsprint Mill'. *The Canadian Geographer* 41, no. 1:41–60.

Simard, Jean-Jacques, et al. 1996. *Tendances Nordiques: Les changements sociaux 1970–1990 chez les Cris et les Inuit du Québec: Une enquête statistique exploratoire*. Québec City: GETIC, Université Laval.

Simeon, Richard. 1988. 'Meech Lake and Shifting Conceptions of Canadian Federalism'. *Canadian Public Policy* XIV:S7–24.

Simpson, Jeffrey. 1993. *Faultlines: Struggling for a Canadian Vision*. Toronto: HarperCollins.

Stanford, Quentin H., ed. 1998. *Canadian Oxford World Atlas*, 4th edn. Toronto: Oxford University Press.

Statistics Canada. 1982. *Census Metropolitan Areas and Census Agglomerations with Components*.

Catalogue no. 95-903. Ottawa: Minister of Industry.

_____. 1992. *Census Division and Subdivisions.* Catalogue no. 93-304. Ottawa: Minister of Supply and Services.

_____. 1996. *Labour Force Annual Averages 1995.* Catalogue no. 71-220-XPB. Ottawa: Statistics Canada.

_____. 1997a. *A National Overview: Population and Dwelling Counts.* Catalogue no. 93-357-XPB. Ottawa: Industry Canada.

_____. 1997b. 1996 Census: Nation Tables—Population by Mother Tongue, Showing Age Groups, for Canada, Provinces and Territories, 1996 Census—20% Sample Data, 2 December 1997 [online database], Ottawa. Searched 15 July 1998; <URL:http://www.statcan.ca/english/census96/>:3 pp.

_____. 1997c. The Daily—1996 Census: Mother Tongue, Home Language and Knowledge of Languages, 2 December 1997 [online database], Ottawa. Searched 14 July 1998; <URL: http://www.statcan.ca/Daily/English/>:15 pp.

_____. 1998a. *Canadian Economic Observer.* Catalogue no. 11-010-XPB. Ottawa: Statistics Canada.

_____. 1998b. The Daily—1996 Census: Aboriginal Data, 13 January 1998 [online database], Ottawa. Searched 14 July 1998; <URL:http://www.statcan.ca/Daily/English/>:10 pp.

_____. 1998c. The Daily—1996 Census: Ethnic Origin, Visible Minorities, 17 February 1998 [online database], Ottawa. Searched 16 July 1998; <URL:http://www.statcan.ca/Daily/English/>:21 pp.

Thibodeau, Jean-Claude, and Yvon Martineau. 1996. 'Essaimage technologique en région périphérique: étude de cas'. *Revue Canadienne des sciences régionales* XIX, no. 1:49–64.

Yakabuski, Konrad. 1997. 'Norsk to Expand Quebec Plant'. *The Globe and Mail* (12 June):B6.

Further Reading

Beauregard, Ludger, ed. 1980. 'Numéro spécial: La problématique géopolitique du Québec'. *Cahiers de géographie du Québec* 24, no. 61: 1–185.

Coffey, W.J. 1996. 'Make or Buy: Internalization and Externalization of Producer Service Inputs in the Montreal Metropolitan Area'. *Canadian Journal of Regional Science* XIX, no. 1:25–48.

Villeneuve, Paul. 1992. 'Un Québec en révolution tranquille'. In *Géographie universelle*, edited by Roger Brunet, 357–73. Paris: Hachette and Reclus.

_____. 1997. 'Le Québec et l'intégration continentale: Un processus à plusieurs vitesses et à directions multiples'. *Cahiers de géographie du Québec* 41, no. 114:337–47.

Chapter 7

Overview

Until 1998, when the Asian economy suffered a setback, British Columbia was the fastest-growing province in Canada—both its economic and population growth outstripped those of the other regions. This rapid expansion altered BC's regional position within Canada, thereby generating tensions between British Columbia and Ottawa. These tensions are described in this chapter as the centralist/decentralist faultline. Much of BC's economic growth is based on its natural resources and more recently, its service industries. This rapid economic growth has, in turn, sparked an influx of Canadians and immigrants into the province.

BC has two distinct subregions—its heavily populated southwest corner is an economic core, while the rest of the province has the characteristics of a resource hinterland. While the economy shows signs of maturing into a core industrial region, it remains dependent on resource industries. The forest industry is of central importance to BC's economy and is discussed in detail in the *Key Topic*. Also, the magnitude of Aboriginal land claims marks the ownership of BC as a critical issue that must be resolved through negotiations with Aboriginal peoples.

Objectives

- Describe British Columbia's physical geography and history.
- Present the basic elements of BC's economy and population.
- Examine these elements within BC's physical setting and the core/periphery model.
- Present arguments for two faultlines in BC—an Aboriginal/non-Aboriginal faultline and a centralist/decentralist faultline.
- Focus on the forest industry and its changing role in BC's economy.
- Ask whether BC is a resource hinterland or an industrial core.

British Columbia

Introduction

British Columbia is an emerging economic and political force within Canada. Since the Second World War, high-technology and service industries have expanded, thereby diversifying the province's economy. The tourism industry, by capitalizing on BC's varied and attractive physical geography, has sparked new investment and generated more service jobs. Foreign trade, especially with Japan and other Pacific Rim countries, has also stimulated BC's economy. From the perspective of British Columbians, BC is part of the economic heartland of Canada and North America, yet BC's economy retains many characteristics of a resource hinterland that is subject to boom-and-bust cycles. The forest industry, the largest resource industry in the province, is the *Key Topic* in this chapter.

Until 1998, the combination of a rapidly growing economy and a mild climate made British Columbia the fastest-growing province in Canada. Much of its population increase was due to the arrival of immigrants and the relocation of Canadians from other provinces into BC. Since population size is a measure of political power, British Columbia can now demand more political respect and attention from Ottawa. Indeed, BC's demands for more political power underscore the struggle between the province and Ottawa, which is described in this chapter as the centralist/decentralist faultline. Within BC there is another faultline—the Aboriginal/non-Aboriginal faultline—whereby Aboriginal peoples are demanding more power through land-claim agreements.

British Columbia Within Canada

British Columbia has undergone rapid economic growth. In the past, the region's economy was propelled by its vast array of natural wealth—fish, forests, and minerals. Though exploitation of these resources is still very important, BC's economy is now powered by other sectors, especially high technology and producer services.

BC's location on the Pacific Rim gives the region access to another part of the world. In fact, British Columbia's economy benefited from the rapidly expanding Asian economy during the 1980s and early 1990s. A large number of Hong Kong immigrants to BC (especially to Vancouver), brought with them skills, capital, and business connections to

Asia, thereby stimulating the region's economy. However, by 1997, the Asian economy had slowed, thus reducing its demand for imports from British Columbia. Also, the influx of wealthy Hong Kong immigrants had diminished due to the transfer of political power in Hong Kong, which became part of China in 1997.

As in Ontario and Québec, there is a core/periphery arrangement of the population in BC, though it is not as prominent in this province. This geographic dichotomy is best expressed by British Columbia's population distribution. There is a population core in the southwest corner of the province, where over 60 per cent of BC's residents are concentrated. The major urban centres, including Vancouver and Victoria, are within this core. Beyond this population core lies more than 90 per cent of the province's territory. Much of this rugged terrain is sparsely populated, though there are growing population clusters in the Okanagan and Thompson valleys and on Vancouver Island north of Victoria. As well, population centres are located along the north coast around Prince Rupert, in the northern interior centred at Prince George, and in the Peace River country. Finally, east of the Okanagan Valley, a few towns, such as Trail, are located near the US border.

In British Columbia, natural resources remain important but no longer dominate the province's economy. For instance, fifty years ago forestry alone accounted for 50 per cent of the provincial economy and for most of the employment. By the late 1980s, forestry comprised only 17 per cent of BC's economy and 14 per cent of employment (Barnes and

Hayter 1997:5). Still, forestry, mining, and fishing generate most of the manufacturing jobs in BC. However, competition for land and water by a growing recreation industry is increasing, and herein lies a dilemma: economic growth depends on the resource industries, but, at the same time, BC's outdoor lifestyle and tourist industry rely on its pristine wilderness. The challenge will be to balance the needs of these two sectors in the coming years.

British Columbia's Physical Geography

The physical geography of British Columbia is far more varied than that of any other region of Canada and even extends offshore to a number of islands and a narrow continental shelf. Unlike in Atlantic Canada, the continental shelf in BC extends only a short distance from the coast. Within this narrow zone there are many islands, the largest being Vancouver Island followed by the Queen Charlotte Islands. The riches of the sea include salmon, which return to the rivers, such as the Fraser and the Skeena, to complete their life cycle. Most of BC's natural wealth, however, is not in the sea but in the province's diversified physical geography, which provides valuable resources, particularly forests, minerals, and rivers.

Most of British Columbia lies in the physiographic region known as the Cordillera, which is a combination of mountains, plateaux, and valleys. A small portion in the northeast of British Columbia, known as the Peace River country, extends into the Interior Plains region (Figure 2.1). The Cordillera is a complex physiographic region. Extending from southern British Columbia to the Yukon, this region extends over 16 per cent of Canada's territory. As explained in Chapter 2, the Cordillera was formed by severe folding and faulting of sedimentary rocks. This coastal zone along the West Coast is subject to earthquakes because of tectonic movement.

The Cordillera has at least ten mountain ranges, the most prominent of which are the Coast Mountains, which extend northward from Vancouver to the Alaskan Panhandle, and the Rocky Mountains, which stretch from the Canada–US border almost to the Yukon (Figure 7.1). Other north-south mountain ranges include the Insular Mountains that rise above the sea to form the Queen Charlotte and

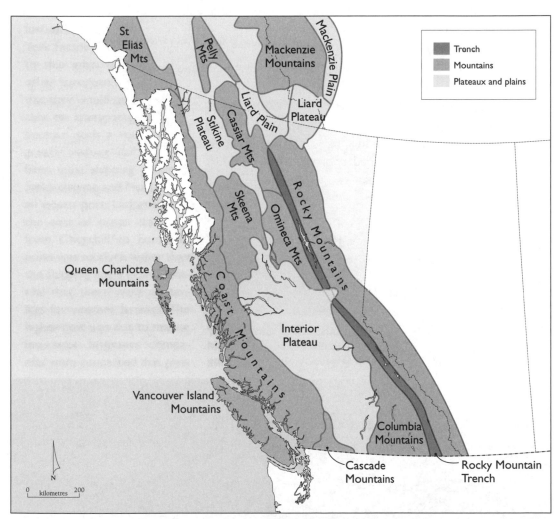

Figure 7.1 Physiography of the Cordillera. British Columbia's complex physiography is evident in the physiographic subregions in the Cordillera. The difficulty of constructing east-west transportation routes can be appreciated if one considers the number of north-south mountain ranges that must be traversed.

Vancouver islands. Two more mountain ranges, which lie in the southern interior of BC, are the Cascade Mountains and the Columbia Mountains. The Columbia Mountains consist of three parallel, north-south mountain ranges—the Purcell, Selkirk, and Monashee—and a fourth range, the Cariboo Mountains, form the Columbia's northern extension. Further north are four mountain ranges—the Hazelton, Skeena, Omineca, and Cassiar mountains.

The Interior Plateau separates the Coast Mountains from the interior mountains (Figure 7.1). North of the Interior Plateau is the Stikine Plateau. The topography of these plateaux consists of gently undulating land with occasional deeply entrenched river val-

leys. The Fraser Canyon, one of BC's most well-known landforms, gouges about 300 to 600 m into the Interior Plateau and extends from just south of Quesnel to Hope. Just south of Lytton, the canyon walls rise about 1000 m above the river. This rocky gorge is called Hell's Gate. The Fraser Canyon is a major fault zone that separates the Cascades from the Coast Mountains.

The Cordillera exerts a strong influence on climate. For that reason, British Columbia has two climatic zones, the Pacific and the Cordillera. (See Figure 2.6 and 'Climatic Zones' in Chapter 2 for more details.) Because of the extremely high elevations in the Coast Mountains and the Rocky Mountains, the Cordillera often prevents the warm, moist

Some areas along the **Coast Mountains**, which stretch 1600 km from the Fraser River Lowlands near Vancouver to the Yukon, have the most rain in British Columbia. (*Corel Photos*)

Pacific air masses from reaching the interior of Canada. At the same time, there is heavy orographic precipitation along the western slopes of the Coast mountain range where 2000 cm of rain falls annually (see Chapter 2). In sharp contrast, southern Alberta and Saskatchewan receive less than 400 cm per year (Figure 2.5).

The natural vegetation of British Columbia is far from homogeneous (Figure 2.7). The Coast mountain range, which rises abruptly from the Pacific Ocean, is covered by a green rainforest that contrasts with the sagebrush and yellow grasses in the lower elevations of the southern Interior Plateau, where higher temperatures occur. At higher elevations and in more northerly areas, where cooler temperatures prevail, forest covers the Interior Plateau. The series of mountain ranges, aligned north to south, are also forested.

Because of the rugged nature of the Cordillera, there is little arable land. Only about 2 per cent of the province's land is classified as arable. British Columbia's largest area of cropland lies outside of the Cordillera physiographic region in the Peace River country. Within the Cordillera, most arable land is in the Fraser Valley, while a smaller amount can be found in the interior, especially the Okanagan and Thompson valleys. This shortage of arable land poses a serious problem for British Columbia. With urban developments spreading onto agricultural land, British Columbia is losing some of its most productive farmland.

British Columbia's Historical Geography

Indians lived along the Pacific Coast of British Columbia for over 10,000 years before European explorers reached the northern Pacific Coast in the mid-eighteenth century. The Spanish had already sailed northward from Mexico to California, but the Russians were the first to reach Alaska and establish fur-trading posts along its coast. In 1778, Captain James Cook established Britain's interest in

Captain James Cook's ships moored in Nootka Sound, as depicted in a watercolour by M.B. Messer. (*Courtesy National Archives of Canada C11201*)

this region by sailing into Vancouver Island's Nootka Sound, where he and his sailors found the Nootka village of Yuquot. The Nootka, now known as Nuu-Chah-Nulth, fished for salmon and hunted the sea otter. Upon landing, Cook engaged in trade for sea otter pelts, then he sailed to China where he and his men sold the pelts for a handsome profit. After the Royal Navy published Cook's record of his voyage, British and American traders came to the Pacific Northwest to seek the highly valued sea otters. Russian fur traders, based in Alaska, also harvested sea otters. Spain, which considered these lands Spanish territory, was disturbed by these interlopers and sent a fleet northward from Mexico in 1789. At Nootka Sound on the west coast of Vancouver Island, the Spanish seized several ships and built a fort to defend their claim to these lands. In 1792, Captain George Vancouver of Britain's Royal Navy sailed around Vancouver Island. In the following year, Sir Alexander Mackenzie travelled overland from Fort Chipewyan to just south of Prince George and then to the Pacific Coast near Bella Coola. Under the Nootka Convention (1794), the Spanish surrendered their claim to the Pacific Coast north of 42° N, leaving the British and Russians in control.

In the early-nineteenth century, the North West Company established a series of fur-trading posts along the Columbia River. From 1805 to 1808, Simon Fraser, a fur trader and explorer, explored the interior of British Columbia on behalf of the North West Company. He travelled by canoe from the Peace River to the mouth of the Fraser River. As elsewhere, the strategy of the North West Company was to develop a working relationship with local Indian tribes based on bartering manufactured goods for furs. After 1821, the North West Company merged with its rival, the Hudson's Bay Company. The Hudson's Bay Company then took charge of the Oregon Territories, which extended from the mouth of the Columbia River to Russia's Alaska.

In 1843, American settlers began to arrive on the coast from the eastern part of the United States. In the same year, the Hudson's Bay Company relocated its main trading post from Fort Vancouver at the mouth of the Columbia River to Fort Victoria at the southern tip of Vancouver Island. The increasing number of American settlers who came west along the Oregon Trail represented a challenge to the authority of the Hudson's Bay Company. A few years later, the United States claimed all the Pacific Coast northward to Alaska, where Russian fur-trading posts existed. In 1846, Britain and the United States agreed to place the boundary between the two nations at 49° N and then to follow the channel that separates Vancouver Island from the mainland of the United States. While the loss of the Oregon Territories was substantial, Britain was fortunate to hold onto the remaining lands administered by the Hudson's Bay Company. Britain recognized that its hold on these lands through the Hudson's Bay Company was tenuous and could not withstand the political weight of the growing number of American settlers. Indeed, without the presence of the Hudson's Bay Company and Britain's negotiating skills, Canada might well have lost its entire Pacific coastline.

The gold rush of 1858 brought about 25,000 prospectors from California to the

Fraser River. Prospectors walked upstream along the sandbanks and sand bars of the Fraser River and its tributaries, panning for placer gold (small particles of gold in sand and gravel deposits). The major finds were made in the interior of British Columbia, where the town of Barkerville was built near the town of Quesnel. By 1863, Barkerville had a population of about 10,000, making it the largest town in British Columbia. In order to ensure British sovereignty over territory north of 49° N, the British government established the mainland colony of British Columbia in 1858 under the authority of Sir James Douglas, who was also governor of Vancouver Island. In 1866, the two colonies were united.

Confederation

By the 1860s, the British government was actively encouraging its colonies in North America to unite into one country. Once the first four colonies were united in 1867, Ottawa adopted the British strategy to create a transcontinental nation. An important part of that strategy was to lure British Columbia into the 'national fabric'. The Canadian Pacific Railway was the first expression of this national policy.[1] Ottawa promised to build a railway to the Pacific Ocean within ten years after British Columbia joined Confederation. In Fort Victoria, however, some wanted to join the United States. By the middle of the nineteenth century, British Columbia had developed significant commercial ties with Americans along the Pacific Coast. San Francisco was the closest metropolis and, with its railway to New York, offered the simplest and quickest route to London. In 1859, Oregon became a state

and Washington was soon to follow. Commercial links with the United States were growing stronger. But the majority of people in Fort Victoria wanted to remain British. In 1871, British Columbia chose to become a province of Canada (Figure 3.5). British Columbians, however, had to wait fourteen years for the Canadian Pacific Railway to reach the Pacific Coast at Port Moody, near Vancouver, on Burrard Inlet. Later, the railway was extended 20 km westward to the small sawmill town of Vancouver, where there was a better harbour and terminal site for the railway (Figure 7.2).

When British Columbia joined Confederation in 1871, the official population of British Columbia was 36,247 (McVey and Kalbach 1995:35). This figure likely underestimated the number of Indians and prospectors who were living in the more remote areas of the province.[2] At the time of the first comprehensive census of British Columbia in 1881, there were approximately 4,200 Chinese, 19,000 white settlers (American, Canadian, and British), and about 30,000 Indians. Most British settlers lived around Fort Victoria. Beyond Fort Victoria, the vast majority of the inhabitants were Aboriginal peoples. In the previous decade, about 25,000 prospectors, mainly Americans, had been scattered in small camps along the Fraser River or at Barkerville. Most Americans had left at the end of the gold rush.

Post-Confederation Growth

At first, Confederation had little effect on British Columbia. The province was isolated from the rest of Canada, Canada's fledgling factories, and even the halls of power in Ot-

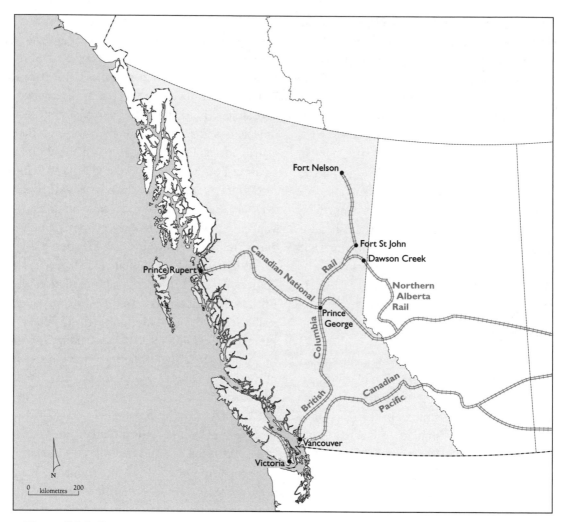

Figure 7.2 Railways in British Columbia. The first railway to cross the mountains of the Cordillera was the Canadian Pacific Railway in 1885. To open the northern interior of the province, Victoria built the Pacific Great Eastern Railway (now known as British Columbia Rail). After several extensions, British Columbia Rail joined North Vancouver to Fort Nelson by 1971. An eastern link from this railway joins Dawson Creek to the Northern Alberta Railway.

tawa. Goods still had to come by ship from San Francisco or London. When the Canadian Pacific Railway was completed in 1885, British Columbia truly became part of the Dominion, and BC's role as a gateway to the world began.

The main line of the Canadian Pacific Railway and its many branch lines were responsible for the formation of many of the province's towns and cities and for providing access to its forest and mineral wealth. The Esquimalt and Nanaimo Railway, the Canadian National Railway, and the British Columbia

Rail (BC Rail) added to the rail network in British Columbia. In 1886, the Esquimalt and Nanaimo Railway was built, connecting the coalfields of Nanaimo with the capital city of Victoria, and stimulated logging and saw-milling along its rail route. With the closure of the coal mines in the early 1950s, the railway lost its main function and now plays a minor role in the transportation system of Vancouver Island. By 1914, the Grand Trunk Pacific Railway (later the Canadian National Railway) provided a trans-Canada rail route to Prince Rupert and an alternative rail service to Vancouver. The Pacific Great Eastern Railway (later the British Columbia Rail) was incorporated in 1912 but laid few rails until the early 1950s, when the provincial government made

a commitment to complete the railway in order to facilitate resource development in the interior of British Columbia. By 1956, this rail line extended from North Vancouver to Prince George. Two years later, British Columbia Rail extended its rail system to Dawson Creek and Fort St John in the Peace River country; and by 1971, this railway reached Fort Nelson in BC's forested northeast.

After the completion of the first railway, Canadian Pacific Railway, in 1885, Vancouver grew quickly and soon became the major centre on the West Coast. By 1901, Vancouver had a population of 27,000 compared to Victoria's 24,000. As the terminus of the trans-continental railway, Vancouver became the transshipment point for goods produced in

Vancouver has a well-sheltered harbour, which contributed to the city's development as a major port on the Pacific Coast. (*Corel Photos*)

the interior of BC and Western Canada. As coal, lumber, and grain were transported by rail from the interior of British Columbia and the Canadian Prairies, the port of Vancouver spearheaded economic growth in the southwest part of the province. It was then possible to tap the vast natural resources of BC and ship them to world markets through the new port of Vancouver. By the twentieth century, Vancouver had become one of Canada's major ports. Vancouver, located on Burrard Inlet, has an excellent harbour. Unlike Montréal, Vancouver has an ice-free harbour, thanks to the warm Pacific Ocean. As Canada's major Pacific port, Vancouver became the natural transportation link to Pacific nations. With the opening of the Panama Canal in 1914, British Columbia's resources were more accessible to the markets of the United Kingdom and western Europe.

Between the completion of the Canadian Pacific Railway in 1885 and the end of the First World War in 1918, British Columbia underwent a demographic explosion. By 1921, British Columbia had over half a million inhabitants. British Columbia had been gradually transformed from a fragile political entity in 1871 into a self-confident political and social entity. During this time, the populations of both European and Asian origin increased by about tenfold. The European population had reached nearly half a million, while the Asian population was about 40,000. At the same time, a combination of disease and social dislocation caused the number of Indians to decline sharply, perhaps from 40,000 to 20,000. This magnitude of demographic decline for Aboriginal peoples was matched in other re-gions of Canada. While there were a number of factors at work, smallpox, tuberculosis, and other communicable diseases caused the greatest losses. (See Chapter 4 for more details.)

While BC's economy and population continued to grow in the 1920s, the Great Depression of the 1930s caused the province's economy to stall and unemployment to rise sharply. Economic disaster struck British Columbia in 1929. Exports of Canadian products from British Columbia, so necessary for its economic well-being, slowed and prices dropped. In the Prairies, the collapse in agricultural prices was accompanied by prolonged drought, turning the land into a dust bowl. Many Prairie farmers abandoned their farms and fled to British Columbia, adding to the burden of unemployment in that province.

The Second World War and Beyond

The Second World War called for full production in Canada, thereby pulling British Columbia's depressed resource economy out of the doldrums. Military production, including aircraft manufacturing, greatly expanded BC's industrial output. As well, resource industries based on forestry and mining (especially coal and copper) were producing at full capacity.

When the war ended in 1945, BC's resource boom continued. With world demand for forest and mineral products remaining high, the provincial government focused its efforts on developing the resources of its hinterland, the central interior of British Columbia. The first step was to create a transportation system from Vancouver to Prince George, the major city in the central interior. The highway system was improved and extended from

Prince George to Dawson Creek in 1952. More important for economic development was the completion of the Pacific Great Eastern Railway (now the British Columbia Rail) to Prince George and then to Dawson Creek in 1958. With rail access to Vancouver, forestry, as well as other resource industries, expanded rapidly, thereby leading to the integration of this hinterland into the BC and global industrial core.

British Columbia Today

Is British Columbia a core or a periphery? No doubt this Pacific province is an emerging powerhouse within Canada, but its economy still has a relatively small manufacturing base. Approximately one-third of BC's manufacturing activities involve the processing of wood or wood fibre. BC exhibits one of the classic characteristics of a resource hinterland: close to 90 per cent of its exports by value fall into the category of natural resources, particularly forest products (Wilkinson 1997:138). Most exported forest products, for instance, undergo little processing. On the other hand, few workers are actually employed in the primary sector; rather, most are involved in the tertiary sector, which may point to a new type of core region. For example, BC's economy now boasts a core of high-technology and **producer services** industries, such as advertising, computer, engineering, and management firms. Most of these are concentrated in Vancouver and Victoria (as they are in major cities in Ontario and Québec). Often these industries are located near research centres and universities, where highly skilled labour is avail-

able and innovative research takes place. As well, BC's natural beauty forms the basis of an expanding tourist industry that accounts for most new jobs created in British Columbia.

High-technology companies, producer services firms, and to a lesser degree, an expanding tourist industry, play a key role in developing BC's diversified economy. Clearly, British Columbia is in a state of change, the leading edge of which is in the Vancouver/Victoria urban cluster, where high-technology and producer services are concentrated. Beyond that cluster, the resource economy remains dominant.

Population and the Centralist/Decentralist Faultline

One measure of BC's expanding power lies in its rapid population growth. In 1969, Premier W.A.C. Bennett boasted that 'With the population of British Columbia growing at twice the rate of the rest of Canada, the presence of British Columbia as an economic region of its own is more obvious as each day passes' (Francis, Jones, and Smith 1996:428). In this case, the numbers support the politician's statement. In 1901, the population of this Pacific province amounted to only 3 per cent of the Canadian population. By 1996, British Columbia's 3.7 million residents constituted 12.7 per cent of Canada's population. By the start of the next century, the population of British Columbia should surpass 4 million, which will be 13 per cent of the national population.

British Columbia is the only province that has made substantial population gains in each decade since Confederation. In comparison to Ontario, British Columbia has had a higher rate of population increase from 1951 to 1996. During this period, British Columbia's rate of growth surpassed Ontario's, reaching an average annual rate of population increase of 4 per cent compared to 3.6 per cent for Ontario. Since then, British Columbia has maintained its demographic momentum over Ontario. From 1991 to 1996, BC's annual rate of increase was 2.7 per cent, well above Ontario's rate of 1.3 per cent and almost two and a half times greater than the national average of 1.1 per cent.

Most of BC's population increase comes from immigration and interprovincial migration. Since the 1970s, more Canadians have resettled in British Columbia than in any other province. Many immigrants also select British Columbia (after Ontario and Québec) as their new home. From 1990 to 1996, many Hong Kong immigrants came to British Columbia before Britain turned its colony over to China. This significant influx of immigrants to BC has changed the cultural makeup of the province. In 1996, for instance, 8.5 per cent of all British Columbians declared Chinese as their ethnic origin, while another 4.5 per cent had South Asian origins (Statistics Canada 1998c:19 of 21). In comparison, 9.2 per cent said that they were of French ancestry, though only 1.5 per cent of British Columbia's residents declared that French was their mother tongue.

Many immigrants have brought considerable wealth and economic connections to invest in British Columbia. Asian investment, particularly from Japan, has had an important impact on British Columbia's economy (Edginton 1994:32). Since the 1960s, Japanese multinational firms have invested heavily in min-

ing, forestry, and tourism in BC. Japanese companies made a massive investment in the Northeast Coal Project in Tumbler Ridge. This project was designed to supply coal to the steel plants in Japan (Bone 1992:139–40). As well, Japanese investors have made many real estate purchases in the Vancouver area rather than in larger centres like Toronto and Montréal because of the large influx of Japanese tourists to Vancouver (Edginton 1996). In the early 1990s, Japanese investors (Nippon Cable) developed a ski resort just north of Kamloops, turning it into the largest ski resort in British Columbia after the Whistler ski resort just north of Vancouver (Jim Miller, personal communications, 28 April 1998).[3]

BC's growing population has not only affected its economic status and cultural diversity; it has also bolstered its political might. The political importance of British Columbia's rapid population increase lies in the increased number of seats held by BC members in the House of Commons. However, since the redistribution of the number of seats assigned to each province occurs after the last census of Canada's population, there is a lag effect in translating BC's increased population size into more seats in the House of Commons. In 1872, British Columbia had six representatives in the House of Commons, which was 3 per cent of the members. By 1996, BC had thirty-two members, forming nearly 11 per cent of the membership of the House of Commons. In 1997, British Columbia held only 11.3 per cent of the seats in the House of Commons, although its population size warranted nearly 12.7 per cent. This lag effect poses a serious irritant in the province's relations with Ottawa, which reinforces the centralist/decentralist faultline.

Economic Growth in the 1990s

From 1990 to 1997, BC's increasing economic strength, partly driven by the Asian economy, outpaced that of all other regions in Canada. In 1994, British Columbia's robust economy accounted for over 13 per cent of Canada's GDP, up several percentage points from a decade earlier. Several factors account for the province's rapid economic ascent. The principal force behind BC's growth is a rich and varied resource base accompanied by high commodity prices. Like the Prairie economy, this Pacific region relies heavily on its natural resources for its economic well-being. While mining and fishing are important industries, forestry remains the centrepiece of British Columbia's vibrant resource economy.

Although British Columbia has only 14.5 per cent of the forest lands in Canada, its share of the area of productive forest land amounts to 21 per cent (Statistics Canada 1998a:19). Moreover, of Canada's productive forest land, British Columbia has the most valuable timber stands because of the size, type, and density of the trees in its rainforest. Along the coastal communities and in the interior centres, logging and wood pro-

cessing have provided employment for workers, wealth for its forest product companies, and valuable primary products for domestic and foreign customers. Just under 2 per cent of British Columbia's labour force is employed in logging and related primary forestry jobs (British Columbia 1998c:4 of 7). Another 10 per cent are involved in wood-processing industries, including sawmilling, plywood manufacturing, and paper-producing activities.

Trade is another dynamic force propelling British Columbia's economy. Vancouver and, to a lesser degree, Prince Rupert and other Pacific ports serve as trade outlets for coal, lumber, potash, and grain from the interior of

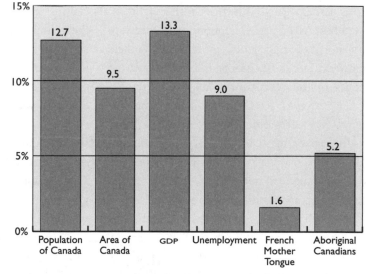

Figure 7.3 British Columbia, 1996. BC's strong place in Canada is revealed by its share of Canada's GDP and population. The Aboriginal population and French by mother tongue show the weak position of Aboriginal peoples and French Canadians in the province. The small percentage of Aboriginal peoples, however, does not reflect the extent of First Nations' potential ownership of land in BC through ongoing land-claim negotiations.

Sources: Statistics Canada, 1997a:Table 1; 1998d:Table 39; 1997b:1 to 3 of 3; 1997c:4 and 5 of 15; 1998b:1 and 2 of 10; 1998c:15 to 20 of 21; Stanford, 1998:185.

British Columbia and the Prairie provinces to reach world markets. In BC's export-oriented economy, it is usually much less expensive in transportation costs to ship the raw material than the finished product. This has led to trade focused on resources in BC. Also, BC has higher labour costs and more stringent environmental regulations than most countries importing BC's products. Those factors make BC a high-cost processing region, which has inhibited the development of manufacturing.

During the last twenty years, China and other Pacific Rim countries have led the world in economic growth, creating new markets for Canadian products shipped through Vancouver. One measure of the potential for trade with Pacific Rim countries is the eighteen-member Asia Pacific Economic Cooperation Forum. Trade opportunities are almost endless between British Columbia and the population of 2.2 billion people in the nations on the Pacific. However, in 1998, Asia's economy faltered. Asia's economic problems quickly led to a drop in demand for primary products, including BC's forest products. This decrease in demand and a fall in commodity prices caused BC's economy to slide into a recession.

British Columbia does have certain natural assets to attract more industrial investment. Low-cost hydroelectric power is one such advantageous feature. Electrical energy generated in British Columbia is relatively inexpensive. While energy is not a critical cost factor for most manufacturing industries, it is crucial for metallurgical industries such as aluminum smelting and fabricating plants. In such cases, 20 per cent or more of the costs of operating smelters are related to the consump-

tion of electrical power. Québec developed an industrial strategy based on providing low-cost electrical power to metallurgical firms (see Chapter 6). BC's approach is discussed later in this chapter in the subsection 'Hydro-electric Power'.

Land Claims and the Aboriginal/Non-Aboriginal Faultline

Who owns British Columbia? This question goes to the heart of the land-claims issue. Indian bands are pressing to settle their claims to lands that they occupied prior to the arrival of Europeans. The overlapping land claims of BC's 197 Indian bands equal 110 per cent of the province (Howard 1996:A5), but land-claim settlements rarely exceed 10 per cent of the claimed territory. Nevertheless, the stakes are very high. Canadians recognize their obligation to settle outstanding Aboriginal land claims. The settlement process has been slow, but now all parties are willing to push forward. Once land claims have been addressed, the tension underlying the Aboriginal/non-Aboriginal faultline in BC will be reduced, and First Nations will be able to focus their energies on improving the economic and social well-being of their members.

Almost 200,000 Aboriginal peoples live in British Columbia. A small number of Indian bands have signed treaties: some fall under Treaty 8, and others under the Douglas and Vancouver Island treaties. These treaties extinguished Aboriginal rights to the land in exchange for various benefits for the Indians. Treaty 8, propelled by the Klondike gold rush of 1898, had two purposes: (1) to open access to miners and prospectors heading to Yukon across the Canadian northwest, and (2) to allow the Indians to continue their traditional way of life and receive annual payments. The fourteen Douglas treaties, on the other hand, focused on providing land reserves for Indians along with their right to hunt and fish on Crown land.

Most of the First Nations in BC have still not negotiated a settlement with Ottawa and Victoria. However, a combination of blockades, legal action, and public protests have made this a central political issue in the province. In 1990, Victoria made a commitment to participate in resolving land claims. Much has happened since then, including the establishment

of a treaty commissioner, whose role is to co-ordinate negotiations between the Aboriginal leaders and the federal and provincial government representatives. By November 1994, forty-two First Nations groups representing about 70 per cent of the Aboriginal population had entered the process of negotiations. By 1996, one claim, that of the Nisga'a, had been approved in principle by the Aboriginal leaders and the public officials representing Ottawa and Victoria. By 1998, the Nisga'a had approved this agreement. The next step requires the legislatures of Ottawa and Victoria to pass a bill incorporating this agreement into law. In exchange for surrendering their Abori-

Totem poles in Vancouver's Stanley Park. Aboriginal peoples' land claims are based on their ancestors' occupation of BC lands for thousands of years. (Corel Photos)

ginal land claims, the Nisga'a obtain forest lands, a share in the commercial fishing catch, and a substantial cash settlement. This land-claim settlement, which is viewed as a precedent for the remaining land claims in British Columbia, will transfer considerable resources, capital, and political power (self-government) to the Nisga'a in exchange for the surrender of their Aboriginal land claim. The Nisga'a will then have an opportunity to put politics behind them and move forward on their economic and social agenda. The political dimensions of this agreement are so critical that Frank Cassidy (1992:11) described it as 'a centrepiece in the historical development of the province of British Columbia'. The remaining land claims in BC should be resolved

in due course, thus completing BC's obligation to its Aboriginal citizens.

Industrial Structure

The economy of British Columbia is based on its resource industries, but these primary industries employ only 4.7 per cent of the workers in the province (Table 7.1). Paradoxically, the industrial structure of this resource-rich province is more oriented to service activities and, to a lesser degree, manufacturing activities rather than to primary employment. Over the years, the percentage of workers in the primary sector has declined, not because resource extraction efforts were decreasing but because firms increased labour productivity by substituting machines for labour.

Table 7.1
Comparison of Employment: British Columbia and Canada, 1995

Industrial Sector	British Columbia (000)	British Columbia (per cent)	Canada (per cent)
Primary	83	4.7	5.4
Forestry	36	2.0	0.7
Secondary	318	18.1	21.2
Tertiary	1,361	77.2	73.4
Transport	140	7.9	7.5
Trade	317	18.0	17.0
Finance	111	6.3	5.8
Service	694	39.4	37.1
Public	99	5.6	5.9
Total	1,762	100.0	100.0

Source: Adapted from Statistics Canada, 1996:Table 40.

Table 7.2
Manufacturing Shipments by Industry, 1997

Industry	$ Millions	Per cent
Wood	11,438	33.0
Paper and allied products	5,412	15.6
Food	3,871	11.2
Refined petroleum and coal	1,753	5.1
Fabricated metal products	1,687	4.9
Primary metals	1,314	3.8
Transportation equipment	1,296	3.7
Machinery and equipment	1,281	3.7
Printing and publishing	1,198	3.5
Electrical and electronic products	1,000	2.9
Chemicals	994	2.9
Non-metallic mineral products	879	2.5
Beverages	765	2.2
Plastics	691	2.0
Clothing	278	0.8
Furniture and fixtures	215	0.6
Textile products	132	0.4
Other	468	1.3
Total	34,672	100.0

Source: British Columbia, 1998a:8–9.

In 1995, BC's secondary sector comprised 18.1 per cent of employment (Table 7.1). Manufacturing is an important but small element of British Columbia's economy; it lags far behind manufacturing in Ontario and Québec. Manufacturing in British Columbia is based on the processing of forest and other resources. By value, just over half of all exports from British Columbia are forest products, while food items contribute another 11 per cent. Coal, grain, and lumber exports have transformed the port of Vancouver into

Vignette 7.1

Heartland or Hinterland?

Resource enterprise and a unique physical environment, above all else, have shaped the geographic character of Canada's west coast province. British Columbia is a staple hinterland for industrial markets throughout the world, yet it is also a prosperous region. This is so because there is a diversity of mineral, forest, and marine resources that have supported continuous economic growth, in turn fuelled by large infusions of people, capital, and technical expertise from metropolitan economies.

The physical environment is of further importance because topographic control has dictated a settlement and communication pattern that concentrates human activity in the southwestern corner of the province around Vancouver. The role of Vancouver as the metropolis of the region—the natural intermediary between world markets and a vast resource hinterland—mirrors much that is characteristic of core and periphery in the world economy.

Source: John Bradbury, 'British Columbia: Metropolis and Hinterland in Microcosm', in *Heartland and Hinterland*, edited by L.D. McCann (Toronto: Prentice-Hall, 1982):339. Reprinted by permission of Prentice-Hall Canada, Inc.

Canada's major trade centre. Unfortunately, most resources are exported in a semi-processed form. The key to expanding the manufacturing sector in British Columbia lies in greater processing of its resources, especially to the final product stage. Such diversification would greatly increase both the value of production and the size of the labour force. In 1997, manufacturing firms produced almost $35 billion worth of goods (Table 7.2).

In 1995, the tertiary sector accounted for 77.2 per cent of employment in BC (Table 7.1). Compared to the national employment percentages, not only does service employment exceed the national average by 4 percentage points but each subsection (except public administration) outstrips the national figures. Employment in transport, trade, finance, and service indicate the relative strength of the service industry. The service

sector owes its strength to several factors, including British Columbia's growing domestic market, strong participation in world trade, a vibrant tourist industry, and its producer services firms. Tourism is a growing industry with many overseas visitors coming to Vancouver. In 1996, $7.0 billion was spent by 28 million overnight visitors to the province (British Columbia 1998a:2). Producer services are also playing a greater role in BC's economy. These firms have been pivotal in transforming Vancouver from 'its traditional role as a provincial higher-order service centre to a place in a network of increasingly interdependent cities of the Pacific Rim' (Davis and Hutton 1994:18).

The evidence is clear—British Columbia is changing, but it is still not an important manufacturing region. In 1981, a leading Canadian geographer, John Bradbury, saw BC as a resource hinterland that also had several characteristics common to a core region (Vignette 7.1). Since 1981, however, BC has experienced growth in its high technology and producer services industries as well as in its tourist industry—the tertiary sector, not the secondary sector, is leading BC to a new level of economic development.

British Columbia's Wealth

The key to British Columbia's past economic prosperity has been the province's natural resources. The industries which have historically led the province's growth and continue to contribute to its economic well-being include: fishing, mining, hydroelectric power, and especially, forestry. These activities have not only fuelled the primary sector, but also the secondary sector. Moreover, a portion of the tertiary sector, the tourism industry, relies almost entirely on the province's natural beauty. Like Ontario and Quebec, BC can be divided into two parts—a core and a periphery—though the core, centred in Vancouver, is much smaller than the industrial cores found in Ontario and Québec. Furthermore, hinterland activities continue to play a pivotal role outside of Vancouver and Victoria, and there is less of a sense of a declining resource hinterland than is found in Ontario and Québec.

Fishing Industry

British Columbia, like Atlantic Canada, has an important fishery. In BC more than eighty species of fish and marine animals are harvested from the ocean, freshwater bodies, and aquaculture areas. Salmon is the most valuable

species, followed by herring, shellfish, groundfish, and halibut. In recent years, excessive harvesting has reduced fish stocks and, in turn, the number of landings (fish caught) (Table 7.3). At the same time, production from fish farms has depressed the price for fish.

Unfortunately, BC shares another similarity with Atlantic Canada—overexploitation of fish stocks, particularly the very valuable salmon stocks. The results of years of overfishing are: overexploitation by excessively large fishing fleets; lower returns per fisher; and longer idle periods for fishers, fishing vessels, and handling/processing facilities.

Ottawa, which is responsible for managing salmon stocks within Canada and for international fishing agreeements, must deal with four perplexing and interrelated issues: (1) the natural cycle (up to five years) of salmon to spawn in rivers, to migrate to the sea, and then return to the rivers to spawn again; (2) the harmful effects of the forestry and hydroelectric industries on salmon spawning grounds; (3) the division of the salmon catch among commercial, Native, and sports fishers; and (4) the harvesting of 'Canadian' salmon in international waters.

Management of fish resources is based on the estimated size of the fish stock. The problem of determining the size of the stock, and therefore the size of a catch each year, is complicated by the natural population cycle of the fish. The natural fluctuation is illustrated by the pink salmon catches in the past. The size of catch has varied from a low in 1975 of 38 000 tonnes to a high of 108 000 tonnes in 1985. In 1994, an estimated 2.3 million fish did not return to BC rivers to spawn. What

Table 7.3
Fishery Statistics, 1996 (preliminary)

Species	Wholesale Value	
	($ millions)	(per cent)
Salmon	448	47.6
Herring	178	18.9
Shellfish	163	17.3
Groundfish	99	10.5
Halibut	51	5.4
Other	3	0.3
Total wholesale value	942	100.0
Total landed value	572	

Source: British Columbia, 1998a:5.

caused this decline is unknown, but a similar drop in 1995 in the number of salmon returning to the Fraser River forced Ottawa to close fishing on the Fraser River. One of the possible reasons for this decline is pollution of fish habitat. Forestry companies, for example, have affected salmon habitat through their logging practices, which have blocked streams and thereby interfered with migration routes to spawning areas. Pulp and paper plants also discharged toxic wastes into rivers and the ocean. Another possibility is a rise in ocean temperature above levels suitable for salmon— the so-called El Niño effect. El Niño's warming of the Pacific waters could provide a natural explanation for the declining salmon stocks, but so far the El Niño effect on ocean temperatures and then on salmon is unsubstantiated by scientific evidence.

A more likely explanation for the apparent decline in salmon stocks is a combination of a larger fishing fleet and the application of new technology (radar and sonar equipment), which both lead to overfishing. Overfishing is generally regarded as the leading cause of declining fish stocks. Fishers have simply taken more fish than the fish populations can sustain. For example, the Federal Department of Fisheries and Oceans, which is responsible for fish management, announced that there was a decline in the number of salmon returning to the Fraser River to lay eggs in 1994 and 1995 (Department of Fisheries and Oceans 1996:4). The fear is that future runs will decline even more.

Salmon are a migratory fish. Like other fish, salmon are common property until caught. This principle is based on the 'rule of capture'. Fishers therefore try to maximize their share of a harvest so no one else will take 'their' fish. The problem is complicated further because the Canadian government cannot regulate the 'Canadian' salmon stocks because they migrate to American waters, where the American fishing fleet harvests them.

The net result is the salmon stocks are threatened. This problem is commonly referred to as the **Tragedy of the Commons**. (See Chapter 9 for an account of the overexploitation of the northern cod.) However, Ottawa is taking action to alleviate the pressure on salmon stocks. It plans to reduce the 4,500 fishing vessels by about one-third. At the same time, Ottawa is allowing Aboriginal fishers, who have treaty rights to harvest fish for subsistence purposes, a share of the commercial stock. Ottawa is also negotiating with Washington to reduce the American catch of the Fraser River salmon. Ottawa is able to exert some management of fish stocks in the Pacific Ocean because of its 200-mile fishing zone (Figure 7.1) and because of its role in the Pacific Salmon Commission. Salmon fishing on the Pacific coast is regulated, based on the 1985 Pacific Salmon Treaty between Canada and the United States (the treaty determines the size of the catch taken by each nation).[4] Unfortunately, Canada has been unable to convince the US to reduce each nation's salmon take. If each country does not lower its

West Coast fish canneries, like this one in Delta, BC, may eventually suffer the same fate as their Atlantic counterparts if BC's salmon stocks continue to decline. (*Corel Photos*)

catch through a new treaty, salmon stocks may decline to dangerous levels similar to those experienced by the northern cod in Atlantic Canada.

Salmon catches by fishing fleets decreased in the 1990s, not because of conservation measures that restrict the allowable amount of catches but simply because the fish were not present in large numbers. Salmon stocks may be failing. Such a collapse spells disaster not only for the fish but also for the people who rely on the fisheries for their livelihood, including the many who work in the fish-processing plants. In fact, there are several large companies, including BC Packers and the Canadian Fishing Company, which control the processing of fish. They have strategically placed their canneries at the mouths of the Skeena and Fraser rivers. Such large companies process about 80 per cent of all the salmon caught each year. The repercussions of declining fish stocks for the 25,000 full-time and part-time employees of these fish-processing plants are most serious.

Mining Industry

The mineral wealth of British Columbia is found in both of the province's physiographic regions—the Cordillera and the Interior Plains. In 1996, the value of total mineral production was almost $4.4 billion (Table 7.4). By value, coal accounted for 24.8 per cent, followed by natural gas at 19.6 percent, and copper at 16.6 per cent. Mining and mineral processing employ about 3 per cent of the labour force, but yield nearly 20 per cent of the value of primary production. Because nearly all of the mineral production is consumed outside of

the province, the fortunes of the mining industry are largely determined by external markets and global prices. This pattern of development is common to all resource development in British Columbia and again emphasizes both the strength and the vulnerability of BC's hinterland economy.

The importance of the export market for mineral products has shaped BC's transportation system—first by requiring a network that would allow mining operations to reach more isolated areas and ship ore back to its markets, and now by adapting to the cost and needs of mining operations. For instance, competition from other resource hinterlands, such as Australia, has forced Canadian producers to reduce their costs of operation and marketing. Transportation costs have therefore been reduced by unit trains and bulk-loading facilities. Unit trains consist of a large number of ore cars, sometimes over 100, pulled by one or more locomotives. The coal-loading terminal at Robert's Point is a large bulk-loading facility. It is able to dump each ore car of coal and then move the coal by conveyor belts to the ship's hold. These ships must be moored in deep water, which necessitated building a long causeway to reach the ships.

As with other resource development, the capital investment in mining is enormous. For that reason, only very large companies, often multinational firms, can venture into such developments. The Northeast Coal Project is one such example. In the late 1970s Denison Mines and Teck Corporation were the major mining companies behind this $4.5 billion construction project (Bone 1992:139–40). With a fourteen-year contract to sell coal

Table 7.4
Value of Mineral Production in British Columbia, 1996

Mineral Product	$ Millions	Per cent
Metals		
Copper	730	16.6
Gold	317	7.2
Zinc	222	5.0
Silver	109	2.5
Molybdenum	94	2.1
Lead	58	1.3
Other	12	0.3
Total metals	**1,542**	**35.1**
Fuels		
Coal	1,090	24.8
Natural gas	861	19.6
Other fuels	443	10.1
Total fuels	**2,394**	**54.3**
Structural materials	**419**	**9.5**
Industrial minerals	**44**	**1.0**
Total mineral production	**4,399**	**100.0**

Source: British Columbia, 1998a.

to Japanese steel companies, banks and other financial institutions were happy to supply most of the capital (about $2.5 billion) to Denison and Teck. The federal and provincial governments provided $1.5 billion of the $4.5 billion to upgrade the CNR line to Prince Rupert, to extend a British Columbia Rail line to Tumbler Ridge, and to build a coal terminal to load ships that would carry the coal to Japan. The two mining companies contributed about $500 million. Both governments provided another $1 billion to build the coal town of Tumbler Ridge. These agreements were signed in the late 1970s, when the world economy was expanding and the demand for coal was increasing. The Northeast Coal Project comprised Quintette Coal Ltd at Tumbler Ridge, Teck Corporation at Bullmoose, and Gregg River Coal Ltd at Gregg River. Like all large-scale construction projects, several years were required to complete the undertaking. Production began in 1984. By then, the world demand and price for coal had peaked and a slow but steady decline set in as the global economy slipped into a recession. At that point, the Japanese steel mills no longer needed so much coal and the price set with the Northeast Coal Project mines was no longer attractive. By 1984, when coal was first produced at the Northeast Coal Project, the world spot price (a price charged for goods available for immediate delivery) for coal was 10 per cent lower than the agreed-upon, contract price! Eventually, the price for coal was renegotiated, but Denison Mines went bankrupt during the process. Since then, Teck Corporation has taken over the coal-mining operation.

The example of the Northeast Coal Project illustrates the vulnerability of the mining industry, indeed of all resource industries. With a heavy reliance on export markets and the fluctuations in world prices, BC's mining industry faces ongoing challenges. BC's economy will also have to adapt to the new economic order, by increasing secondary activities and relying less on such primary industries.

Hydroelectric Power

British Columbia has many natural hydroelectric sites. Within the Cordillera, a combination of elevation, steep-sided valleys, and steady flowing rivers provides ideal conditions for the construction of hydroelectric dams. The Columbia, Fraser, and Peace rivers offer many excellent hydroelectric sites. In addition, heavy precipitation and melt water from the mountain snowpack ensure a regular and abundant supply of water. Major hydroelectric developments have taken place on the Columbia and Peace rivers as well as on the Nechako River, a tributary of the Fraser River. Minor developments have occurred on Vancouver Island.

In the early days of hydroelectric development, small-scale hydroelectric projects were established near Vancouver and Victoria. These two cities are the major markets in the province for electricity. Hydroelectric dams were also built near smelters that required vast amounts of power to operate. For example, early in this century, power was generated from privately owned dams on the Kootenay River for the lead-zinc smelter at Trail.

After the Second World War, public and private companies began to undertake megaprojects. Among these giant industrial construction efforts, three—one on the Columbia River, another on the Peace River, and the third on the Nechako River—had enormous impacts on the economy. They all involved the harnessing of water-power from the province's rivers to generate low-cost electrical power, but they also flooded valuable farm and Indian lands and led to the loss of salmon spawning grounds. Such construction projects were considered major engineering feats. In 1951, Alcan completed the construction of the Kenney Dam across the Nechako River, impounding the water draining from an area more than twice the size of Prince Edward Island (see 'Alcan in Northwest British Columbia' later in this chapter). In 1968, BC Hydro built the W.A.C. Bennett hydroelectric dam on the Peace River, creating Williston Lake, the largest fresh-water body in the province. From this giant reservoir, vast quantities of water flow through the turbines at the powerplant, generating much of British Columbia's electrical power and creating a surplus of power for sale in the US.

The most ambitious of these early mega-projects was the one developed on the Columbia River. A combination of rising demand for power in the state of Washington and innovations in the transmission of electricity over long distances enabled the development of water-power along the Columbia River in British Columbia. In the 1960s, dams were constructed on the Columbia River to increase the flow of water to the hydroelectric power-plant at Grand Coulee in the United States (Vignette 7.2). Initially, BC Hydro sold its share of the electrical power generated at Grand Coulee to its counterpart in the state of Washington. In 1997, this power was no longer needed in the US and British Columbia was faced with a huge surplus of electrical power. Similar to the Québec government's efforts to stimulate industrial development in that province, Victoria has now begun offering this surplus power at a reduced price to firms that locate in British Columbia. So far, no large firms have expressed an interest, but Victoria has used this surplus as part of a settlement it reached with Alcan over the cancellation of one of its hydroelectric projects (see below).

Alcan in Northwest British Columbia

The Alcan hydroelectric development in British Columbia is an example that illustrates the advantages and pitfalls of megaprojects. The Aluminum Company of Canada (Alcan) is one of the giants in the world aluminum industry. To keep its production costs low, Alcan is attracted to sites where low-cost hydroelec-

---

Vignette 7.2

The Columbia River Treaty

In 1961, Canada and the United States agreed to cooperative development of the Columbia River for the production of hydroelectric power. Canada undertook to construct three dams for water storage in the Canadian portion of the Columbia River Basin and to operate them to produce maximum flood control and power downstream. In return, the United States would return half the power generated to Canada and pay for half the value of the flood protection for property in the United States. The three dams constructed as a result of this agreement are the Duncan (1967) north of Kootenay Lake, the Hugh Keenleyside (1968) on the Columbia River, and the Mica (1973) north of Revelstoke. The major hydro-electric powerplant is located in the state of Washington at Grand Coulee. Before the construction of the Grand Coulee Dam, the Columbia River was one of the world's great spawning grounds for salmon.

tric power can be developed and then used by its smelting plants to reduce bauxite (a clay-like mineral) into alumina (the chief source of aluminum) and then aluminum. In Canada, water resources are under provincial jurisdiction, so Alcan seeks long-term arrangements with provincial governments to develop the water-power for its smelting plant. In turn, provincial governments are attracted to megaprojects because they can develop a region and employ large numbers of people.

In the late 1940s, Alcan proposed to build an aluminum complex in northwest British Columbia. After obtaining the rights to the waters of the Nechako River for fifty years from the provincial government, Alcan began construction of the first phase of this complex in 1951.[5] Three years later, Alcan built a new town called Kitimat, a powerplant at Kemano, and a dam on the Nechako River. To produce hydroelectric power, Alcan dammed the Nechako River, thereby reversing its flow westward to the Pacific Ocean near Kemano. To accomplish this task, a tunnel had to be drilled through the Coast Mountains. At Mount DuBose, a 16-km tunnel allowed waters from the Nechako reservoir to flow to Kemano where the powerplant is located. These waters plunge 860 m from the western end of the tunnel to drive the turbines at Kemano. Electricity is transmitted about 80 km to Alcan's aluminum smelter at Kitimat.

In the late 1940s, concerns about environmental and social impacts of industrial projects were not given much attention. Both the provincial and federal governments bent over backwards to assist Alcan in what was considered one of the great engineering and construction feats of the twentieth century. The flooding of

the Cheslatta Indian reserve, which resulted from the Alcan project, was not considered a significant social impact and the Department of Indian Affairs quickly arranged for a relocation of this tribal settlement. As for the sockeye salmon spawning grounds that were affected, no protest was raised. Three reasons account for the lack of concern: (1) the general public was convinced that industrial growth was 'good' for the province; (2) the public and the fishing industry were not aware that the destruction of the Nechako spawning grounds could have a major impact on salmon stocks; (3) the environmental movement and Aboriginal groups had not yet emerged as forces capable of challenging such projects.

In 1989, Alcan began the second phase of its hydroelectric complex—the Kemano Completion Project. Estimated costs were $1.3 billion. The plan was to enlarge its hydroelectric facility by boring a second tunnel through Mount DuBose, thereby diverting more waters of the Nechako reservoir westward. This time, both the Carrier-Sekani Tribal Council (whose members claim the Nechako River as part of their land claims) and environmental groups spearheaded by the Rivers Defense Coalition openly challenged this project, claiming that the social and environmental costs were too great. Even so, the project was approved by the federal government without an environmental review. By 1991, the project was half completed at a cost of about half a billion dollars, but opposition from environmental groups and Aboriginal leaders had grown so strong that the federal government was forced to order an environmental inquiry. Alcan ceased its construction efforts, hoping to restart work after federal approval.

In 1995, the provincial government cancelled its approval for Alcan's Kemano Completion Project. Alcan threated to sue. Later, the British Columbia government agreed to compensate Alcan for the money spent on construction by selling the company electricity at a very low price (BC was able to make use of the surplus power it had). Alcan claimed to have spent $500 million toward the construction of the $1.3 billion project. In the arrangement reached in 1997, Alcan will receive electrical power for its proposed smelter in two ways (Cernetig 1997:1): (1) 115 MW a year from BC Hydro at a price pegged to the price of aluminum on the London Metal Exchange until 1 January 2024; and (2) 60 MW a year from BC Hydro until 1 January 2024. At 1997 prices for power, the difference between the cost to Alcan and the market value of the 175 MW a year results in a savings of approximately $9 million a year, or $243 million over the course of the 27 years.

The prospects for the construction of new hydroelectric projects in BC in the near future are poor. Environmental and social impacts may well be too great, as was the case with the Kemano Completion Project. Other adverse factors include: BC has a surplus of power; world demand for aluminum is very weak; and Alcan has just announced a hugh expansion of its aluminum operation in Québec, which it perceives as a 'more friendly government'.

Key Topic: Forestry

The forest is British Columbia's greatest natural asset. Forest covers 62 per cent of the territory of British Columbia. Almost all of this forest is coniferous and constitutes about half of the Canadian softwood timber stock. Within Canada, British Columbia harvests just over half of the total logging output. The main species harvested are lodgepole pine, spruce, hemlock, balsam, Douglas fir, and cedar (Table 7.5). The processing of timber into lumber, pulp, newsprint, paper products, shingles, and shakes supports a major manufacturing industry in British Columbia. Within Canada, British Columbia leads in the production of wood products like lumber and plywood, while Ontario and Québec produce a greater proportion of pulp and paper products.

One estimate suggests that half of British Columbia's population is directly or indirectly dependent on forestry for their livelihood. Statistics support this claim. In 1996, timber harvest was a staggering 75.2 million m³, an increase of nearly 5 per cent over the 1991 harvest. The value of this harvest, including processing, was about $18 billion. Led by lumber and pulp, forest exports are valued at over $15 billion (Table 7.6). Wood processing consisted of softwood lumber, pulp, paper, and plywood. Almost 20,000 people are employed in logging operations, 15,000 in wood processing, and 5,000 in pulp and paper making. This does not include the many jobs that rely indirectly on forestry like tree planting and furniture making.

Forest Regions

British Columbia's forest is far from homogeneous, largely because of its climatic zone and the varied relief in the Cordillera. The variations in the type of forest have strong implications for forestry. For example, trees along BC's moist West Coast grow much more quickly than those in the dry interior. BC's for-

Table 7.5
Timber Harvest by Species, 1996

Species	Value (million cubic metres)	Per cent
Lodgepole pine	20.4	27.1
Spruce	14.2	18.9
Hemlock	10.9	14.4
Balsam	10.6	14.2
Douglas fir	8.3	11.0
Cedar	6.5	8.7
Other species	4.3	5.7
Total timber harvest	75.2	100.0

Source: British Columbia, 1998a.

Table 7.6
Forest Product Exports from British Columbia, 1996

Commodity	$ Millions	Per cent
Lumber (softwood)	7,736	51.3
Pulp	3,407	22.6
Newsprint	1,284	8.5
Paper and paperboard	851	5.6
Plywood (softwood)	585	3.9
Cedar shakes and shingles	222	1.5
Selected value-added wood products	230	1.5
Others	758	5.0
Total forest product exports	15,073	100.0

Source: British Columbia, 1998a.

Vignette 7.3

Making Pulp

Wood is reduced to fibres called pulp by mechanical means (grinding wood), by chemical means (cooking wood in a chemical solution), or by a combination of these two methods. All three approaches produce a different type and quality of pulp. They are: groundwood pulp, thermomechanical pulp, and chemical pulp. **Groundwood pulp**, made by grinding wood chips or sawdust into fine fibres, is used to make high-quality newsprint. **Thermomechanical pulp** is a combination of the groundwood and chemical pulp processes. It produces pulp with many of the qualities of groundwood pulp, but with the addition of greater strength such as found in chemical pulps. In both the groundwood and thermomechanical processes, grinding the wood produces very soft pulp and makes effective use of about 95 per cent of the cellulose content of the wood. However, grinding the wood breaks the fibres and shorter fibres produce a pulp with little strength. **Chemical pulp** mills change wood into sulphate pulp by chemical means, thereby retaining the wood's long fibres. However, only about 50 per cent of the cellulose fibre of the wood is used. The chemical pulp is used to make industrial paper, cardboard, and a wide range of quality paper products.

est lands are divided into two major regions: the Coast Forest and the Interior Forest. Within the Interior Forest, there are four subregions: the Northern Forest, the Nechako Forest, the Fraser Plateau Forest, and the Columbia Forest (Figure 7.4). Within each of these subregions there is great variation in the forests due to the differences in local growing conditions, which are affected by precipitation and length of growing season. Other factors are elevation, soil conditions, and topography.

The Coast Forest is the most luxuriant coniferous forest in Canada. With its wet marine mild temperatures and abundant rainfall, this coniferous forest is one of the most densely forested areas of North America. The key species are Douglas fir, western red cedar, and western red hemlock. Under ideal conditions, mature cedar and hemlock trees reach 45 m in height and about 1 m in diameter. The Douglas fir is an even larger tree. Mature stands may average 60 m in height and over 1.5 m in diameter. Logs from the mature stands (known as old growth) are highly valued for lumber and plywood, while logs from immature stands (known as secondary growth) are used for pulp and paper.

Inland from the Coast Forest, the climate changes from a wet marine climate to a semi-arid one. At this point, the Interior Forest begins. However, differences exist within the Interior Forest. The Fraser Plateau subregion is an open woodland with much smaller trees than those in the Coast Forest. The controlling factor is the dry climate. Trees in the Fraser Plateau subregion must be able to cope with drought conditions. Ponderosa pine and lodgepole pine are the most common species. They often attain heights of 25 m and a diameter of 1 m. These trees are usually converted into lumber.

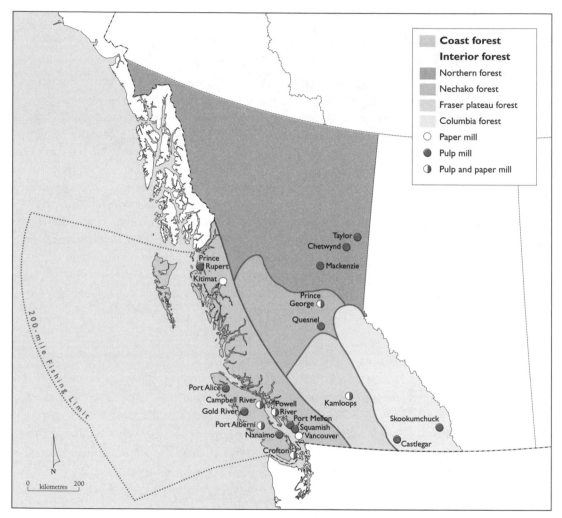

Figure 7.4 Forest regions in British Columbia. The two principal regions are the Coast Forest and the Interior Forest. The Interior Forest is subdivided into four areas that reflect the variations in growing conditions.

The Nechako Forest subregion, which lies to the north of the Fraser Plateau, receives more precipitation and experiences lower summer temperatures. The net result is more moisture for tree growth. Consequently, the forest cover is denser than that in the Fraser Plateau. A common species in the Nechako Forest is the Engelmann spruce, which can at-

tain heights of 40 m and diameters of 60 cm. Timber from this region is processed by pulp and paper mills at Prince George.

The Columbia Forest subregion lies in the easternmost area of southern British Columbia. Again, a dry climate limits tree growth. Because of widely varying terrain, the forest cover is very heterogeneous. West-

Old-growth forest on the Queen Charlotte Islands. (*Corel Photos*)

ern red cedar and western hemlock are common species, though stands of ponderosa pine and even Douglas fir are also found in this region. Large timber is sent to sawmills, while the smaller logs are used for making pulpwood.

The Northern Forest subregion is the most remote of BC's forests. High transportation costs to ship wood to world markets hinder logging. Tree growth is hampered by cool growing conditions and poorly drained land. Large blocks of land that contain muskeg are either devoid of trees or have trees with little commercial value.

Forest Industry

Of all its natural resources, the exploitation of the forest was the mainstay of the British Columbian economy. While other sectors have become equally important, the forest industry remains a cornerstone of BC's economy. British Columbia has always been oriented to **staple** (or primary) **production** to supply its resources to markets in Canada, the United States, and other foreign countries. Such a pattern of development is the mark of a hinterland economy. As with other resource hinterlands, attempts to diversify its resource-based industry by increasing the amount of processed resource products has been slow. Manufacturing increases the value of a product, such as a log, by changing it into a final product, such as a window frame. This process is called **value-added production**. Besides increasing the value of production in British Columbia, this strategy would also provide more employment. In 1997, the British Columbia government announced the Jobs and Timber Accord, an agreement it reached with the forest industry. One aspect of the accord attempts to increase the value added to forest products by requiring lumber producers to sell one-sixth of their output to local manufacturing firms, such as a furniture factory. In this way, the government hopes to increase the size of forestry's manufacturing sector and thereby generate more jobs.

At first, the forest industry consisted of small-scale logging and sawmilling operations. Prior to the Second World War, the forest industry was concentrated along the Pacific Coast. After the war, logging, sawmilling, and pulp mills began to locate in the interior. Sawmills in the interior account for about three-quarters of the lumber production. Pulp and paper mills are mainly along the West Coast and in the Prince George area of the interior. Wood manufacturing plants are heavily concentrated in the Lower Mainland. With the increased demand for forest products, logging intensified during the 1970s. By the 1980s, almost all of British Columbia's vast commercial timber lands were leased to forestry companies.

Today, ten forestry companies account for over half of the timber harvest and three-quarters of the processing of wood products. The pulp and paper companies are controlled by a few multinational corporations. Since the 1970s, Japanese firms have invested heavily in the forest industry, often as minority partners. This shift towards multinational corporations is common in resource hinterlands. The rise in multinational ownership is due to two factors: (1) the huge capital investment required to undertake megaprojects, and (2) resource hinterlands' integration into the global economy with foreign owners securing their source of raw materials.

In 1995, enormous forest reserves supported a major industry with an output valued at nearly $20 billion. Just over 95 per cent of the 60 million ha of forest lands were owned by the British Columbia government and allocated to forestry companies on long-term leases. MacMillan Bloedel, the largest forestry company in British Columbia, held about 15 per cent of the leased forest area and, like other firms, had its headquarters in Vancouver. Despite the forest industry's success, through the 1990s it became more and more difficult to obtain raw material. What was thought to be a limitless supply of wood fibre was a myth.

The Forest Myth

At the time of Confederation, the vast timber lands of British Columbia seemed endless. Today, there is a wood fibre crisis. Few forest leases for mature timber stands are available, and usually include inaccessible timber stands. The major drawback of these stands is the cost of constructing logging roads.

The rapid depletion of wood is revealed in the changes, over a five-year period, in the volume of British Columbia's exports of wood chips. In 1989, nearly 2 million tonnes were exported. By 1994, this amount had fallen by more than half to 630 000 tonnes (Statistics Canada 1998a:36). Imports of wood fibre from Alberta and Alaska, on the other hand, increased during the same period. Several factors account for this change from a surplus of wood chips in BC to a shortage. BC forestry companies' past logging practices were one factor. Driven by profit, companies harvest the more accessible (and more cheaply harvested) stands first. A second factor is the scarcity of accessible timber stands. In the past, logging companies simply obtained another timber lease, which provided them with easy access to commercial timber stands. Now, with the depletion of forest stands, fewer accessible ones are available. The third factor is the intervention of the provincial government. The provincial government, which was pushed by environmental activists over the issue of clear-cutting the ancient rainforest on Vancouver Island's Clayoquot Sound, was forced to 're-assess' its forestry position. In 1995, the provincial government introduced new restrictions on logging and, at the same time, announced the establishment of wilderness parks in areas of mature forests. The combination of these two policies has reduced the amount of Coast Forest available to forestry companies by about 10 per cent and reduced the **annual allowable cut** by 6 per cent (British Columbia Ministry of Forests 1996:5).

Forest Management

In the late 1970s, the provincial government had a system of long-term forest leases that encouraged companies to manage these lands to allow another 'crop' of trees to mature to a commercial state. The assumption was that such a management system would lead to a sustainable harvesting of the forest. Typically, these licences extend for twenty-five years, but are subject to renewal. Companies responded

by cutting the best and most accessible trees. The failure to build additional logging roads to reach new areas often led to overcutting in the more accessible timber areas. Private companies' attempts to plant seedlings have met with mixed success. In 1994, the Forest Renewal Act was passed, nearly doubling the stumpage fees the companies paid for each tree cut. This revenue is designed to finance tree planting through a government corporation, Forest Renewal BC.

Until 1995, neither the province nor companies had approached forestry from a 'serious' sustainable perspective. Now forest management practices are enforceable by law (British Columbia Ministry of Forests 1996:1). The assumption of the 1960s that 'old' timber stands would last into the next century, by which time 'natural' and 'artificial' reforestation of earlier logged areas would develop into commercial forest stands, no longer has any credence. The primary reason is that the increase in logging over the last twenty-five years was much greater than anticipated in the 1960s. This increase has reduced the mature forest stands much more quickly and has left less time for the logged areas to regenerate. By the mid-1990s, the wood fibre shortage has pushed timber prices so high that it has become economical to transport logs from Alberta's foothills to mills in the southern interior of British Columbia and from southern Yukon to mills in northern areas of the province. At some point, however, additional transportation costs will force the closure of such mills. It is hoped that forest management measures will alleviate problems related to the depletion of forest stands.

Free Trade and Forest Exports

One of the challenges facing the forest industry in BC extends beyond the province's borders and relates to trade relations between Canada and the United States. The annual value of Canada–US trade is nearly $400 billion, which is the largest in the world between two countries. The vast majority of this trade takes place without incident, but there are 'hot points' and lumber is one of them.

NAFTA has reduced trade barriers between the United States and Canada, but it has not prevented trade disputes involving lumber to erupt. Forest exports from British Columbia have already been subjected to the heavy hand of Washington. Most Canadians do not understand why NAFTA has not given Canadian products 'free' access to US markets. After all, was not 'free' access to each other's market the purpose of the original Free Trade Agreement and the subsequent NAFTA? The answer to that question is 'freer' but not 'free' trade. For example, when an American producer is adversely affected by imported products, the US company complains to the American government about the lower-priced products, claiming that such lower prices are a form of unfair trade. The company expects Washington to protect it by creating a trade barrier, which it has done on occasion. The trade agreement has a dispute-settlement mechanism, but resolving trade disputes takes time, during which the Canadian exporter loses sales and profits.

Problems over free and fair trade have been particularly frequent in the forest industry. American lumber production, which operates mainly in the Pacific Northwest and Georgia, can produce a maximum of 15 billion board feet per year. Canadian lumber production nearly doubles the American figure. In fact, lumber production from British Columbia is roughly equal to that of the entire United States (Table 7.7). The main difference is that US-produced lumber serves its domestic market, while most Canadian-produced lumber is exported to the United States, Japan, and other foreign customers. Trade disputes over lumber exports can therefore significantly affect BC's forest industry.

When American lumber producers lose market share to their Canadian counterparts, they turn to their lobby organization, the US Coalition for Fair Lumber Imports. The US political system is extremely sensitive to lobby efforts because individual members of the House of Representatives and the Senate have the power to intercede on trade matters. In addition, these elected officials rely on lobby organizations for support at election time. Since 1982, the US Coalition for Fair Lumber Imports has managed to convince the American government to reduce Canadian forest imports into the United States. In 1986, for example, they launched a trade action, claiming that Canadian softwood lumber production was subsidized by low provincial stumpage fees. In the following year, the US Department of Commerce ruled that Canadian stumpage rates constituted a countervailing subsidy. Ottawa agreed to impose a 15 per cent export tax on lumber exported to the US. British Columbia and several other provinces raised their stumpage rates to counter the American claim of low stumpage rates and to replace the fed-

Table 7.7
Lumber Production by Province, 1995

Province	Production (million board feet)	(per cent)
Newfoundland	36.3	0.1
Prince Edward Island	46.7	0.1
Manitoba	130.3	0.5
Saskatchewan	239.6	0.9
Nova Scotia	334.5	1.2
New Brunswick	1,148.2	4.2
Alberta	2,307.6	8.4
Ontario	2,584.6	9.4
Québec	6,535.1	23.7
British Columbia	14,220.4	51.5
Total	27,583.3	100.0

Source: Adapted from Statistics Canada, 1998a: Table 19.

eral export tax. In 1991, Canada terminated the 15 per cent export tax. In response, the US Coalition for Fair Lumber Imports called for another trade action to restrict the flow of Canadian lumber into the United States. In 1992, the United States government imposed a countervailing duty on Canadian lumber. Over a two-year period, the US government collected more than $800 million from Canadian exporters. In 1994, a bilateral trade panel (the dispute mechanism created by NAFTA) declared the tax invalid under NAFTA and ordered the US government to reimburse the Canadian companies. Not to be outdone, the US Coalition for Fair Lumber Imports called for new restrictions. In 1996, the US government insisted on a ceiling for Canadian lumber shipments to the US. In 1996, Ottawa and Washington reached a new agreement: a limit of 14.7 billion board feet would be allowed into the US in exchange for five years of non-interference with the lumber trade by the US government. The agreed-to figure is an average of the volume of Canadian lumber exported to the United States over the past three years. The first 650 million board feet in excess of the figure will face a US tax of $50 per 1,000 board feet. Even with this additional cost, Canadian lumber will still be competitive in the US market. Once the 650 million board feet figure is exceeded, the tax will jump to $100 per 1,000 board feet. At that tax level, Canadian lumber will not be competitive in the US market. All of these efforts by the US lumber lobby and the US government are attempts to bypass NAFTA in order to protect US lumber interests.

In the coming years, the forest industry in BC will continue to face a number of challenges. In addition to trade disputes, BC will have to contend with environmental concerns, dwindling forest resources, and value-added production. The forest industry remains an important part of BC's economy, but in order to protect it, and the economy in general, such issues will have to be addressed.

British Columbia's Urban Geography

The urban geography of British Columbia is dominated by its largest city, Vancouver (Figure 7.5). Similar to the province, Vancouver's population has increased at a rate well above the national average. In fact, it is the fastest-growing large metropolitan area in Canada. Vancouver's spectacular growth rate is closely followed by Oshawa, Toronto, and Ottawa. The next category of cities with high rates of increase are Saskatoon, Calgary, and Victoria.

Smaller urban centres in BC are also growing quickly. Since 1981, Kelowna, Chilliwack, and Nanaimo have exhibited the greatest rates of increase of all cities in British Columbia (Table 7.8). From 1991 to 1996, these three cities recorded rates of population increase of 22.1 per cent, 20.5 per cent, and 16.4 per cent

respectively. In comparison, Vancouver and Victoria had increases of 14.3 per cent and 5.7 per cent respectively.

Another striking aspect of the population geography of British Columbia is the concentration of people in the southwest corner of the province and the dominant position of Greater Vancouver within that population cluster. Greater Vancouver encompasses a number of cities and towns, including Burnaby, Coquitlam, Delta, Langley, New Westminster, North Vancouver, Richmond, Surrey, and West Vancouver. Besides Greater Vancouver, three other metropolitan centres, Victoria, Nanaimo, and Chilliwack, are located in this densely populated area of British Columbia. The Okanagan Valley's cities of Kelowna, Vernon, and Penticton constitute a modest secondary cluster.

Greater Vancouver is the dominant urban centre of British Columbia. Unlike Ontario and Québec, which each have more than one population cluster, BC can only claim the Vancouver area as a veritable population cluster. Defined in terms of a census metropolitan area, Vancouver's 1996 population was 1.8 million. Victoria, the second largest urban centre, has a much smaller population at just over 300,000 inhabitants. As the capital of the province, Victoria is a 'government' town, as well as an important tourist and service centre. Its mild climate has also attracted retired people, especially from the Prairie provinces. Kelowna, with a population of just over 136,000, falls into a distant third place, but has grown rapidly in recent years. After Kelowna, there are four cities with populations over 65,000 but under 86,000:

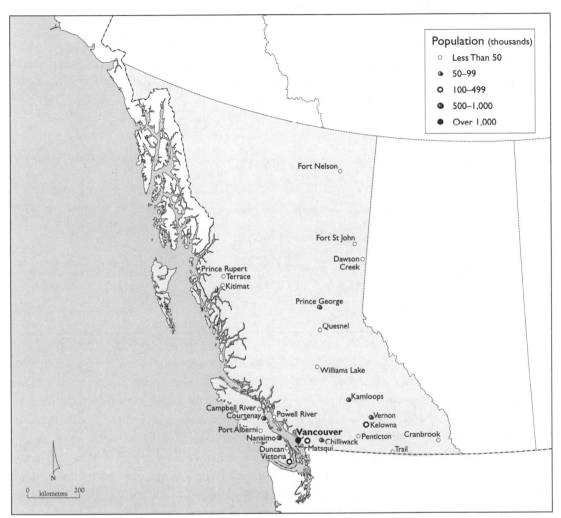

Figure 7.5 Major urban centres in British Columbia. Vancouver dominates BC's urban geography, followed by Victoria and Kelowna. BC's population is clustered in the southwest corner of the province, with smaller centres dotting the northern region.

Nanaimo, Kamloops, Prince George, and Chilliwack. Below the 60,000 mark are about a dozen smaller cities with populations of 10,000 or more (Table 7.9).

Vancouver

Vancouver has a majestic physical setting. The city, located on the shores of Burrard Inlet, lies across the water from the snowcapped peaks of the North Shore mountains. To the west lies the island-studded Strait of Georgia, while the Fraser River and its deltaic islands (flat, low islands composed of silt and clay near the mouth of a river) mark Vancouver's southern edge. Vancouver has a mild, marine climate, though some find the frequent rain and over-

Table 7.8
Major Urban Centres in British Columbia, 1981–1996

Centre	Population 1981	Population 1996	Change (per cent) 1981–96	Change (per cent) 1991–6
Chilliwack	41,471	66,254	59.8	20.5
Prince George	67,559	75,150	11.2	7.9
Kamloops	64,997	84,914	30.6	14.2
Nanaimo	57,694	85,585	48.3	16.4
Kelowna	77,468	136,541	76.2	22.1
Victoria	233,481	304,287	30.3	5.7
Vancouver	1,268,183	1,831,665	44.4	14.3

Sources: Statistics Canada, 1982:Table 1; 1997a.

cast skies unappealing. Vancouverites are rightfully proud of their city, which has one of the most beautiful settings in the world.

Vancouver emerged from its forest wilderness in a remarkably short time to become one of the great cities of North America. Few cities have grown as quickly. In 1870, the frontier settlement of Granville on Burrard Inlet had fifty residents. Little changed until the arrival of the Canadian Pacific Railway in 1886. At that time, the population of Granville had soared to about 1,000 and the residents incorporated their village as the City of Vancouver. By 1891, the newly incorporated city had reached a population of 15,000. The spectacular growth continued, partly as a result of amalgamation with neighbouring municipalities. By 1931, Vancouver's population had reached a quarter of a million. Twenty years later, Vancouver's population had doubled. By 1996,

Vancouver's population had more than tripled compared to what it was in 1951. Assuming that such a rate of population increase continues, Vancouver will exceed the 2 million mark early in the twenty-first century.

Like Montréal, much of Vancouver's commercial strength stems from its role as a trade centre. Most of the world's population is located along the Pacific Rim and Vancouver's economic well-being is closely tied to these countries. Most of BC's exports go to the United States and Japan. In 1997, three-quarters of exports from British Columbia were shipped to those two countries (British Columbia 1998b:3 of 4). Forest products made up approximately half of all 1996 exports, of which softwood lumber comprised 51 per cent, pulp 23 per cent, newsprint 9 per cent, and paper and paperboard 6 per cent. The remaining 20 per cent consisted of manufac-

Table 7.9
Smaller Urban Centres in British Columbia, 1991–1996

Centre	Population 1991	Population 1996	Change (per cent)
Dawson Creek	10,981	11,125	1.3
Kitimat	11,305	11,136	(1.5)
Fort St John	14,156	15,021	6.1
Cranbrook	16,447	18,131	10.2
Powell River	18,477	19,936	7.9
Terrace	18,908	20,941	10.8
Port Alberni	26,601	26,893	1.1
Campbell River	30,860	35,183	14.0
Williams Lake	34,690	38,552	11.1
Penticton	35,823	41,276	15.2
Courtenay	44,523	54,912	23.3
Vernon	48,139	55,359	15.0

Sources: Statistics Canada, 1997a: Table 2.

tured wood products such as doors and window frames (British Columbia 1998a:3 of 4).

Vancouver is also a transshipment point for resource products from the interior of British Columbia, Alberta, Saskatchewan, and Manitoba; a service centre; and a tourist town. Head offices of private companies (especially fishing, forest, and mining companies) and federal and provincial public offices are located in Vancouver. While Vancouver (known in 1867 as Gastown) began as a sawmill centre, these air-polluting mills relocated to other centres or were dismantled. False Creek, in downtown Vancouver, had been the prime site for processing logs, but in 1986, False Creek became the site of Expo '86 and then the site of an upscale residential and farmers' market complex known as Granville Island.

Much like British Columbia itself, Vancouver lies at a crossroads. It continues to rely on the business and trade generated by resource industries for it prosperity. However, in recent years, it has developed into a major financial and cultural centre, much like the major cities of Central Canada. With a growing population and continued growth, Vancouver may come to rival Montréal and Toronto in the coming years.

Victoria's metropolitan population is mostly of British origin despite postwar immigration of other ethnic groups. (*Corel Photos*)

British Columbia's Future

The future poses several challenges for British Columbia. The first will be to regain the high rates of economic growth and population increase experienced prior to 1998. The second challenge will be to avoid environmental deterioration of BC's major cities, parks, and wilderness areas while pursuing economic growth. The third challenge will be to reduce the mounting pressure placed on renewable resources in the province, particularly its agricultural lands, old-growth forest, and salmon stocks. Related to BC's resources is the question of industrial diversification and this re-

source hinterland's position in the Canadian, North American, and global economies. Unless BC's resource industries are able to shift to a more value-added production, additional economic growth in manufacturing is unlikely, and, with continued automation in the resource industries, fewer jobs will exist. The final and most contentious challenge will be settling Aboriginal land claims. Land-claim agreements will benefit both First Nations and BC's economy as First Nations buy land, set up businesses, and purchase goods and services. The negotiations of these agreements, however, will come up against considerable opposition because they call for the sharing of BC's

Vancouver's skyline, looking southward across the Lions Gate Bridge and over the city's famous Stanley Park. (*Corel Photos*)

resources. The idea of sharing resources, particularly fish and forest resources, is not popular among other resource users.

British Columbia's past has been based on resource development; its future may well lie in the expansion of the high technology, producer services, and manufacturing sectors. In the nineteenth century, the sheer extent of BC's natural resources suggested limitless wealth. By the end of the Second World War, harvesting of these resources had increased exponentially, forcing governments and private companies to rethink their approach to resource extraction. Now public policy emphasizes sustainable resource development—a policy which industry seems to support. But do they

support it in practice? Now that the fish and forest resources are dwindling, private companies can no longer count on expanding their production without threatening the capacity of these renewable resources to replenish. Instead, loggers and fishers will pay the price for a declining fish stock and a shrinking volume of old-growth timber. Perhaps the recognition of these resource problems suggests that it will become easier to address them. At the 1998 shareholders' meeting of MacMillan Bloedel, the president and chief executive officer, Tom Stephens, announced that the company *may* halt logging of old-growth forests (Lush 1998:B1).

How can British Columbia adjust to these

new resource constraints? Hayter (1996:101) argues that the forest industry must change its outlook from that of a staple producer to that of a processor. More specifically, forestry companies must shift their efforts from low-value processing of timber, such as lumber, to high-value processing, such as furniture. But can forestry companies make this adjustment? If private forestry companies must maintain their profits, why would they change the nature of their operations from the profitable practice of exporting lumber, pulp, and other semi-processed forest products to the uncertainties of wood manufacturing? Given that scenario, Barnes and Hayter pose two possibilities:

> The pessimistic interpretation is of an industry on the decline, fizzling out as the

resource base itself disappears. The optimistic interpretation, though, is of a kind of industrial renaissance, of a newly fashioned forest products industry that emphasizes high-value products, skilled labour, and leading edge technology (Barnes and Hayter 1997:7).

Herein lies the key to BC's complete transformation from a resource hinterland to an industrial core.

Summary

Since the Second World War, British Columbia's population has grown rapidly and its economy has diversified. Except for the late 1990s, when the downturn in the Asian eco-

Islands in Howe Sound, which extends north from West Vancouver to Squamish. (*Corel Photos*)

nomy dampened BC's exports and stalled its economic growth, British Columbia's economy outperformed the economies of other provinces. In 1996, for example, BC constituted 12.7 per cent of Canada's population, but accounted for 13.3 per cent of Canada's GDP. Both these factors indicate that British Columbia is an emerging economic powerhouse within Canada. High-technology companies, producer services firms, and resource industries combine to drive BC's economy. Trade with Asia, coupled with Asian investments in BC, have further stimulated British Columbia's economic growth. Together these economic forces have propelled BC's economy into Canada's third economic force behind Ontario and Québec.

Before BC becomes an industrial core, however, several important issues must be resolved:

- The resource economy must be based on sustainable harvesting practices.
- In order to break its dependency on the primary sector and its inevitable boom-and-bust cycles, BC must increase the processing of its natural resources to expand its manufacturing base and increase employment opportunities.
- Land-claim agreements must be negotiated with BC's First Nations to settle past obligations and remove uncertainty over land ownership.

Notes

1. Besides the issue of luring British Columbia into Confederation, there were several other political reasons for constructing a transcontinental railway. First, there was the urgent need to exert political control over the newly acquired but sparsely settled lands in Western Canada. As in British Columbia, the perceived threat to these lands was from the United States. Second, there was the need to create a larger market for manufactured goods produced by the firms in southern Ontario and Québec.

2. The exact number of people in British Columbia in 1871 is not known. How many Aboriginal peoples is a guess because their numbers declined sharply as they came into contact with European diseases. Similarly, the number of Americans and people from other countries who remained in the country after the gold rush is unknown. Certainly

most moved on to the next gold rush, but some stayed. The gold rush of 1858 may have attracted about 25,000 Americans who sailed from San Francisco to New Westminster at the mouth of the Fraser River.

3. Professor Jim Miller is a member of the Department of Geography, The University College of the Cariboo, Kamloops BC.

4. On 27 June 1997, negotiations between Canada and the United States over salmon fishing in the Pacific Northwest collapsed. Efforts to resume negotiations have failed, leaving the fish quotas at the levels established in the 1985 Pacific Salmon Treaty.

5. In 1950, Alcan secured the rights to the water in the Nechako River system until 1999. When the

British Columbia government signed this long-term agreement, they saw this industrial development as the key to opening up the province's Pacific Northwest. At that time, Victoria imagined that Kitimat would become an industrial complex with a population quickly reaching 20,000.

Key Terms

annual allowable cut

The amount of timber that can be cut under a publicly controlled timber lease; the amount of allowable cut is based on sustainable harvesting principles.

chemical pulp

A soft, shapeless mass of fine, long wood fibres produced from wood chips boiled in a chemical solution placed under pressure.

groundwood pulp

A soft, shapeless mass of fine but short wood fibres produced from wood chips that are ground by mechanical means.

producer services

Services that have enabled firms and regions to maintain their specialized roles in marketing, advertising, administration, finance, and insurance industries. Producer services are one of several parts of the growing service sector of the economy.

staple production

Resource development results in primary or staple production.

thermomechanical pulp

A soft, shapeless mass of fine, long wood fibres produced by grinding wood chips into a soft mass and then boiling the fibrous mass.

Tragedy of the Commons

The destruction of renewable resources because of the absence of collective control over these resources, which are available to all. The problem lies in individuals maximizing the use of such resources for personal gain; such usage, in total, overwhelms the capacity of renewable resources to maintain and regenerate themselves.

value-added production

Manufacturing that increases the value of primary (staple) goods.

References

Bibliography

Barman, Jean. 1996. *The West Beyond the West: A History of British Columbia*, 2nd edn. Toronto: University of Toronto Press.

Barnes, Trevor, Roger Hayter, and E. Grass. 1990. 'MacMillan Bloedel: Corporate Restructuring and Employment Change'. In *The Corporate Firm in a Changing World Economy*, edited by M. De Smidt and E. Wever, 145–65. London: Routledge.

Bone, Robert M. 1992. *The Geography of the Canadian North*. Toronto: Oxford University Press.

Bradbury, John. 1982. 'British Columbia: Metropolis and Hinterland in Microcosm'. In *Heartland and Hinterland: A Geography of Canada*, edited by L.D. McCann, Chapter 10. Scarborough: Prentice-Hall.

British Columbia. 1998a. *British Columbia Home Page: Quick Facts—the Economy* [online database], Victoria. Searched 31 August 1998;

<URL:http://www.bcstats.gov.bc.ca/data/QF_econo.HTM#for.

_____. 1998b. Quick Facts—Statistical Appendix [online database], Victoria. Searched 31 August 1998; <URL:http://www.bcstats.gov.bc.ca/data/QF_stats.HTM.

_____. 1998c. Quick Facts—People [online database], Victoria. Searched 31 August 1998; <URL:http://www.bcstats.gov.bc.ca/data/QF_peopl.HTM>:4 of 7.

British Columbia Ministry of Forests. 1996. *Forest Practice Code:Timber Supply Analysis*. Victoria: Ministry of Forests.

Canada. Department of Fisheries and Oceans. 1994. *Canadian Fisheries Statistical Highlights 1992*. Ottawa: Department of Fisheries and Oceans.

Cassidy, Frank. 1992. 'Aboriginal Land Claims in British Columbia: A Regional Perspective'. In *Aboriginal Land Claims*, edited by Ken Coates, 10–43. Toronto: Copp Clark.

Cernetig, Miro. 1997. 'BC Makes Peace with Alcan'. *The Globe and Mail* (6 August):A1, A5.

Christensen, Bev. 1995. *Too Good to Be True: Alcan's Kemano Completion Project*. Vancouver: Talonbooks.

Davis, H. Craig, and Thomas A. Hutton. 1994. 'Marketing Vancouver's Services to the Asia Pacific'. *The Canadian Geographer* 38, no. 1: 18–28.

Department of Fisheries and Oceans. 1996. *Fraser River Pink: Report of the Fraser River Action Plan Fisheries Management Group*. Ottawa: Department of Fisheries and Oceans.

Edginton, David W. 1994. 'The New Wave: Patterns of Japanese Direct Foreign Investment in Canada During the 1980s'. *The Canadian Geographer* 38, no. 1:28–36.

_____. 1996. 'Japanese Real Estate Investment in Canadian Cities and Regions, 1985–1993'. *The Canadian Geographer* 40, no. 4:292–305.

Farley, A.L. 1979. *Atlas of British Columbia*. Vancouver: University of British Columbia Press.

Forward, Charles, N., ed. 1987. *British Columbia: Its Resources and People*. Victoria: Department of Geography, University of Victoria.

Francis, R. Douglas, Richard Jones, and Donald B. Smith. 1996. *Destinies: Canadian History Since Confederation*, 3rd edn. Toronto: Harcourt Brace.

Harris, Cole. 1997. *The Resettlement of British Columbia: Essays on Colonialism and Geographical Change*. Vancouver: University of British Columbia Press.

Hayter, Roger. 1992. 'The Little Town That Did: Flexible Accumulation and Community Response in Chemainus, British Columbia'. *Regional Studies* 26:647–63.

_____. 1996. 'Technological Imperatives in Resource Sectors: Forest Products'. In *Canada and the Global Economy: The Geography of Structural and Technological Change*, edited by John N.H. Britton, Chapter 6. Montréal–Kingston: McGill-Queen's University Press.

Howard, Ross. 1996. 'BC Land-Claim Stakes Highest Yet'. *The Globe and Mail* (16 January):A5.

Hutchinson, Bruce. 1942. *The Unknown Country: Canada and Her People*. Toronto: Longmans, Green and Company.

Koroscil, Paul, ed. 1991. *British Columbia: Geographical Essays in Honour of A. McPherson*.

Burnaby: Department of Geography, Simon Fraser University.

Liu, Xiao-Feng, and Glen Norcliffe. 1996. 'Closed Windows, Open Doors: Geopolitics and Post-1949 Mainland Chinese Immigration to Canada'. *The Canadian Geographer* 40, no. 4: 306–19.

Lush, Patricia. 1998. 'MacBlo Posts Profit, Sells MB Paper'. *The Globe and Mail* (24 April):B1.

McKee, Christopher. 1996. *Treaty Talks in British Columbia: Negotiating a Mutually Beneficial Future*. Vancouver: University of British Columbia Press.

McVey, Wayne W., and W.E. Kalbach. 1995. *Canadian Population*. Toronto: Nelson Canada.

Marchak, M. Patricia. 1995. *Logging the Globe*. Montréal–Kingston: McGill-Queen's University Press.

Marsh, James H. ed. 1988. *The Canadian Encyclopedia*, 2nd edn. Edmonton: Hurtig Publishers.

Miller, Jim. 1998. Personal communications, 28 April. University College of the Cariboo, Kamloops, British Columbia.

Robinson, J. Lewis, ed. 1972. *British Columbia*. Studies in Canadian Geography. Toronto: University of Toronto Press.

———, and W.G. Hardwick. 1973. *British Columbia: 100 Years of Geographical Change*. Vancouver: Talonbooks.

Saku, James C., Robert M. Bone, and Gérard Duhaime. 1998. 'Toward an Institutional Understanding of Comprehensive Land Claim Agreements in Canada'. *Études/Inuit/Studies* 22, no. 1:109–21.

Stanford, Quentin H., ed. 1998. *Canadian Oxford World Atlas*, 4th ed. Toronto: Oxford University Press.

Statistics Canada. 1982. *Census Metropolitan Areas and Census Agglomerations with Components*. Catalogue no. 95-903. Ottawa: Ministry of Supply and Services Canada.

———. 1988a. Canadian Statistics—Labour Force, Employed and Unemployed [online database], Ottawa. Searched 2 September 1998; <URL: http://www.statcan.ca/english/Pgdb/People/Labour/labor07a.htm>:2 pp.

———. 1988b. Labour Force Characteristics for Both Sexes, Aged 15 and Over [online database], Ottawa. Searched 12 August 1998; <URL:http://www.statcan.ca/english/econoind/lfsadj.htm>:2 pp.

———. 1988c. 1981–1996 Census: Labour Force Activity [online database], Ottawa. Searched 2 September 1998; <URL:http:www.statcan.ca/english/census96/mar17/labour/table6/t6p59s.htm>:1 p.

———. 1992. *Mother Tongue*. Catalogue no. 93-313. Ottawa: Industry Canada.

———. 1996. *Labour Force Annual Averages 1995*. Catalogue no. 71-220-XPB. Ottawa: Statistics Canada.

———. 1997a. *A National Overview: Population and Dwelling Counts*. Catalogue no. 93-357-XPB. Ottawa: Industry Canada.

———. 1997b. 1996 Census: Nation Tables—Population by Mother Tongue, Showing Age Groups, for Canada, Provinces and Territories, 1996 Census—20% Sample Data, 2 December 1997 [online database], Ottawa. Searched 15 July 1998; <URL:http://www.statcan.ca/english/census96/>:3 pp.

_____. 1997c. The Daily—1996 Census: Mother Tongue, Home Language and Knowledge of Languages, 2 December 1997 [online database], Ottawa. Searched 14 July 1998; <URL: http://www.statcan.ca/Daily/English/>:15 pp.

_____. 1998a. *Canadian Forestry Statistics 1995*. Catalogue no. 25-202-XPB. Ottawa: Ministry of Industry, Science and Technology.

_____. 1998b. The Daily—1996 Census: Aboriginal Data, 13 January 1998 [online database], Ottawa. Searched 14 July 1998; <URL:http://www.statcan.ca./Daily/English/>:10 pp.

_____. 1998c. The Daily—1996 Census: Ethnic Origin, Visible Minorities, 17 February 1998 [online database], Ottawa. Searched 16 July 1998; <URL:http://www.statcan.ca/Daily/English/>:21 pp.

_____. 1998d. *Canadian Economic Observer*. Catalogue no. 11-010-XPB. Ottawa: Statistics Canada.

Wilkinson, Bruce W. 1997. 'Globalization of Canada's Resource Sector: An Innisian Perspective'. In *Troubles in the Rainforest: British Columbia's Forest Economy in Transition*, edited by Trevor J. Barnes and Roger Hayter, 131–47. Canadian Western Geographical Series, vol. 13. Victoria: Western Geographical Press.

Further Reading

Barnes, Trevor, and Roger Hayter. 1997. *Trouble in the Rainforest: British Columbia's Forest Economy in Transition*. Canadian Western Geographical Series, vol. 33. Victoria: Western Geographical Press.

Harris, Cole. 1997. *The Resettlement of British Columbia: Essays on Colonialism and Geographical Change*. Vancouver: University of British Columbia Press.

Wynn, Graeme, and Timothy Oke, eds. 1992. *Vancouver and Its Region*. Vancouver: University of British Columbia Press.

Chapter 8

Overview

Western Canada occupies the vast interior of Canada. Within this geographic region there are two distinct zones: a sparsely populated northern resource hinterland in the Canadian Shield, and a more densely populated southern zone in the Interior Plains. Western Canada's economy is in a state of change, shifting from one anchored in primary production to one that is more involved in the processing of these products. Alberta leads the way in this transformation, powered by its oil and gas industry. Similar changes are underway in other sectors of the economy, especially agriculture. This notion of transition is the *Key Topic* in this chapter.

Objectives

- Describe Western Canada's physical and historical geography.
- Present the basic characteristics of its population and economy.
- Examine Western Canada's population and economy within the context of the region's physical setting and dry continental climate.
- Explore Western Canada's changing position within the core/periphery model—from a resource frontier to an upward transitional region.
- Focus on the transition in Western Canada's agricultural sector, and to a lesser degree, in its resource and manufacturing sectors.

Western Canada

Introduction

Western Canada is at a turning point. Behind it lies its history as Canada's chief exporter of grain, oil, and other primary products. Ahead of it is the prospect of playing a different economic role, one that is much more involved in the processing of Western Canada's agricultural and resource products. This chapter examines the factors underlying this radical departure for Western Canada from its traditional role as a resource hinterland, especially an agricultural hinterland. The *Key Topic*, 'Transition', looks at the economic and political forces propelling Western Canada's economy to diversify and resulting in land-use changes in the Canadian Prairies. In a broad context, this transition is illustrated in Friedmann's core/periphery model as the upward transitional region (see Chapter 1).

Western Canada Within Canada

Western Canada is part of the vast western interior of North America (Figure 1.1). This huge area, comprising the three western provinces—Alberta, Saskatchewan, and Manitoba—contains many natural resources. These resources range from the fertile agricultural lands of the Canadian Prairies to the commercial timber stands in the northern coniferous forest. As well, there is considerable mineral wealth in the Interior Plains and the Canadian Shield.

In terms of economic output, Western Canada falls behind Ontario and Québec. Overall, Western Canada's economy accounted for just over 17 per cent of Canada's GDP in 1996 (Figure 8.1). Western Canada's economic performance translates into a $116 billion economy. An important measure of its economic well-being is Western Canada's low unemployment rate. In 1996, Western Canada had, at 7.1 per cent, the lowest unemployment rate of the six regions, well below the national average of 10 per cent (Figure 8.1).

Since the Second World War, most of the region's economic and demographic growth has occurred in oil-rich Alberta. While oil and gas are the region's most valuable non-renewable assets, agriculture remains its key renewable resource. A basic problem confronting farmers and resource companies in Western Canada is the high cost of transportation, particularly the long rail distance their products must travel to reach ocean ports, from where they are then shipped to world markets. Once

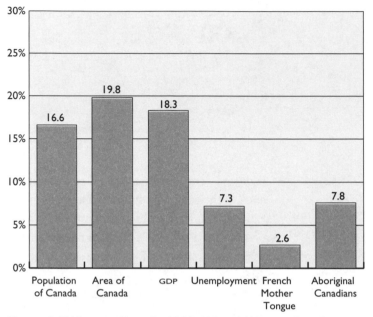

Figure 8.1 Western Canada, 1996. Although Western Canada covers a large area of the country, the region's economic and population strength are more modest. The region has fewer French Canadians than other parts of the country, but a significant number of Aboriginal Canadians.

Sources: Statistics Canada, 1997a:Table 1; 1997b:1 to 3 of 3; 1997c:4 and 5 of 15; 1998a:Table 39; 1998b:1 and 2 of 10; 1998c:15 to 20 of 21; and Stanford, 1998:185.

products are aboard ocean vessels, the transportation cost per kilometre drops drastically. Overcoming the cost of shipping products to ocean ports has been a persistent struggle in Western Canada's development of its agricultural, forest, and mineral resources (Vignette 8.1).

The population of Western Canada totals nearly 5 million. The vast majority lives in the **regional core**, an area of major cities, towns, and villages, located in the southern half of the region. Agricultural, industrial, and service activities are well established within this population core. Edmonton, Calgary, Winnipeg, Saskatoon, and Regina are the leading CMAs. In

the northern half of Western Canada lies its regional hinterland, where most settlements are linked to resource developments, with some consisting of Aboriginal reserves. Western Canada's northern hinterland has less than 5 per cent of the region's population, most of whom live in resource towns, such as Fort McMurray and Thompson. Most of the small Aboriginal settlements in this hinterland owe their origins to the fur trade, but have no economic function today.

Western Canada's Physical Geography

Western Canada, which extends over nearly 20 per cent of Canada, has two physiographic regions—the Interior Plains and the Canadian Shield (Figure 2.1). In addition, a thin portion of the Cordillera is found along the western edge of this region, where the Rocky Mountains form a natural and political border between southern Alberta and British Columbia. Each physiographic region has a particular set of geological conditions, physical landscapes, and natural resources. (See also 'The Interior Plains' and 'The Canadian Shield' in Chapter 2.)

The tiny section of the Cordillera, along the eastern flank of the Rocky Mountains in Alberta, provides logging and mining opportunities for Western Canada. However,

Vignette 8.1

The Hudson Bay Railway

In the 1920s, Prairie farmers thought they could solve their high transportation costs by constructing a rail line from The Pas to York Factory on Hudson Bay, thereby minimizing the length of rail transport. Centuries before, the Hudson's Bay Company had loaded its furs on ships destined for London, England, at York Factory, so why not use the same route to ship grain to the United Kingdom and other European countries? Farmers assumed that they would save considerably on transportation costs because such a route would greatly reduce rail distance from most shipping points in Saskatchewan and Manitoba to an ocean port. Unfortunately, the cost of ocean transport from Churchill to European ports was so much higher than the shipping costs from Montréal that there were no savings for western farmers. The higher cost was due to marine insurance. Insurance companies were concerned that grain ships travelling through the icy waters of the North Atlantic, Hudson Strait, and Hudson Bay might be damaged by ice or sunk as a result of a collision with an iceberg. Insurance companies therefore set much higher marine insurance rates for grain ships travelling through Hudson Strait and Hudson Bay than for those ships departing for Europe from the port of Montréal. High transportation costs for Prairie farmers would remain a problem.

York Factory, situated at the mouth of Hayes River in Manitoba, was the historic centre of the Hudson's Bay Company's fur trade. (Courtesy Bill Barr)

the main attraction of this slice of the Cordillera is its spectacular mountain landscape. This mountainous terrain has two internationally acclaimed parks, Banff National Park and Jasper National Park, which attract visitors from around the world. Calgarians are especially fortunate in having easy access to the Kananaskis Country Provincial Park, where a number of mountain recreational activities—camping, hiking, or skiing—are available.

In the Interior Plains region the sedimentary rocks contain valuable deposits of fossil fuels. By value, the four leading mineral resources are oil, gas, coal, and potash. Most petroleum production occurs in a geological structure known as the **Western Sedimentary Basin**, which underlies most of Alberta and

Lake Louise is one of Banff National Park's most stunning features. The lake's famous turquoise colour is caused by fine rock particles that are carried down by melt water from glaciers. (*Corel Photos*)

thermal electricity. In northeastern Alberta, the huge petroleum reserves in the Athabasca tar sands are exploited by surface mining techniques, such as hydro-transport in which the oil sands are mixed with extremely hot water and transported to the upgrader plant by pipeline.

In the Canadian Shield, which extends into Manitoba and Saskatchewan, rocky terrain makes cultivation virtually impossible. While forestry does take place along the southern edge of the Canadian Shield, particularly in southeastern Manitoba, large-scale commercial enterprises are generally limited to mining and the production of hydroelectricity. In northern Saskatchewan, uranium companies produce most of Canada's uranium from open-pit and underground mines. Large-scale hydroelectric dams and generators are located on Manitoba's northern rivers, particularly the Nelson River.

All of Western Canada has a continental climate characterized by extreme daily and seasonal fluctuations in temperature and low annual precipitation (Figure 2.5). The region is dry not only because of its distance from the ocean, but also because the mountain ranges in the Cordillera block the eastward movement of mild, moist Pacific air masses. During the winter, the Arctic air masses often dominate

portions of British Columbia, Saskatchewan, and Manitoba. However, not all mineral extraction takes place deep in these sedimentary rocks. A few deposits are exposed at the surface of the ground. In southeastern Saskatchewan, brown coal is extracted through open-pit mining and then burned to produce

weather conditions in the Prairies, placing the region in an Arctic 'deep-freeze' (Figure 2.3). A combination of strong winds and sub-zero temperatures can produce blizzard-like weather. (See 'Prairies Zone' in Chapter 2 for further details.)

Precipitation is a critical climatic element in Western Canada. Not only is the annual precipitation in Western Canada low, it also varies widely from a high of 550 mm in southern Manitoba to less than 400 mm in the dry lands of southern Alberta (Figure 2.5). In addition to this spatial variation, precipitation varies from year to year, causing so-called wet and dry years. For instance, in the 1930s, a series of dry years resulted in the disastrous Dust Bowl, a period of severe drought that devastated Prairie farming and led to extensive wind erosion of the topsoil. The Dust Bowl drove thousands of homesteaders off the land in the short-grass Prairies. Annual variations in precipitation have led Prairie farmers to coin the term 'Next Year Country', a term that refers to the bad harvest years, which prompt farmers to hope for a better crop the following year. (See 'Next Year Country' later in this chapter for further information.)

In Western Canada the natural vegetation and the soil, and therefore the success of commercial crops, are largely determined by the

Dust storm southwest of Lakenheath near Assiniboia, Saskatchewan in 1934. The drought that struck Western Canada in the 1930s was so severe that some grain farmers abandoned their homesteads. (Courtesy Saskatchewan Archives Board R-A4665)

evapotranspiration rate, which is a combination of precipitation and temperature (See 'Prairies Zone' in Chapter 2). The evapotranspiration rate varies within Western Canada and therefore creates differences in the region's soil and vegetation. In the northern parts of Western Canada, in the Canadian Shield, cooler weather leads to a lower evaporation rate, resulting in sufficient moisture for tree growth. In this area the Boreal Forest grows in a podzolic soil (Figures 2.7 and 2.8). The only exception is in the Peace River country, a gently rolling plain where black chernozemic soils are found beneath an aspen parkland. Further south, where the evaporation rate is much higher, the

podzolic soil gives way to chernozemic soils, which are more favourable for agriculture.

Indeed, the soil and natural vegetation of the southern part of Western Canada have had a significant impact on the region's role as an agricultural heartland. This southern portion, in the Interior Plains, has two natural vegetation zones (Parkland and Grassland) and three chernozemic soil zones (black, dark brown, and brown) (Figures 2.7 and 8.2). Just south of the Boreal Forest, podzolic soil gives way to the black chernozemic soil that supports a parkland vegetation. This Parkland is a transition zone between the Boreal Forest and the Grassland natural vegetation zone, which is

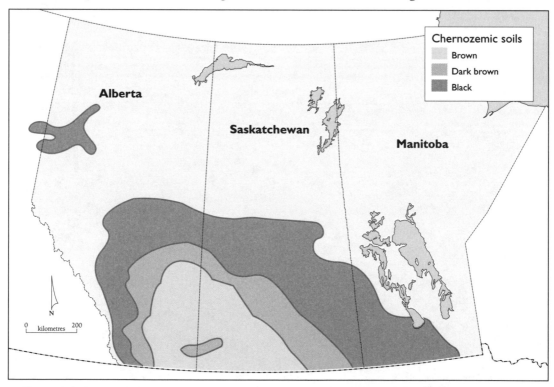

Figure 8.2 Chernozemic soils in Western Canada. There are three types of chernozemic soils in the Canadian Prairies: black, dark brown, and brown. The differences in colour are due to varying amounts of humus in the soil. The Peace River country's soil, formed under an aspen forest, is 'degraded' black soil.

located a little further south and is supported by dark brown and brown chernozemic soils. Within the Grassland, the evaporation rate increases toward the American border, causing this mid-latitude grassland to change from tall grass to short grass as one moves southward. Tall-grass natural vegetation occurs in a wide arc from southwestern Manitoba to Saskatoon to Edmonton. Here there are dark-brown chernozemic soils, except in southern Manitoba, where black soils occur because of higher annual precipitation. This tall-grass natural vegetation zone and the parkland to its north comprise the area known as the **Fertile Belt**, where farming has traditionally been more successful. South of this fertile arc is the area of short-grass natural vegetation and brown chernozemic soils known as the **Dry Belt** (Vignette 8.2). This semi-arid area has presented the highest risk for grain farming, so much of the land is now worked in fallow

rotation (a crop is planted every second year) or used for pasture.

Western Canada's Historical Geography

Western Canada's history began long before the three Prairie provinces became part of Canada. In fact, Western Canada's recorded history goes back to the fur-trading days. Beginning in 1670, the Hudson's Bay Company administered for 200 years much of Canada's western interior. This area was part of Rupert's Land (all the land draining into Hudson Bay). In 1821, when the company merged with its rival, the North West Company, the Hudson's Bay company acquired control over more land, known as the North-Western Territory (lands draining into the Arctic Ocean). Before 1870, when Canada purchased these lands from the British Government (Figure 3.4), the

---

Vignette 8.2

Palliser's Triangle

In 1857, the Palliser Expedition set out from England to assess the potential of Western Canada for settlement. This expedition spanned three years (1857–60). In his report to the British government, Palliser identified two natural zones in the Canadian Prairies. The first zone was described as a sub-humid area of tall grasses and parkland vegetation, while the second, located further south, was described as a semi-arid area with short-grass vegetation. Palliser considered the area of tall grasses and parkland to be suitable for agricultural settlement. He

named this area, extending from central Alberta to southern Manitoba, the Fertile Belt. Palliser believed that the semi-arid zone, located in southern Alberta and Saskatchewan, was a northern extension of the Great American Desert. According to Palliser, these semi-arid lands were unsuitable for agricultural settlement. The semi-arid area described by Captain Palliser became known as Palliser's Triangle—its geographic extent corresponds closely with the Dry Belt, as shown in Figure 8.4.

Hudson's Bay Company used these lands exclusively for the fur trade.

The land in Western Canada began to be used for purposes other than fur trading at the beginning of the nineteenth century. In 1810, Lord Selkirk, a Scots nobleman who was concerned with the plight of poor Scottish crofters (peasants) evicted from their small holdings, acquired land in the Red River Valley from the Hudson's Bay Company. The first Scottish settlers arrived in 1812 to form an agricultural settlement near Fort Garry, the principal Hudson's Bay trading post in the region. This settlement became known as the Red River Colony. Selkirk's settlers, however, faced an unfamiliar and harsh environment and had great trouble establishing an agricultural colony. Over the years, many gave up and left for Upper Canada and the United States.

At the same time, many former officers and servants of the Hudson's Bay Company, along with their Indian wives and children, settled at Fort Garry. In addition to these English-speaking people were the French-speaking Métis who had worked for the North West Company. Because the Métis were Catholic and spoke French, they formed a separate cultural group within the Red River Colony. After the consolidation of the Hudson's Bay Company and the North West Company in 1821, many who worked for the North West Company were no longer employed by the new company. Many Métis, particularly those who were French-speaking, settled in the Red River Colony, where they turned their attention to subsistence farming, freighting, and buffalo hunting.

Agricultural Potential

Despite the modest settlement near Fort Garry and the thriving fur-trade, little was known about the geography of the Canadian West in the mid-nineteenth century. To the south, in the United States, American settlers had begun to occupy land west of the Mississippi River. By 1854, a railway stretched across the United States from New York through Chicago to St Paul on the Mississippi River. In 1858, Minnesota had a sufficiently large population to warrant statehood. By then, American homesteaders began moving beyond the Mississippi into the northern Great Plains, including the Red River Valley. West of the Red River Valley, however, were the dry lands of the Great Plains, which limited further settlement. In fact, about thirty years earlier American explorers had dubbed Montana part of 'the Great American Desert'. Blocked to the west by these dry lands of North Dakota and Montana, settlers began to look northward to the unoccupied lands of British North America.

In 1857, the British government and the Royal Geographical Society sponsored an expedition into the Canadian West. Their central task was to determine the suitability of the Canadian West for agricultural settlement. John Palliser, an explorer, led the British North American Exploring Expedition. After his party arrived from England, they quickly travelled by rail from New York to St Paul and then by steamboat to the Red River Colony. They travelled by horseback from Fort Garry across the Canadian Prairies to the Rocky Mountains. Palliser reported that there was fertile land in much of the western interior, but that the land in southern Alberta and Saskatchewan near the border with Montana was far too dry for farming (Vignette 8.2). Palliser believed the Great American Desert that American explorers had already identified extended into the grasslands of southern Alberta and Saskatchewan. Named after him, this semi-arid area is now known as **Palliser's Triangle**. Another expedition in 1858, led by Henry Hind, a geologist and naturalist, confirmed that the parkland (the natural vegetation zone between the grasslands and the boreal forest) offered the best land for agricultural settlement.

Opening the West

By the middle of the nineteenth century, arable land in Upper Canada was in short supply. Some settlers began to look to the Canadian West for new land. During the negotiations with Britain over Confederation, the subject of the annexation of Rupert's Land into the Dominion of Canada arose, and provision was made in the British North America Act for its admission into Canada. In 1869, the Hudson's Bay Company signed the deed of transfer, surrendering to Great Britain its chartered territory for £300,000—with the exception of the lands surrounding its posts and about 1 133 160 ha of farmland. In 1870, Great Britain transferred Rupert's Land to Canada.

In 1869, Canada's government sent surveyors into the Red River Colony to prepare a land registry system for the expected influx of settlers. The surveyors employed a grid system known as a township survey. A township con-

sisted of 93 km², or thirty-six sections. Each section was subdivided into four quarter sections of 65 ha each. Each settler would receive a quarter section and would be required to till the land and build a house on that section. The opening of Western Canada for agricultural settlement officially began in 1870 and continued until 1914, when the First World War halted the influx of European immigrants into Western Canada. Following the First World War, veterans were encouraged to establish homesteads, especially in the Peace River country, where there remained some arable land.

Original Inhabitants

When the Canadian government sent surveyors to the west in 1869, it was not uninhabited. Many Aboriginal peoples lived there, but the new agrarian economy would quickly marginalize them and overtake their lands. In 1867, the Red River Colony had a population of nearly 12,000 people, mostly Métis. The arrival of land surveyors and settlers led the Métis, under the leadership of Louis Riel, to mount the Red River Rebellion in 1869. (See 'The Red River Rebellion' in Chapter 3 for more details.) The Métis wanted to negotiate the terms of entry into Canada from a position of strength; that is, as a government. The Métis obtained major concessions from Ottawa— they were guaranteed ownership of land, recognition of the French language, and permission to maintain Roman Catholic schools. In 1870, the fur-trading district of Assiniboia became the province of Manitoba. But the rebels' victory was hollow. As the newcomers poured into Manitoba, the Métis society was overwhelmed.

Many Métis left the colony to search for a new place to settle in the Canadian West. Such a place was Batoche, just north of the site where the city of Saskatoon now stands. Within fifteen years, however, settlers would again encroach upon the Métis' agricultural settlement. In 1885, as before, the Métis

The execution of Métis leader Louis Riel remains a subject of much debate today. *(National Archives)*

staged a rebellion led by Louis Riel. This time, the Canadian army defeated the Métis at the Battle of Batoche and Louis Riel was captured, found guilty of treason, and hanged. To this day, Riel remains a controversial figure in Canadian history—a traitor to some, he remains a hero to the Métis.

The experiences of other Aboriginal peoples during the early period of western settlement were somewhat different. Indian tribes, such as the Blackfoot, had roamed across the Canadian Prairies and the northern Great Plains of the United States long before the arrival of European explorers, fur traders, and settlers. The tribes were nomadic and hunted buffalo. By the 1870s, the buffalo had virtually disappeared from the Prairies, leaving the Prairie Indians destitute. By then, life for the Blackfoot, Blood, Plains Cree, Peigan, and Saulteaux tribes became almost unbearable. They had little choice but to sign treaties with the federal government. Between 1873 and 1877, all the tribes (except the Cree) signed numbered treaties in exchange for reservations, cash gratuities, annual payments in perpetuity, the promise of educational and agricultural assistance, and the right to hunt and fish on Crown land until such land was required for other purposes. In 1882, impending starvation for his people also forced the Cree leader, Big Bear, to accept treaty conditions. Over the next few years, however, the Cree sought other concessions from the federal government. When these efforts failed, Cree warriors supported the doomed Métis rebellion in 1885 by attacking several settlements, including Fort Pitt, the Hudson's Bay post on the North Saskatchewan River near the Alberta–Saskatchewan border.[1]

Treaties with Ottawa offered the Indians prospects for survival and time to find a place in a new economy, but the treaties also made them wards of the Crown. Living on reserves, Indians were isolated from the evolving Canadian society and became increasingly dependent on the federal government. Further north, the Woodland Crees and Dene (Chipewyan) tribes who lived in the boreal forest were not as affected by the encroachment of western settlers. Although they too signed treaties, these northern Indians continued their migratory hunting and trapping lifestyle well into the next century. In the 1950s, their dependency on Ottawa grew with the demise of the fur trade and their subsequent relocation to settlements.[2]

Canadian Pacific Railway

Once treaties ensured the availability of land for homesteading, the Canadian government needed to make this land more accessible for new settlers. The next step in the plan to open Western Canada to agricultural settlement was a transcontinental railway. Macdonald's vision of Canada extending from the Atlantic to the Pacific hinged on a transcontinental railway. There were already three transcontinental railways in the United States. Without a Canadian counterpart, Ottawa feared the worst; namely,

that the West would be lost to the Americans. Even if Canada could retain its western territories in the absence of a Canadian transcontinental railway, the north-south transportation pull exerted by the American railways would prevent Ontario's fledgling industrial core from reaching the market in Western Canada, and Western settlers would be unable to ship their products to eastern markets. For instance, in 1870, the new province of Manitoba was linked by steamboat to the rail centre of Fargo, North Dakota. Only from Fargo could passengers and freight from Manitoba quickly reach Toronto, Ottawa, and Montréal.

British and Canadian companies were not interested in a risky railway construction project across Canada unless they could obtain substantial financial assistance from Ottawa. Two reasons accounted for their lack of interest: the Canadian Shield and the Cordillera were two formidable (and therefore costly) barriers to overcome in building a railroad. Indeed, physical geography posed a much greater challenge to Canadian railway builders than to their American counterparts. In 1881, Ottawa announced generous terms: the Canadian Pacific Railway Company was awarded a charter, whereby the company received $25 million from the federal government, 1000 km of existing railway lines in eastern Canada owned by the federal government, and over 10 million ha of Prairie land in alternate square-mile sections on both sides of the railway to a maximum depth of 39 km. The terms were successful—the Canadian Pacific Railway was completed in 1885.

Settlement of the Land

The settling of Western Canada marks one of the world's great migrations and the transformation of the region into an agricultural resource frontier. Under the Dominion Land Act that Ottawa had passed in 1872, homesteaders were promised 'cheap' land in Manitoba—by building a house and cultivating some of the land, they could obtain 65 ha of land for only $10. Following 1872, an influx of prospective homesteaders began arriving, most coming from Ontario and, to a lesser degree, the Maritimes, Québec, and the United States. When the Canadian Pacific Railway was completed, the settlement of Saskatchewan and Alberta began. Many homesteaders now came from Great Britain.

By 1896, the federal government sought to increase immigration to Western Canada by promoting Western Canada in Great Britain and Europe as the last agricultural frontier in North America. The Canadian government initiated an aggressive campaign, administered by Clifford Sifton, Minister of the Interior, to lure more settlers to the Canadian west. Thousands of posters, pamphlets, and advertisements were sent and distributed in Europe and the United States to promote free homesteads and assisted passages. Prior to 1896, most immigrants came from the British Isles or the United States—these were 'desirable' immigrants. Sifton's campaign, however, cast a wider net to areas of central and eastern Europe that were not English-speaking and therefore provided 'less desirable' immigrants. The strategy generated considerable controversy among some English-speaking Canadians who believed in the racial superiority of British people.

Nevertheless, Clifford Sifton's efforts paid off. At the end of the last century, the Canadian Prairies had few settlers beyond Manitoba, and most of them had taken land near the Canadian Pacific Railway. Following the recruitment campaign, a flood of settlers arrived and the land was quickly occupied. Thus began the great migration to Western Canada. After 1896, the majority of settlers—about 2 million—were central or eastern European from countries like Germany, Russia, and the Ukraine. This large influx of primarily non-English-speaking immigrants led to a quite different cultural makeup in Western Canada from that in Central Canada, where a French/English composition had developed. By 1905, Alberta and Saskatchewan had sufficient numbers to warrant provincial status. By the outbreak of the First World War, the region of Western Canada was settled.

Life was not easy for homesteaders. Many were ill prepared for farming, let alone farming in a dry continental environment. Securing supplies of wood and water often posed a problem. Those settlers who could not afford to import lumber were forced to live in sod houses and burn buffalo chips and cow dung for heat. While many members of ethnic and religious groups settled together in the same area, forming communities, isolation still posed a problem for many. The land survey system encouraged a dispersed rural population; that is, individual farmsteads rather than rural villages. As a result, farm families sometimes did not visit the town or see their neighbours for weeks or even months. Such isolation was particularly hard on farm wives. In spite of these difficulties, the land was set-

Vignette 8.3

The People of Western Canada

In 1996, nearly 97 per cent of the residents in Western Canada declared that English was spoken at home, while the remainder spoke French. Yet, unlike other regions of Canada, those declaring English or French ancestry constituted barely half of the population. This ethnic composition is a result of the influx of non-English-speaking settlers from Europe and Russia that arrived in Western Canada about 100 years ago. Today, German and Ukrainian are still spoken, but by new immigrants from these countries and within cultural organizations, which, among their many activities, conduct language classes. Within a remarkably short span of time, cultural and linguistic acculturation has created an English-speaking society in which the multicultural aspects are still evident.

tled, towns sprang up, and institutions were created to meet the local and regional needs. In short, a new society was in the making.

By 1921, there were over 250,000 farms in Western Canada. Homesteaders now had to turn to the Peace River country for arable land. The Peace River country, part of the high Alberta Plain, is much further north and its short growing season makes agriculture risky. Even so, the Peace River country has a climate and soils that allow for mixed farming (a combination of grain and hay crops with livestock). The section of the Rocky Mountains located to the west of the Peace River country is considerably lower and thus allows more rainfall from Pacific air masses to reach the Peace River country. The final settlement of the Peace River country took place after the First World War, when returning soldiers were encouraged to settle there.

Emergence of an Agricultural Economy

By the early-twentieth century a Prairie agricultural economy had emerged. Based on a single staple—grain—this economy had to contend with fundamental geographic weaknesses. The first of these was its geographic position within the interior of Canada. Western Canada's geographic isolation from the rest of the world was partly overcome by the building of a railway system across the Prairies. The purpose of the CPR was twofold: (1) to bring settlers to the West, and (2) to allow these settlers to export their farm products to markets in Central Canada, Britain, and other European countries. However, once the railway system was in place, farmers faced a new problem—the high cost of shipping grain long distances by rail. (See Vignette 8.4 and 'Grain Transportation Subsidy' later in this chapter.)

Railway expansion within the Prairies solved the second geographic problem, namely, the difficulty of transporting grain from the farm to the grain elevator. In the late-nineteenth century most farmers transported their grain by horse-drawn wagons to loading points (grain elevators) along the two east-west railway lines. Under the best conditions, farmers operating 15 km from a grain elevator were fortunate if they could haul their grain to an elevator within one day. For that reason, farmers did not cultivate much land beyond 15 km of a grain elevator. By building many branch lines to the main east-west railway, railway companies expanded their rail systems and made grain farming commercially viable in almost all areas of the Prairies.

A third geographic challenge was the region's short growing season. In 1910, a new strain of wheat, Marquis wheat, was developed. Its shorter maturation period overcame the threat of frost.[3] Marquis wheat also extended the growing area for wheat in the Prairies. By 1920, Marquis Wheat was the most popular spring wheat in Western Canada and the adjoining Great Plains states.

Drought posed the fourth geographic problem for farmers, especially those farming in the semi-arid Dry Belt of southern Alberta and Saskatchewan. While drought remained a threat, the technique of dry land farming reduced the risk of crop failure. In dry land farming, part of the land is left in **summer fallow** each year. In this way, sufficient soil moisture is accumulated over several years, allowing for the seeding of the land every other year.[4]

From Intensive to Extensive Agriculture

Over the last century, agriculture in Western Canada underwent significant changes. The first change was the shift in the farm economy from a labour-intensive operation to a capital-intensive one. At the turn of the nineteenth century, many hands were required to successfully deal with the sowing, growing, and harvesting of grain. The introduction of machinery, such as self-propelled steam tractors and threshing machines, changed the way farms were run and reduced the need for labour. Further technological changes continued to affect the size of farm labour. For instance, the development of the combine harvester, which can cut and harvest a swath of grain as wide as 15 metres, allowed farmers to harvest hundreds of hectares in a single day.

After the mechanization of the farm economy, a trend of consolidating farms into larger and larger units occurred. The number of farms declined while the size of farms increased (Tables 8.1 and 8.2). In the Canadian West, this shift began after the Second World War but is most readily apparent in the twenty-year period from 1971 to 1991. Over this period, the number of farms declined by 18 per cent, while the average farm size increased by about 30 per cent, and nearly 31,000 farms disappeared. Less efficient farmers were forced off the land, allowing others to expand their landholdings. For the surviving grain farmers, larger farms increased productivity and thereby lowered per unit costs.

The move from intensive to extensive agriculture transformed the grain economy of Western Canada. With fewer people engaged in agriculture, the rural landscape lost much of its farm population to towns and cities. The ripple effect was that many villages, small service centres, were abandoned and disappeared from the landscape.

Economic Diversification

While Western Canada remains an important agricultural region, development of its forests,

Table 8.1

Number of Farms in Western Canada, 1971–1996

Year	Alberta	Saskatchewan	Manitoba	Western Canada
1971	62,702	76,970	34,981	174,653
1981	58,056	67,318	29,442	154,816
1991	57,245	60,840	25,706	143,791
1996	58,990	56,979	24,341	140,310
Difference 1971–96	(3,712)	(19,991)	(10,640)	(34,343)
Difference (per cent)	(5.9)	(26.0)	(30.4)	(19.7)

Sources: Statistics Canada, 1992b: Table 10; 1998d: 1 of 2.

Table 8.2

Average Size of Farms in Western Canada, 1971–1996

Year	Alberta (acres)	Saskatchewan (acres)	Manitoba (acres)
1971	790	845	543
1981	813	952	639
1991	898	1,091	743
1996	881	1,152	785
Difference 1971–96	91	307	242
Difference (per cent)	11.5	36.3	44.6

Sources: Statistics Canada, 1992b: Table 1; 1998d: 1 of 2.

minerals, and petroleum deposits has diversified the region's economy, particularly in Alberta. Following the Second World War, rising world prices for primary products led to a resource boom in Western Canada. American demand for oil and gas from Alberta rose sharply, while Saskatchewan saw major resource developments in potash and uranium. At the same time, Manitoba became a major producer of nickel and hydroelectric power. Furthermore, the forest industry in all three provinces expanded due to a rising demand for lumber, pulp, and paper in the United States and other industrial countries.

While these resource developments helped diversify the economy of Western Canada, rising oil prices in the 1970s triggered a major resource boom that had a most dramatic impact on the economy of Alberta. This boom had important spin-offs for the Albertan economy including jobs and royalties, technological advances that allowed for the mining of the vast tar sands in northern Alberta, pipeline construction projects to supply the large markets of Ontario and the United States, and the emergence of Calgary as the headquarters for the offices of major oil companies.

By the 1980s, Alberta's economy had not only grown rapidly, it had also diversified, transforming Alberta into a have province. While Alberta's economy and population expanded rapidly during the 1970s and 1980s, Manitoba and Saskatchewan followed at a much slower pace. In fact, during the decade that followed, low prices for primary products had a greater impact on Manitoba and Saskatchewan's economies because they continued to rely heavily on agriculture and resources. Even today, Manitoba and Saskatchewan remain more vulnerable to a resource-driven economic bust than Alberta.

The Prairie Psyche and Western Alienation

Western Canada's geography and its role as a resource hinterland forged a 'Prairie psyche' based on a sense of collectivity among farmers and a feeling of alienation caused by the lack of control over their economy. Before the First World War, farmers felt frustrated by external forces—world prices for grain remained low, while prices for agricultural machinery made in Ontario increased. Furthermore, dealers at the Winnipeg Grain Exchange bought low from the farmers and sold high to the grain buyers, and operators of private grain elevators along the railways assigned the farmers a low grade for their wheat (resulting therefore in a low price). These conditions led farmers to form new institutions or movements. For example, in 1913 farmers banded together to create the United Grain Growers, to provide an alternative means of selling their grain and thus protect themselves from the low grades and prices offered by private grain companies.

The deep sense of alienation expressed by farmers has been an ongoing theme in the history of Western Canada and can now be found in all sectors of society in Western Canada. This negative feeling stems from the peripheral position of Western Canada within Canada and the global economy. Ottawa is often seen as either an uncaring government that ignores western grievances or a manipulative state power that places the interests of Central Canada over those of Western Canada.

Over half the wheat grown in Canada is from Saskatchewan—wheat has played such an important role in that province that it figures prominently on Saskatchewan symbols like the province's flag and coat of arms. *(Corel Photos)*

For instance, at the time that Alberta and Saskatchewan joined Confederation in 1905, the two western provinces were denied control over natural resources while Ontario, Québec, Nova Scotia, and New Brunswick had obtained this control (and taxing power) when they united to form the Dominion of Canada in 1867. A more recent example is the federal initiative known as the National Energy Program, which was in effect in the early 1980s. In this case, Ottawa exerted its control over oil prices and levied taxes on oil production. In Alberta's eyes, the National Energy Program was both a 'tax grab' and a political means by which Ottawa favoured energy-deficient Ontario over Alberta, securing low oil prices for Central Canada's manufacturing industry while interfering with Western Canada's resource revenue.

Sometimes western alienation has led to the formation of new political movements, particularly political parties, as a means of combating the apparent inequities of centralist governments. One such example was the Co-operative Commonwealth Federation (CCF), a political party formed in 1932 by a coalition of labour and farming interests in an effort to combat the destitution people were experiencing during the Great Depression. (The CCF later became the NDP.) It met with considerable success, and is attributed with leading the way for the creation of a variety of social programs in Canada. Social Credit, another Western-based political party, came into existence at about the same time. The Reform Party of Canada provides the most recent example of a Western protest party.

In the 1960s, Howard Richards, who founded the Department of Geography at the University of Saskatchewan in 1960, argued that isolation from other populated regions was at the root of western alienation in the early part of this century. This isolation was worsened by the physical barriers of the Cordillera and the Canadian Shield. As a consequence, westerners felt exploited by Central Canada's businesses and politicians, who wielded the economic and political power that ultimately affected those living in the western hinterland. For farmers, the targets of their resentment were the banks and railways.[5] Later, for oil producers and provincial governments, the target would become Ottawa's centralist policies. Western alienation represents both a psychological version of the core/periphery model and an example of the centralist/decentralist faultline.

Western Canada Today

Western Canada has evolved from a narrowly based agrarian economy to a more diversified one. Increased exploitation of its natural resources, a growing trend towards processing these resources, and a fundamental shift in its agricultural sector are the factors contributing to the region's transformation. Alberta's economy, driven by the oil and gas industry, has become more diversified. Saskatchewan and Manitoba have also diversified but, lacking huge petroleum reserves and still relying heavily on agriculture, they lag behind Alberta's economy. Of particular concern to the agricultural sector of the Prairie provinces is the 1995 cancellation of the Crow Benefit, a transportation subsidy that allowed farmers to ship their products by rail at a reduced cost (Vignette 8.4). In 1995, the cancellation of the Crow Benefit initiated massive changes in the agricultural sector. Whether or not the changes being experienced in the agricultural sector will generate rapid economic growth remains uncertain, but the initial shift to greater processing of agricultural products within Western Canada augurs well for the region's future.

Alberta is the economic giant of the three provinces. It has over half of the population in Western Canada and produces about 63 per cent of the region's GDP (Table 8.3). In Alberta the 1996 GDP per capita amounted to just over $39,000. Both Saskatchewan and Manitoba fall well below this level. What accounts for this variation? Each province has much natural wealth. Saskatchewan, for example, has most of the cropland and is the leading producer of potash and uranium. In addition to having the richest agricultural land in the West, Manitoba produces vast amounts of hydroelectric power from the Nelson River. Even so, Alberta holds the trump resource card—oil and gas.[6]

Vignette 8.4

The Origin and End of the Crow Benefit

When the Canadian Pacific Railway was built, the railway company required public financial assistance to cover the costs of the railroads. In exchange for that assistance, the CPR agreed to lower its shipping rates to Thunder Bay, thus reducing the cost of marketing grain in Canada's major market, Great Britain. Signed in 1897, the Crow's Nest Pass Agreement between the Canadian Pacific Railway and the federal government called for the CPR to lower the rate of moving grain to Thunder Bay by 3 cents per hundredweight in return for a $3 million federal subsidy to extend the rail line from Lethbridge to the Kootenay Valley in British Columbia. The purpose of this subsidy was to ensure that the rail rates for grain were low; that is, below actual shipping costs. This would allow the agricultural sector in Western Canada to thrive despite geographic distance from markets.

Over time, the transportation subsidy for Canada's two national railways increased, reaching $550 million in 1994. It increased for two reasons: shipping costs increased, and the subsidy was extended to include West Coast ports. On 1 August 1995, the Western Grain Transportation Act ended and so did the railway subsidy known as the Crow Benefit. Now farmers would bear the full burden of shipping grain by rail to Canadian ports.

Table 8.3

Basic Statistics for Western Canada by Province, 1996

Province	Population (000)	Population Density (per km²)	GDP (per cent)	GDP/ Capita ($)	Percentage of Canada's cropland
Alberta	2,697	4.2	11.4	39,245	27.3
Saskatchewan	990	1.6	3.4	28,342	41.2
Manitoba	1,114	1.8	3.5	25,429	13.5
Western Canada	4,801	2.5	18.3	31,244	82.0
Canada	28,847	2.9	100.0	28,437	100.0

Sources: Statistics Canada, 1997a: Table 1; 1998a: Table 39; 1998d: 2 of 2.

Back in 1972, when a barrel of oil was worth $2.00 on the world spot market, Alberta oil was not valuable enough to dominate the western economy. Now that oil is far more valuable, the same oil reserves have greatly appreciated in value and thus economic importance. The annual value of natural gas and oil production in Alberta is close to $20 billion.

The expansion of the oil and natural gas industry in Alberta, coupled with high demand for this resource, quickly transformed Alberta from a have-not province into a have province. (Corel Photos)

In comparison, Saskatchewan's annual value of petroleum production is about $2.5 billion, while Manitoba's amounts to less than $100 million.

Though Western Canada exports both primary and processed products to other countries—recently increasing its exports to Pacific Rim countries—the petroleum industry is more closely tied to the North American market. Most of the natural gas and oil go to markets in Ontario and the United States. After the Second World War, a network of oil and gas pipelines was constructed to serve both the Canadian and American markets.

Energy, unlike manufactured goods, had easy access to American markets before the Free Trade Agreement.

World prices hold the key to economic growth for the resource industries, especially oil. Imagine the impact on the Prairie economy if other **primary prices** (prices for primary resources) followed the upward path of oil prices! Unfortunately, Prairie farmers, miners, and loggers are all too well acquainted with a long-term downward trend in commodity prices, as well as sudden fluctuations in the prices for their primary resources—all caused by changes in world demand. While a rise in wheat prices is highly unlikely, dramatic land-use changes that might turn around the agricultural sector are taking place in Western Canada.

Population

Western Canada's population has undergone significant changes since settlers first came to this region, and these changes have affected and been affected by many aspects of Prairie life. In 1921, Western Canada had a population of nearly 2 million. At that time, it comprised 22 per cent of Canada's population. By 1996, its population had increased to 4.8 million, but accounted for less than 17 per cent.

Two migrations transformed Western Canada. Just 100 years ago, land-hungry homesteaders poured into the Prairies, creating a rural landscape of small farms and villages. No one could have foreseen that this rural landscape would lose most of its population in a remarkably short time. In the second migration, the so-called rural-to-urban migration, rural people moved from the countryside to towns and cities in Western Canada and to urban centres in British Columbia, Ontario, and the United States. The push factors of this migration were the mechanization of the agricultural sector, the consolidation of farms, and the shrinking need for farm labour that re-

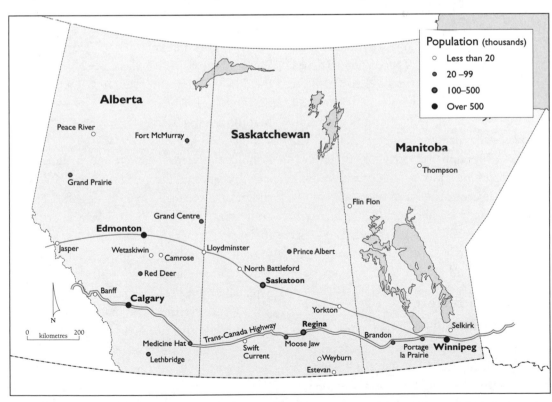

Figure 8.3 Major urban centres in Western Canada. Edmonton, Calgary, Winnipeg, Saskatoon, and Regina are the major population centres in the region.

sulted. Pull factors included the employment opportunities and greater amenities that existed in towns and cities. The rural-to-urban migration resulted in a redistribution of people within Western Canada. Today, most people live in the five major cities of the region: Edmonton, Calgary, Winnipeg, Saskatoon, and Regina (Figure 8.3).

Within Western Canada, Alberta has the largest population with 2.5 million residents. Manitoba and Saskatchewan each have about 1 million. Over the last twenty years, the rate of population increase has been highest in Alberta followed by Manitoba. Recent population changes between the 1991 and 1996 censuses confirm that this trend continues with Alberta growing at a rate of 5.9 per cent over this five-year period followed by Manitoba at 2 per cent. Saskatchewan's population increased by only 0.1 per cent (Statistics Canada 1997a: Table 1).

Another demographic feature of Western Canada is the size of the Aboriginal population. By 1996, approximately 400,000 Aboriginal peoples lived in Western Canada, forming about 8 per cent of Western Canada's population. A growing number reside in urban centres, largely because Indian reserves and Métis communities have insufficient employment opportunities and limited urban amenities.

Industrial Structure

Employment by industrial sector in Western Canada reveals the prominence of primary economic activities. While Table 8.4 is only a generalized picture of the western economy, it illustrates two important aspects. One is the importance of the primary sector. The percentage

of people employed in this sector (13 per cent) is nearly triple the national figure of 5.4 per cent. The second is the small size of the secondary sector. Employment in the secondary sector is well below the national average.

The employment structure of Western Canada reveals the importance of two subsectors, agriculture and transport. Relatively high employment in agriculture in Western Canada is an indication of its importance in the western economy. Employment in agriculture accounts for 8.8 per cent of those working in Western Canada. High employment in the transportation industries reflects the distance factor, namely, the need for a transportation system to overcome the great distances between places in Western Canada and the region's export markets. Those employed by transportation firms make up 8.2 per cent of the total workforce in Western Canada.

Manufacturing

Though mechanization of Western Canada's resource industries has created a more balanced economy, the manufacturing sector is still relatively small. In fact, in 1984 Blackbourn and Putnam (1984:160) were surprised to learn that such a vast area as Western Canada accounted for only 8 per cent of Canada's manufacturing employment. Their explanation for the weak state of manufacturing in the Prairies was that manufactured goods could be produced in Ontario and Québec and sold in Western Canada at less cost than manufacturing the same goods in Western Canada; that is, market size and economies of scale are necessary in order for manufacturing to flourish. From 1981 to

Table 8.4
Employment by Industrial Sector in Western Canada by Province, 1995

Industrial Sector	Western Canada (000)	Western Canada (per cent)	Canada (per cent)
Primary	306	13.0	5.4
Agriculture	207	8.8	3.2
Secondary	341	14.5	21.2
Tertiary	1,707	72.5	73.4
Transport	193	8.2	7.5
Trade	398	16.9	17.0
Finance	117	5.0	5.8
Service	865	36.7	37.1
Public	134	5.7	5.9
Total	2,354	100.0	100.0

Source: Adapted from Statistics Canada, 1996b: Table 40.

1996, however, there was a significant increase in Western Canada's domestic market as the region's population went from 4.2 million to 4.8 million. Western Canada's portion of Canada's manufacturing also increased to 9.2 per cent by 1995. While much of this increase stems from the petrochemical industry, small-scale manufacturing aimed at particular market niches in Western Canada is expanding through the production of specialized products for the agricultural, forestry, and mining industries. In a few cases, these specialized products are being purchased by customers in the United States. Perhaps the highly specialized manufacturing, the increase in the domestic market, and access to the United States'

market explain this modest rise in Western Canada's share of the country's manufacturing industry over the past fifteen years.

Agriculture

Agriculture was the driving force behind the settlement and development of Western Canada. Today it remains a staple of the region's economy. Within Western Canada, there are three distinct agricultural regions, which parallel natural vegetation zones. They are: (1) the Fertile Belt (parkland and long-grass natural vegetation), (2) the Dry Belt (short-grass natural vegetation), and (3) the **agricultural fringe** (southern edge of the boreal forest) and the Peace River country (Figure 8.4). These subregions each have very different growing conditions. The major factors controlling those conditions are the number of frost-free days and the soil moisture. The Fertile Belt provides the best environment for crop agriculture. The Dry Belt occupies semi-arid lands and has become a grain/livestock area. The agricultural fringe is along the southern edge of the boreal forest, and the Peace River country forms a pocket in the northwestern area of this forest. In the agricultural fringe, the short growing season encourages farmers to grow feed grains and raise livestock, while in the Peace River country, farmers grow both grain and feed grain for raising livestock.

The Fertile Belt

The Fertile Belt extends from southern Manitoba to the foothills of the Rocky Mountains west of Edmonton (Figure 8.4). The higher levels of soil moisture, an adequate frost-free period, and rich soils make this belt ideal for a variety of crops and livestock. The most popular crop, since farmers first arrived in the West, has been wheat. In recent years, however, the acreage in grain has declined, while the planting of oil-seed and specialty crops, such as beans, field peas, and sunflower plants has increased. This change was fuelled by rising prices for these crops and declining prices for wheat. In 1995, the area seeded for wheat

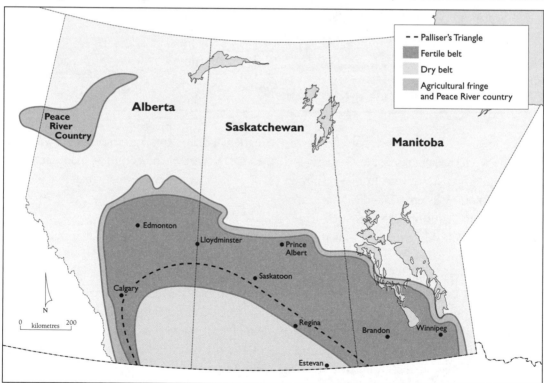

Figure 8.4 Agricultural regions in Western Canada. Farming in the Prairies can be divided into three areas: the Fertile Belt, the Dry Belt, and the agricultural fringe and Peace River country. Each region has a different type of agriculture because of variations in physical geography.

was approximately 11 331 600 ha, the area for canola was 5 261 000 ha, and the area for field peas was over 809 000 ha. In fact, field peas have become a major crop, with the annual acreage sown for field peas increasing since 1990.

The predominance of wheat in the Fertile Belt is most evident in western Saskatchewan and eastern Alberta. In southern Manitoba and the adjacent parts of eastern Saskatchewan, there is more annual precipitation so farmers there can grow a wider variety of crops as well as keep livestock. As a result, mixed farming is common. Grain and specialty crops (canola, flax, sunflowers, and lentils) are combined with beef, pork, and poultry livestock produc-

tion. Near Winnipeg, for instance, a livestock industry has developed where cattle are fattened before shipment to meat-packing plants in Winnipeg and Ontario. Feed lots and nearby meat-packing plants are located in other parts of Western Canada, particularly in Brandon, Calgary, Edmonton, Lethbridge, and Saskatoon. Since NAFTA, meat production has increased, especially pork, and most meat products are being exported to markets in the United States.

As the urban markets grow, market gardens, dairy farms, and other specialized forms of intensive agriculture are developing near major cities in the Fertile Belt. The demand for specialty-crop products has spawned a num-

Grain is stored, cleaned, and weighed in grain elevators before being transported by railway cars and trucks to markets. *(Corel Photos)*

An area of the Fertile Belt near Saskatoon. Ideal conditions in this zone—from moist, rich soils to a long growing season—allow for a range of agricultural production. *(Courtesy Bill Archibold)*

The Dry Belt

The Dry Belt contains both cattle ranches and large grain farms. It extends from the Saskatchewan–Manitoba boundary to the southern foothills of the Rockies and north nearly to Saskatoon (Figure 8.4). However, the driest area, or heart of the Dry Belt, occupies a much smaller area, stretching southward from the South Saskatchewan River to the US border. The arid nature of the Dry Belt is not due to low annual precipitation (it is comparable to other areas of western Canada) but to longer summers and higher evaporation rates. Within the Dry Belt, feed grain and hay crops are grown to supply winter feed for the cattle farming that dominates in this area. Prior to settlement, the Dry Belt's short-grass vegetation provided a natural grazing area for buffalo. Later, cattle replaced the buffalo. Cattle ranching began in this area in the 1880s. Today, ranches are large, often many times the size of grain farms, because of the lower productivity of the dry land and the need for huge grazing areas to support a rotational grazing system.

Along the northern edge of the Dry Belt, grain farming is pursued, but the risk of crop failure is high. To conserve soil moisture, summer fallowing is practised—crops are only planted on parcels of land in alternate years. Until the 1980s, farmers planted wheat in narrow rows separated by a band of

ber of smaller but very intense production units, including nurseries and greenhouses, to produce flowers and vegetables for local sale. For that reason, farm sizes are much smaller around the major cities.

fallow land. This method of farming is called **strip farming**. Strip farming is not really practised now but summer fallowing remains common. The hope is that two years of precipitation will accumulate sufficient soil moisture to germinate the seed and sustain the young wheat plant. The crop will still require summer rainfall to reach maturity. On average, half of the arable land in the Dry Belt is kept in summer fallow each year.

However, land left to fallow has been declining in favour of the growing trend towards continuous cropping. Though summer fallow was practised partly to conserve moisture and partly to control weeds, a drawback of this technique was that the ploughed (fallow) land was subject to water and wind erosion. In a dry spring, windy weather could result in extensive loss of topsoil with huge clouds of dust stretching for many kilometres across the Prairies. Advances in technology have allowed for 'one-pass' seeding, spraying, and fertilizing. The expense and time not spent on repeated tilling of a field reduces a farmer's costs and conserves moisture, which in turn, allows for continuous cropping.

Irrigation on these semi-arid lands has provided another solution to dry conditions. The most extensive irrigation systems are in southern Alberta. In fact, nearly two-thirds of the 750 000 ha of irrigated

land in Canada is located in Alberta. In the 1950s and 1960s, two major irrigation projects were developed in the dry lands of Alberta and Saskatchewan: the St Mary River Irrigation District is based on the internal storage reservoirs of the St Mary and Waterton

An area of the Dry Belt near Swift Current, Saskatchewan. Short-grass vegetation, longer summers, and a higher rate of evaporation characterize this semi-arid area of Western Canada. *(Courtesy Bill Archibold)*

dams in southern Alberta, and Lake Diefen-baker serves as a massive reservoir on the South Saskatchewan River. However, development of irrigated land has been hampered by the problem of finding crops that will provide sufficient revenue to offset the high cost of irrigation water. Another problem, faced by Saskatchewan farmers in particular, is that with a shorter growing season, their selection of crops is more limited compared to the corn, sugar beets, and other specialty crops that farmers in southern Alberta are able to grow.

The Agricultural Fringe and Peace River Country

After the First World War, much new land was brought under cultivation in the Peace River country in northern Alberta and along the southern edge of the boreal forest that constitutes the agricultural fringe of the three Prairie provinces. Settlers moved into the agricultural fringe after the more promising lands of Western Canada were occupied. Unfortunately, much of the agricultural fringe was ill-suited for crops. In these higher latitudes, the threat of frost increased and, in the boreal forest, soil quality decreased. These two natural factors, plus the increased cost of shipping agricultural products to local and world markets at greater distances, made farming in these areas even more risky and marginal than in the rest of Western Canada.

During the Second World War, agriculture in these areas declined—some farmers joined the armed forces, while others migrated to the cities to find jobs. Farmers in the Peace River country fared better—particularly grain

farmers—but they had to contend with a circuitous rail route in getting their products to markets. Abandoned farms reverted to bush while remaining farmers turned to mixed farming. In 1955, however, farmers in the Peace River country benefited from the construction of a rail line from Prince George to Dawson Creek. This shorter link to grain terminals in North Vancouver and then to Pacific Rim customers reduced shipping costs for these farmers.

Next Year Country

The dry continental climate in Western Canada makes farming a risky business. The threat of crop failure is greatest along the margins—the northern agricultural fringe is subject to frost, while grain farmers in the Dry Belt constantly face the possibility of drought. Prairie farmers often describe grain farming as 'Next Year Country'. The meaning behind these words is simple: Our crops did poorly this year, but we hope that they will do better next year. Often the reason for a bad year with low yields is insufficient precipitation during the growing period. However, farmers face a whole range of natural hazards including: summer frosts, which occur when a cold air mass slides unimpeded from the Canadian Arctic to Western Canada; hail and early snowfall; and pests, such as grasshoppers, and diseases, such as stem rust. All these hazards can have devastating effects—from delaying a harvest or lowering the grade of wheat, to reducing a yield or destroying an entire crop. Nevertheless, when compared to the days of the pioneer farmers, who did not have the advantages of modern farm technology and improved strains of wheat, farmers now have a better chance of dealing with adverse weather conditions.

However, if another prolonged dry spell were to occur, similar to that of the Dust Bowl of the 1930s, grain farmers in the Dry Belt of Alberta and Saskatchewan would face several years of crop failure, thereby threatening their survival. Some fear that global warming could create such arid conditions again. Others argue that since weather records indicate that Prairie weather follows a cycle of wet and dry years, it is only a matter of time before another dry cycle occurs. Successive hot, dry summers can quickly dry up soil moisture, bringing back the spectre of drought, dust storms, and crop failure in the semi-arid areas. Perhaps Palliser's original assessment of the limited agricultural potential of the semi-arid areas of Western Canada was correct. In any case, a rise in temperature, without a corresponding increase in precipitation, would threaten grain growing in the Dry Belt and increase the risk of crop failure in the Fertile Belt.

The Canadian Wheat Board

Besides the constant struggle against natural elements, grain farmers have no control over the prices for their commodities. Unlike dairy and poultry farmers, whose output and prices are set by a marketing board, grain farmers live with prices that are set by the market forces of supply and demand. In order to assist Western Canadian farmers to market their grains and to obtain the best possible prices, the federal government established the Cana-

dian Wheat Board in 1935. Today, this Crown corporation remains the grain handling and marketing agency for wheat, barley, and oats destined for export or use by the food industry in Canada. The twin goals of the Canadian Wheat Board are to sell as much grain as possible at the best possible prices, and to ensure that each producer gets a 'fair' share of the market.

To achieve these goals, the Canadian Wheat Board issues delivery quotas. These quotas tell farmers that the board is willing to accept a certain volume of a specified grain from each farmer at their local elevators in a defined geographic region. Quotas, based on the board's sales commitments and stocks on hand in the elevator system, are issued so as to maintain a relatively even flow of grain from farms through the primary (local) elevator system. The Canadian Wheat Board may sell directly to foreign governments' buying agencies, to commercial interests in foreign countries, or to private grain trading companies, which then resell to foreign buyers.

Over the years, farmers have had mixed reactions towards the Canadian Wheat Board. At times, it has provided much-needed protection from an unstable market and low prices. At other times, farmers have resented having to sell their grain through the Board at a Canadian price that is lower than the American price. In recent years, some farmers have challenged the authority of the Canadian Wheat Board, trucking their grain across the border to the US market, where prices have been higher for durum. The Canadian Wheat Board retaliated, taking the issue to court—in 1999, the matter remained unresolved.

Key Topic: Transition

In the 1990s, global forces were unleashed that are reshaping Western Canada's economy. While this transformation affects all sectors of the western economy, the changes are most visible in the agricultural sector. Therefore the focus of this key topic will be recent and pending changes in agriculture. For the most part, these changes are being driven by three factors: global forces such as trade liberalization and fluctuating world prices; national policies such as government programs and subsidies; and regional conditions such as the high costs of transporting products to distant markets.

The liberalization of international trade, with agreements like NAFTA and GATT, have opened up foreign markets for Western Canada, but this has made resource sectors more vulnerable to downturns in these foreign markets. For Western Canadian farmers, trade liberalization is a double-edged sword. It has provided exports opportunities, but it has also led to the elimination of farm subsidies, such as the transportation subsidy for grain, and may now make farm marketing boards and farm organizations, such as the Canadian Wheat Board, unnecessary.[7]

Grain Transportation Subsidy

The greatest change affecting agriculture in Western Canada is the loss of the Crow Benefit in 1995. Since then, farmers have had to pay the full cost of rail transportation to ship their grain to port. The magnitude of the new transportation costs is illustrated by the size of the former subsidy. In 1994, the Crow Benefit subsidy amounted to approximately half the cost of shipping grain. Once this subsidy was cancelled, most agricultural economists expected that the loss of the annual $550-million rail subsidies would slash farm income by a similar amount. For farmers in Saskatchewan, their share of the rail cost of shipping spring wheat to port doubled. For instance, a Swift Current farmer's grain shipping costs to Vancouver were $13.82 per tonne under the Crow Benefit, but rose to $29.39 per tonne immediately after August 1995 (Budhia 1995b:43). Manitoba farmers face the largest increases because they are the farthest from ocean ports. Spring wheat shipments from Winnipeg to Montréal via Thunder Bay could triple rail costs for Manitoba farmers, causing them to lose money growing spring wheat. At the same time, however, the cost of shipping durum wheat by rail from Winnipeg to Montréal is about half the cost of shipping spring wheat (Budhia 1995a:38–9). What accounts for the variation? The rail freight rates for durum were lower in the mid-1990s because large quantities of durum were sold in the US market, where the price for durum was high. Consequently, Canadian railways were forced to lower their rates to remain competitive.

The changes in rail transportation costs for grain farmers are changing the agricultural land-use pattern in Western Canada.[8] The speed with which these changes occur will depend on the world price for spring wheat. If the price is high, as in 1994/5, the pace of change will be slow; but if the price declines sharply, as in 1997/8, farmers will lose money growing spring wheat and will have to grow alternative crops, thus accelerating the pace of change. These are some of the changes farmers might make in order to keep their farms profitable:

- Farmers in eastern Saskatchewan and in Manitoba may grow less grain but more feed grains and specialty crops, as the climate here is both wetter and warmer than most other areas of the Canadian West, allowing for a wider variety of crops. Feed grains can be sold to new beef and hog producers in these provinces, while specialty crops, such as canary seed, can be grown under contract to larger firms.
- Farmers in western Saskatchewan and in Alberta may not decrease their wheat acreage because they have the lowest rail costs for shipping spring wheat to Vancouver. On the other hand, access to the Asian and California markets for pork and beef products is encouraging an expansion of the livestock industry and the production of fodder crops.

The Prairie Staple

Grain production, particularly spring wheat, has been the prairie staple for nearly 100 years. Recent developments, however, such as low world prices for spring wheat and the loss of the rail transportation subsidy for grain, have led grain farmers to seek alternative crops for spring wheat.[9] Another development affecting grain agriculture are the two trade agreements—the North American Free Trade Agreement and the General Agreement on Tariffs and Trade. These agreements have opened American and other foreign markets to farmers, thereby further encouraging a diversifica-

tion in crops. Access to the American and global markets has encouraged:

- growing more high-priced durum wheat for sale in the US market
- producing and processing more beef and pork in Western Canada, thereby providing a local market for feed grains
- growing and locally crushing seed crops, such as canola, and shipping the oil product to the US, Japan, and other Asian countries
- growing specialty crops, such as sunflowers and field peas, under contract to US firms
- producing alfalfa pellets from hay for shipment to the US and Japan for feed in their dairy industry

In Saskatchewan, for example, the proportion of land planted for wheat, durum, and canola from 1986/7–1995/6 reflects a shift in the choice of crops. Wheat is down by 35 per cent and durum is up by 26 per cent. Canola, a seed oil, made the greatest gains and was up 144 per cent in 1995/6 over its 1986/7 figure (Table 8.5). Cash receipts from the sale of these crops also reflect this change. Cash receipts for wheat were $1.3 billion in 1987, which was 58 per cent of all crop sales. By 1995, wheat sales had increased to $1.5 billion, but accounted for only 37 per cent of all crop sales. Canola jumped from 13 per cent of crop sales to 20 per cent, and durum wheat from 11 per cent to 19 per cent. The cash receipts from these two crops combined ($1.7 billion) was slightly above that of wheat (Saskatchewan Agriculture and Food 1996:tables 28, 29, and 34).

Table 8.5
Shift in Acreage: Wheat, Durum, and Canola, 1986/7–1995/6

Crop	1986/7 (million acres)	1995/6 (million acres)	Change (per cent)
Spring wheat	17.8	11.5	(35)
Canola	2.5	6.1	144
Durum	3.5	4.4	26

Source: Saskatchewan Agriculture and Food, 1996:tables 28, 29, and 34.

Table 8.6
Shift in Price: Wheat, Durum, and Canola, 1986/7–1994/5

Crop	1986/7 ($/tonne)	1994/5 ($/tonne)	Change (per cent)
Spring wheat	105	167	59
Canola	199	348	75
Durum	122	237	94

Source: Saskatchewan Agriculture and Food, 1996:tables 28, 29, and 34.

Faced with increasing rail rates for grain products, farmers will be discouraged from shipping grain to foreign markets.[10] They have several alternatives. One is to grow more durum wheat, which is ideal for making pasta products. As a more specialized wheat, durum commands a higher price and therefore can better withstand the higher rail transportation costs. For instance, in the 1994/5 crop year, the price of durum was 42 per cent higher than that of spring wheat (Table 8.6) (Saskatchewan Agriculture and Food 1996: tables 28 and 29). Another attractive feature of durum is the potential market for it in the United States.

Farmers are also growing more oil-seed crops, such as canola and flax, and more specialty crops, such as peas, lentils, and canary seed. The primary reason for the viability of these alternative crops is that their prices have risen steadily over the last twenty years, while the price for spring wheat has not changed appreciably. In fact the price of spring wheat has dropped so low that in 1998/9, farmers were losing money. Another factor is the lower transportation costs for the farmer because oil-seed and specialty crops only need to be trucked to local processing plants. Major food companies have built large canola-crushing plants in Western Canada over the past ten years, encouraging farmers to grow more canola. By shipping their products to local plants by truck, farmers' transportation costs are minimized. However, canola has its limitations. Canola cannot be grown in the more arid areas of Western Canada, and the crop is prone to disease and weed infestations if cultivated repeatedly.

With the end of the Crow Benefit in 1995, livestock producers in southern Ontario can no longer afford to import feed grains from Western Canada. However, a growing livestock industry in Western Canada now supplies beef and pork products to Ontario and Québec. As well, the demand for pork is increasing, opening more markets for Western Canadian producers in both the United States and Pacific Rim countries. To meet the demand, huge hog farms, similar to those in the United States, have sprung up across Western Canada that take advantage of modern agrotechnology, mass production techniques, and economies of scale. Because of the smell and the risk of contaminating groundwater, large hog barns are located in rural settings far from settlements. Here, the hog barns are providing employment for residents of rural communities, though the long-term viability of mass hog production and the stability of demand remain uncertain.

Western Canada's Resource Base

Western Canada's resource economy began to diversify in the 1970s, when energy developments began in Alberta. Since then, some of the same factors that have been affecting the region's agricultural sector, have also been fuelling further growth and changes in its resource sectors. The significance of resource exports to the economy of Western Canada is evident in the 1996 export figures. In that year, the leading exports from Western Canada were grain, petroleum, potash, and uranium. By value, these exports amounted to $37 billion, about 12 per cent of Canada's total exports (Statistics Canada 1998f:1 of 1). Besides its fertile soils, Western Canada's main assets are its minerals, fuels, and forests. (Figure 8.5).

Mining Industry

The mining industry has helped to diversify the economy of Western Canada. The variety and value of mineral production in Western Canada is enormous. Mining companies produce a full range of mineral products, including metals, non-metals, structural materials, and fuels. In 1996, the three Prairie provinces produced almost $32.6 billion of the energy and mineral production in Canada (Table 8.7). This production amounted to 66 per cent of the national output (Statistics Canada 1998e: 9 of 11).

In Western Canada the geology of each province differs sufficiently to produce three distinct types of mining. Alberta contains rich coal reserves along the eastern slopes of the Rocky Mountains. Coal mining began in 1872 just west of Lethbridge. By the 1990s, the province of Alberta produced more coal than any other province within Canada, which contributed to Canada's position as the fourth largest coal exporter in the world. Today, coal

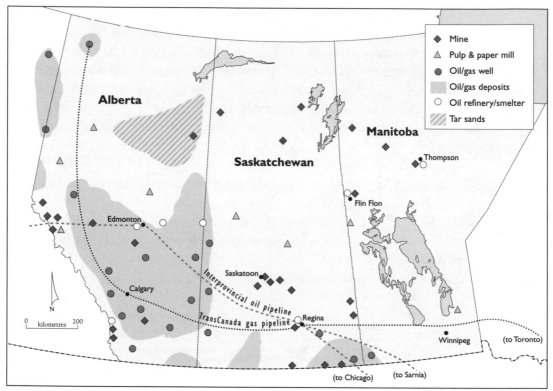

Figure 8.5: Western Canada's resource base, 1996. The number of mines, pulp and paper mills, oil and gas wells, and oil refineries and smelters that dot Western Canada's landscape attest to the region's range of natural wealth.

Table 8.7
Mining Production by Value in Western Canada, 1996

Mineral Product	Alberta ($000)	Saskatchewan ($000)	Manitoba ($000)	Western Canada ($000)
Metals	304	627,834	825,385	1,453,523
Non-metals	129,846	1,116,924	19,095	1,265,865
Structural materials	310,246	26,614	50,110	386,970
Fuels	25,777,635	3,559,482	107,355	29,444,472
Total	26,218,031	5,330,854	1,001,945	32,550,830

Source: Adapted from Statistics Canada, 1998e:1, 3, 5, and 7 of 11.

mining takes place in the East Kootenay and Peace River coalfields. Like other resource industries, these Alberta mines depend on world demand and prices. In 1997, the Asian economic slump in Japan and other Asian countries resulted in a drop in demand and price for Alberta coal.

Potash and uranium are the major mineral deposits in Saskatchewan. The potash deposit lies approximately 1 km below the surface of the earth, reaching its thickest extent around Saskatoon, where six of the nine potash mines are located. Canada is the world's largest producer and exporter of potash. Saskatchewan has the largest and highest quality deposit in the world, and except for a potash mine in New Brunswick, all of Canada's potash production comes from Saskatchewan. Potash is used to produce potassium-based fertilizers, and is sold to firms in the United States and Asia, especially China and Japan. Even though China and Japan have been suffering from the Asian economic downturn, they continue to import Canadian potash to ensure high yields for their crops. Still, world demand varies, forcing potash mines to adjust the volume of their production.

Uranium mining also takes place in Saskatchewan. Production first took place on the northern shores of Lake Athabasca near Uranium City. Since the late 1970s, uranium mining has shifted south to the geological area of the Canadian Shield known as the Athabasca Basin. Three mines are currently operating (Cluff Lake, Key Lake, and Rabbit Lake). These older open pit mines will soon cease production and, subject to the approval of the federal environmental agency, four new

mines (Cigar Lake, McArthur River, McClean Lake, and Midwest) are expected to begin production. Saskatoon is the major supply centre for both potash and uranium mines as well as their corporate headquarters.

Manitoba has two major mineral deposits—copper-zinc and nickel—both located in the Canadian Shield. Copper and zinc ore bodies are found near Flin Flon, a mining and smelter town in northern Manitoba that began production in 1930, shortly after a rail link to the Pas was completed in 1928. At one time, Flin Flon produced most of Canada's copper and zinc but today it is an aging resource town, with a declining population. Thompson, located some 740 km north of Winnipeg, is a nickel mining town. In 1957, after a rail link to the Hudson Bay Railway was completed, the mine facility, smelter, and town were constructed. Unlike Flin Flon, Thompson was a specially designed resource town that had a complete array of urban amenities. The first nickel was produced in 1961. During the 1960s, Thompson's population soared to 20,000. Since then, export demand for nickel has become more competitive, resulting in a population decline to 14,000 by 1996. Both the copper and nickel deposits in Manitoba are relatively expensive mining operations. As well, the high cost of shipping the processed product to market adds to their economic disadvantage.

Fuels: Oil and Natural Gas

What has proven to be Western Canada's most valuable mineral deposit is its vast reserves of oil and natural gas (Vignette 8.5). In 1996, Canada's production of fuels (including coal, oil, and natural gas) was valued at nearly $30 billion. Western Canada, but mainly Alberta and Saskatchewan, accounted for approximately 85 per cent of this production. Western Canada contains over 70 per cent of Canada's oil reserves. These oil deposits are located in the southern half of the Western Canadian Sedimentary Basin. Alberta has the largest oil reserves (55 per cent of Canada's reserves) followed by Saskatchewan (17 per cent). Manitoba has less than 1 per cent. In addition to oil and natural gas deposits, vast amounts of oil are contained in the tar sands—oil mixed with sand, known as **bitumen**—of northern Alberta. The two major oil sands mining sites are

situated near Fort McMurray and Cold Lake. Alberta has the major oil and gas industry, so it leads the three Western provinces in the value of mineral production.

The economic impact of petroleum on Western Canada, especially Alberta, was caused by the sudden rise in oil prices in the 1970s. During the 1970s, the sale of large quantities of oil and natural gas to Ontario at much higher prices than in the previous decade resulted in a transfer of enormous wealth and economic power from the consuming provinces (mainly Ontario) to the producing provinces (mainly Alberta). Alberta was transformed into a 'rich' province. Oil was

---

Vignette 8.5

Westward Expansion of the Oil Industry

While oil was first discovered in the southwestern corner of southern Ontario in 1858, the search for petroleum soon switched to southern Alberta. In 1883, Canadian Pacific Railway workers found gas while drilling for water near Medicine Hat. In 1890, gas was again found in the area by miners drilling for coal. Natural gas was soon used as a fuel for heating and lighting the homes and offices of Medicine Hat. The streets of the young Prairie city were lit by gas lamps; natural gas was so cheap that the lamps were never turned off. Soon major natural gas deposits were found throughout southern Alberta. In 1912, a gas pipeline was built, linking Calgary to the Bow Island gas fields near Medicine Hat, 275 km away.

In 1902, oil was discovered in southwestern Alberta. However, Alberta did not experience a true oil boom until ten years later, when oil was discovered at Turner Valley. Drilling companies also discovered natural gas, but because it was so abundant, it was considered a nuisance and burned at the site of the well. Turner Valley was Canada's largest and most important oil field in the years before the Second World War. In 1947, when Imperial Oil discovered a major oilfield near the small town of Leduc, just outside of Edmonton, the Alberta oil boom began in earnest. This oilfield lies in a geological formation of the Western Sedimentary Basin. Since such basin formations have potential for oil and natural gas deposits, they were the focus of more oil and gas exploration.

Source: Adapted from Stan Garrod, *Economics in Society: Canadian Case Studies* 192. Copyright © 1984 Addison Wesley Longman Limited. All rights reserved. Reprinted by permission.

not only a source of economic power but also political power. In terms of the core/periphery model, the export of oil reversed a century-old economic relationship—Central Canada was dependent on the Western provinces. Ottawa interfered with this new relationship by introducing the National Energy Program. The federal government was seen by Alberta as intruding on provincial rights and attempting to prevent the transfer of wealth. The results of this federal program were to widen the political gap between Alberta and Ottawa and to deepen the sense of Western alienation and mistrust towards Ottawa and Central Canada.

Forest Industry

The boreal forest stretches across the northern part of Western Canada. Within the boreal forest, a few coniferous species—spruce, fir, pine, and tamarack—predominate. The commercial forest zone lies in the Interior Plains where soil conditions are more favourable for tree growth than in the thin, rocky soils of the Canadian Shield. In 1995, logging in Western Canada amounted to over 5 per cent of Canada's total timber harvest (Table 8.8). Approximately 70 per cent of this production takes place in Alberta.[11] Alberta's advantage comes from its large northern forest stands. The forest-covered Interior Plains region, which extends beyond the North Saskatchewan River to the border with the Northwest Territories, is the largest forest stand in Western Canada. This, together with the timber in the foothills of the Rocky Mountains, means that Alberta has by far the largest commercial forest stands of the three Prairie provinces.

Each of Western Canada's three provinces has a diversified timber operation, ranging from sawmills to pulp and paper plants. Most pulp and paper mills are located on the southern edge of the forest. In Saskatchewan these mills are located at Prince Albert and Meadow Lake. In Alberta they are in Athabasca, Grande

Table 8.8
Forest Area and Production in Western Canada, 1995

Province	Area (million of ha)	Productive Forest Area (million of ha)	Primary Production (million of m³)
Manitoba	26.3	13.2	826
Saskatchewan	28.8	10.6	2954
Alberta	38.2	20.9	9936
Western Canada	93.3	44.7	13 716
Per cent	22.3	20.9	5.1

Source: Adapted from Statistics Canada, 1998g: Table 1.

Prairie, Hinton, and Peace River. In Manitoba The Pas is the site of a major pulp and paper mill.

By the end of the 1980s, virtually all the commercial forest stands had been leased, signalling the end of available virgin timber. In fact, several mills were unable to obtain additional Crown timber leases and have suffered from a shortage of wood fibre. Now that the original boreal forest has been harvested, the second growth is best suited for the pulp and paper industry. These trees, small and large, known as pulpwood, are cut and then trucked to a pulp mill. Each mill has a forest lease, giving it the right to log on Crown land. While a small amount of lumber production does occur, pulp and paper products predominate. Distance from markets also affects the focus of the forest industry. Pulp and paper products, which have a much higher value per tonne than lumber, are better able to withstand rail transportation costs from Western Canada to distant markets.

In the 1980s, it was discovered that aspen forests in Alberta, Saskatchewan, and Manitoba provided an excellent source of pulpwood for high-quality paper production. With considerable provincial financial support, mills were established at Hinton, Grande Prairie, Prince Albert, and The Pas. The Alberta government, seeking to diversify its economy by encouraging more processing of resources, tried to lure pulp and paper companies to northern Alberta by offering them vast northern hardwood timber leases. Two Japanese firms, Daishowa Canada Company Ltd and Alberta-Pacific Forest Industries Inc., established mills at Peace River and Athabasca in northern Alberta. By the end of the 1980s, virtually all Alberta timber had been leased to five pulp and paper companies. Similar developments occurred in Saskatchewan and Manitoba. One Alberta-based firm, Millar-Western Ltd, turned to Saskatchewan where it arranged to purchase pulp logs from Norsask, a local company that holds the timber lease to lands north of Meadow Lake. With an assured supply of pulp wood, Millar-Western built a mill at Meadow Lake. The Millar-Western pulp mill is a unique industrial development because: (1) the mill's internal circulating system means that no toxic wastes are released into the local rivers and lakes, and (2) it has a business arrangement with Norsask Forest Products, which is jointly owned by the Meadow Lake Tribal Council and Techfor Services, a company owned by the employees of the local sawmill (Anderson and Bone 1995:127). This business venture represents a new approach for resource development, combining the financing and expertise of an established pulp and paper company with the commitment by Aboriginal peoples to supply the raw materials.

Western Canada's Urban Geography

The process of urbanization in Western Canada has lagged behind that of Ontario, British Columbia, and Québec. Even so, nearly three-quarters of Western Canada's residents live in urban centres, and most of these urbanites reside in the five major cities— Regina, Saskatoon, Winnipeg, Calgary, and Edmonton. A second order of urban centres includes Lethbridge, Red Deer, Medicine Hat,

Brandon, Fort McMurray, Prince Albert, Moose Jaw, and Swift Current. In 1996, these thirteen major cities and urban centres had over 3 million people.

An important urban corridor is emerging within Alberta, linking Edmonton with Calgary. Nearly 2 million people live and work within this corridor, so it not only forms a major market within Western Canada but also has the potential for the establishment of high-tech and producer services industries. Two anchors—Calgary and Edmonton—each contribute special functions to the corridor. Calgary is the headquarters of the oil and gas industry, while Edmonton, besides being the provincial capital, is known as the 'Gateway to the North'. Over the past fifteen years, this urban corridor has had a high rate of population growth, exceeding 20 per cent.

From 1981 to 1996, the rate of urban growth varied considerably for the towns and cities of Western Canada. This variation reflects differences in local economic growth and in the pace of consolidating populations into regional centres. From 1981 to 1996, the fastest-growing cities in Western Canada were in Alberta and Saskatchewan: Medicine Hat, Prince Albert, Calgary, Red Deer, and Saskatoon all exceeded a growth rate of 25 per cent. The slowest-growing cities were Moose Jaw, Brandon, Regina, Winnipeg, and Fort McMurray—all below 15 per cent (Table 8.9).

Saskatoon, situated alongside the South Saskatchewan River, is Saskatchewan's largest city. *(Courtesy R.M. Bone)*

Table 8.9
Major Urban Centres in Western Canada, 1981-1996

Centre	Population 1981	Population 1996	Difference (per cent)
Moose Jaw	33,943	34,829	0.3
Fort McMurray	31,000	35,213	13.6
Brandon	36,320	40,581	11.7
Prince Albert	30,995	41,706	34.6
Medicine Hat	40,700	56,570	39.0
Red Deer	46,393	60,075	29.5
Lethbridge	54,558	63,053	15.6
Regina	173,226	193,652	11.8
Saskatoon	175,058	219,056	25.1
Winnipeg	592,061	667,209	12.7
Calgary	625,966	821,628	31.3
Edmonton	740,882	862,597	16.4
Total	2,581,102	3,096,169	20.0

Sources: Statistics Canada, 1997a:Table 2; 1982:Table 1.

Western Canada has, in effect, two types of urban settlement patterns. These two urban patterns reveal a north-south split that is based on a regional core and a hinterland. In the south, centres are aligned close to the southern border, whereas an oasis-like pattern exists in the northern hinterland; that is, settlements are far apart. Such a pattern is common in resource hinterlands where a single industry town is located near an ore body. Three examples are: Flin Flon (copper), Fort McMurray (heavy oil), and Thompson (nickel).

Across Western Canada, the trend for over sixty years has been a rural-to-urban migration with many of the smallest centres disappearing. At the turn of the twentieth century, these small communities formed the heart of the local farming area, but today they have lost their function. The main reasons why these centres lost their function are complex and interconnected:

• Farms consolidated, thereby reducing the size of the rural population.
• Highways were built, facilitating shopping in larger centres, where the variety and prices of goods were superior to

Vignette 8.6

Winnipeg

Winnipeg, the capital and the largest city in Manitoba, is located at the confluence of the Red River and the Assiniboine River. With the building of the Canadian Pacific Railway, Winnipeg became known as the 'Gateway to the West'. Because of its strategic location, Winnipeg was and remains a transportation centre. By the turn of the twentieth century, Winnipeg was the principal city in Western Canada, controlling the grain trade and also acting as the administrative, financial, and wholesale hub for Western Canada.

Winnipeg's role in Western Canada has gradually weakened over time. After the completion of the Panama Canal in 1914, Alberta and some Saskatchewan farmers began to ship their wheat through Vancouver instead of through Winnipeg. Next, service industries began to emerge in Edmonton, Calgary, Saskatoon, and Regina. Each city captured some of the trade previously held by Winnipeg merchants. Winnipeg's stranglehold over the sale and distribution of agricultural machinery and products to the farmers of Western Canada was broken.

After the Second World War, Winnipeg still remained the largest city in Western Canada. In 1951, Winnipeg's population was 357,000, while Edmonton and Calgary were much smaller at 177,000 and 142,000 respectively. However, Calgary and Edmonton grew rapidly in the following years, partly because of developments in the oil and gas industries. By 1981, both these cities surpassed Winnipeg in population size. At that time, Winnipeg had a population of 652,000, while Edmonton's population was 840,000 and Calgary's was 754,000. Today, Winnipeg dominates the economy of Manitoba, but its role within Western Canada has been considerably diminished by Calgary and Edmonton and, to a lesser degree, by Saskatoon and Regina.

those in smaller centres; eventually, stores in smaller communities closed.

- Schools, hospitals, and other public services were concentrated in larger centres, attracting the people who use these services.
- Railways abandoned branch lines, causing grain elevator companies to close their operations; for smaller towns, the closure of the grain elevator and the loss of their rail service meant the loss of the town's last function.
- Grain companies built larger grain elevators on main lines, expecting that the lower freight rates offered at the larger grain elevators would attract farmers, drawing business away from elevators in small centres.
- Farmers moved to larger centres, where more urban amenities were available, and commuted to their farms.

Western Canada's Future

Western Canada began this century as an agricultural hinterland. This hinterland slowly began to diversify into other forms of resource development. After the Second World War, the process of change accelerated, particularly in Alberta, where petroleum diversified the resource economy and provided the basis for a strong petrochemical industry. Two events, trade liberalization and the termination of the Crow Benefit, launched further economic changes in the 1990s, especially in agriculture.

In Western Canada, agriculture is undergoing most of the economic restructuring caused by free trade and the loss of the transportation subsidy. As farmers shift to more high-value crops and crops that can be sold locally, the importance of spring wheat will diminish. The structural changes to transportation rates will also encourage local processing of agricultural products, thereby expanding the meat-processing industry and stimulating the development of new food-processing industries. Agricultural land use is also changing across Western Canada, with an expansion in the growing of speciality crops. For instance, farmers in Manitoba pay more for shipping their wheat to Thunder Bay than Alberta farmers pay for sending their grain to Vancouver—this spatial price differential is encouraging a shift in agricultural land use with more spring wheat being grown in Alberta than in Manitoba. Manitoba farmers now sell more feed grain to the province's expanding livestock industry, which then ships its meat products to markets in Ontario, Québec, and the United States.

The greatest impact of economic diversification within Western Canada has been in Alberta. In the 1970s, soaring oil and gas prices transformed Alberta into a wealthy province. Yet resource industries, while they add greatly to the value of provincial produc-

tion, required only a small labour force and therefore had a limited effect on the workforce. The indirect effects of oil, however, were substantial in the construction, pipeline, and service sectors of Alberta's economy. Saskatchewan and Manitoba, while also diversifying but without the benefit of a large oil and gas industry, have proceeded at a much slower rate. With the basic changes now underway in agriculture, these two provinces may hasten their pace towards more diversified economies.

Summary

Western Canada occupies the western interior of Canada. The region has a range of resources, including arable land, vast forests, and mineral wealth. The geography of Western Canada has posed several challenges to economic development. Two major challenges are distance to ocean ports and world markets, and a dry climate for agriculture. Initial attempts to overcome distance took the form of railway construction and a subsidy for shipping grain by rail. To overcome the dry climate, farmers grew wheat, which can withstand the semi-arid conditions and thus became the staple crop.

As Western Canada enters the twenty-first century, its role as a resource hinterland is changing through economic diversification. The expansion of the petroleum industry, the liberalization of world trade, and the loss of the Crow Benefit have accelerated this trend. While the region is still far from becoming a major industrial core, Western Canada has become an upward transitional region as described in Friedmann's core/periphery model.

The process of diversification has included strong growth in the manufacturing sector, agricultural transition, and the greater processing of natural resources (mineral, forest, and agricultural).

Alberta has led the way in diversification with its vast reserves of oil and gas. Manitoba and Saskatchewan, while lagging behind Alberta's economic performance, are heading in the same direction by engaging in more processing of their resources. While the Asian economic slow-down in the late 1990s has hurt certain sectors of the resource economy, this economic blow was softened by the emergence of a more broadly based economy. The direction of economic change in Western Canada is clear—more processing and a more diversified economy—but the final shape of this new economy will not be revealed until well into the twenty-first century.

Notes

1. The Cree attacked the outpost at Frog Lake, laying siege to Fort Battleford and defeating the Northwest Mounted Police at Fort Pitt and Cut Knife Hill. With the arrival of the Canadian army from eastern Canada, the Cree were defeated.

2. It was only after the Second World War that the federal government extended the rights and privileges accorded to Canadian citizens to its Aboriginal peoples. Until then, Indians could not vote in provincial and federal elections. The right to vote marked the start of a long journey to find a place in Canadian society. This journey is far from over.

3. The search for a quicker maturing wheat began in 1892 when a cross was made between Red Fife, a popular wheat grown on the Prairies, and an earlier maturing wheat. After a decade of trials, Marquis wheat was tested at the Dominion Experimental Farms at Indian Head, Saskatchewan. By 1910, this variety of wheat made Canada famous for producing an exceptionally high-quality, hard spring wheat.

4. Summer fallowing accomplished two goals: it conserved moisture and controlled weeds. In the 1990s, farmers began to abandon this technique because advances in technology enabled them to accomplish these goals without having to resort to summer fallowing.

5. Western alienation is based on past experience of real or perceived 'abuse' by big business, such as the CPR, and big government, meaning government controlled by eastern interests. Such feelings were deeply felt by homesteaders who settled the West. By the beginning of the twentieth century, western alienation was particularly strong and the subsequent resentment was often aimed at the Canadian Pacific Railway. In fact, farmers regarded this railway company as the most rapacious agent of eastern Canadian interests. Not only had the company obtained millions of acres of fertile land, but its real estate offices often manipulated station sites to ensure their location on Canadian Pacific Railway property. Since farmland increased in value with proximity to rail-loading sites, such Canadian Pacific Railway land increased in value and could be sold to settlers at a higher price. Even more galling to westerners was the fact that CPR landholdings could not be taxed for twenty years. To extend this tax-free period, the CPR sometimes delayed selecting land, which meant that large areas were not avail-

able for homesteading because the railway could opt to select such lands as part of its grant.

6.   Alberta, Manitoba, and Saskatchewan did not gain full control over Crown lands, and therefore over natural resources, until 1930. At that time, Ottawa transferred federal property to these provinces, giving them access to lucrative sources of taxation associated with natural resource developments. Until then, taxes from these lands went to Ottawa, supposedly to pay for railway building in the West. At first, the revenue from natural resources was small, but with the discovery of oil at Leduc, Alberta, in 1947, royalties from the production of petroleum provided most of the revenue for Alberta. Equally important, the exploitation of the vast oil and gas deposits in Alberta grew rapidly and eventually resulted in a variety of petroleum-processing plants and pipeline-construction firms in Alberta. Later, more modest energy and mineral developments were found in Saskatchewan, British Columbia, and Manitoba.

7.   Changes in the structure of farming are also anticipated. With expansion of the livestock and meat-processing industries, the number of farm workers may also increase. Large-scale hog farms, for instance, require farm workers. Also, the increase in specialty crops is causing more farmers to undertake contract farming; that is, they grow a certain quantity of canary seed, for example, for a set price. Some of these crops are now processed locally, a trend that is expected to intensify.

8.   Farmers are becoming less sheltered from economic forces. Through contract farming, agribusiness—the sector of the economy that provides inputs to farms, such as chemical fertilizers, and procures agricultural products from farms for processing and distribution to consumers—is increasingly affecting the daily lives of farmers. Some fear that in the next severe economic downturn or drought, family farming operations will be replaced by corporate farms. In fact, the trend towards contract farming by agribusiness may be the first sign of such an ownership shift. While not yet apparent in the grain industry, agribusiness has established itself in specialty crops, poultry, and hog farming.

9.   Farmers' decisions are affected by a complex set of variables, but ultimately the cost of doing business is the determining factor. For instance, the increased production of canola may affect the price of livestock feed. A protein by-product derived from the crushing of canola seeds is suitable for livestock. Instead of importing a similar protein supplement and paying for the built-in freight costs, using a local canola-based product to feed livestock may reduce costs.

10.   The impact of higher rail costs for grain shipments has not yet occurred because prices of wheat were at a twenty-year high in 1995 and 1996. However, grain farmers are sensitive to price. As already stated, it was continuing low prices for wheat and higher prices for canola and specialty crops in the early 1990s that resulted in a shift from wheat to a more diversified crop production.

11.   Since the early 1990s, sawmills in the interior of British Columbia have exhausted local supplies of timber. They have had to purchase logs from Alberta ranchers who own timber lands in the foothills around Calgary.

Key Terms

agricultural fringe
Agriculture at its physical limits. Along the southern edge of the boreal forest and in the Peace River country, farmers clear the land, but the short growing season prevents most crops from maturing, so many farmers turn to cattle.

bitumen
A tar-like mixture of sand and oil.

Dry Belt
An agricultural area in the semi-arid parts of Alberta and Saskatchewan that is primarily devoted to grain farms and cattle ranches. Crop failures due to drought are more common.

Fertile Belt
A mixed farming area where crop failures due to drought are less common. This area of long grass and parkland natural vegetation is associated with black and dark-brown chernozemic soils.

Palliser's Triangle
Captain Palliser conducted a survey of the Canadian West in 1857–8. He concluded that this short-grass natural vegetation area in southern Alberta and Saskatchewan was a northern extension of the Great American Desert and was therefore unsuitable for agricultural settlement.

primary prices
The prices for commodities such as foodstuffs, raw materials, and other primary products.

regional core
Within the core/periphery model, cores can occur at different geographic levels. A regional core is an area (often a large city) that dominates trade and stimulates economic growth in the region.

strip farming
An old farming practice in which alternating strips of land were cultivated and left fallow. The purpose was to conserve soil moisture and reduce wind and water erosion. Due to advances in agricultural technology, farmers can now maximize soil moisture conservation while practising continuous cultivation.

summer fallow
The farming practice of leaving land idle for a year or more in order to accumulate sufficient soil moisture to produce a crop; also practised to restore soil fertility; being replaced by continuous cropping.

Western Sedimentary Basin
Within the geological structure of the Interior Plains, the normally flat sedimentary strata are bent into a basin-like shape. These basins often contain petroleum deposits.

References

Bibliography

Anderson, Robert B., and Robert M. Bone. 1995. 'First Nations Economic Development: A Contingency Perspective'. *The Canadian Geographer* 39, no. 2:120–30.

Barr, Brenton M., and John C. Lehr. 1982. 'The Western Interior: Transformation of a Hinterland Region'. In *A Geography of Canada: Heartland and Hinterland*, 1st edn, edited by L.D. McCann, Chapter 8.

Blackbourn, Anthony, and Robert G. Putnam. 1984. *The Industrial Geography of Canada*. London: Croom Helm.

Budhia, Narendru. 1995a. 'Manitoba—Outlook for 1995 Improves'. *Provincial Outlook: Economic Forecast* (Spring) 10, no. 2:38–41.

_____. 1995b. 'Saskatchewan—As the Crow Flies'. *Provincial Outlook: Economic Forecast* (Spring) 10, no. 2:42–5.

Garrod, Stan. 1984. *Economics in Society: Canadian Case Studies*. Don Mills: Addison-Wesley Publishers.

Marsh, James H., ed. 1988. *The Canadian Encyclopedia*, 2nd edn. Edmonton: Hurtig Publishers.

Neatby, L.H. 1979. *Chronicle of a Pioneer Prairie Family*. Saskatoon: Western Producer Prairie Books.

Richards, J. Howard. 1968. 'The Prairie Region'. In *A Geographical Interpretation*, edited by John Warkentin, Chapter 12. Toronto: Methuen.

Saskatchewan Agriculture and Food. 1997. *Agricultural Statistics 1996*. Regina: Statistical Branch.

Smith, P.J., ed. 1972. *The Prairie Provinces: Studies in Canadian Geography*. Toronto: University of Toronto Press.

Spry, Irene M. 1963. *The Palliser Expedition: An Account of John Palliser's British North American Expedition 1857–1860*. Toronto: Macmillan.

_____, M.R. Olfert, and Murray Fulton. 1992. *The Changing Role of Rural Communities in an Urbanizing World: Saskatchewan 1961–1990*. Canadian Plains Report no. 8. Regina: Canadian Plains Institute.

Stanford, Quentin H., ed. 1998. *Canadian Oxford World Atlas*, 4th edn. Toronto: Oxford University Press.

Statistics Canada. 1982. *Census Metropolitan Areas and Census Agglomerations with Components*. Catalogue no. 95-903. Ottawa: Minister of Supply and Services Canada.

_____. 1992a. *Census Division and Subdivisions*. Catalogue no. 93-304. Ottawa: Minister of Supply and Services.

_____. 1992b. *Census Overview of Canadian Agriculture 1971–1991*. Catalogue no. 93-348. Ottawa: Ministry of Industry, Science and Technology.

_____. 1993. *Canada Year Book 1994*. Ottawa: Ministry of Industry, Science and Technology.

_____. 1994a. *Canada's Aboriginal Population by Census Subdivision and Census Metropolitan Area*. Catalogue no. 94-326. Ottawa: Ministry of Industry, Science and Technology.

_____. 1994b. *Canadian Forestry Statistics 1993*. Catalogue no. 25-202-XPB. Ottawa: Ministry of Industry, Science and Technology.

_____. 1996a. *Canada's Mineral Production*. Catalogue no. 26-202-XPB. Ottawa: Mining Sector, Natural Resources Canada.

_____. 1996b. *Labour Force Annual Averages 1995*. Catalogue no. 71-220-XPB. Ottawa: Statistics Canada.

_____. 1997a. *A National Overview: Population and Dwelling Counts*. Catalogue no. 93-357-XPB. Ottawa: Industry Canada.

_____. 1997b. 1996 Census: Nation Tables—Population by Mother Tongue, Showing Age Groups, for Canada, Provinces and Territories, 1996 Census—20% Sample Data, 2 December 1997 [online database], Ottawa. Searched 15 July 1998; <URL:http://www.statcan.ca/english/census96/>:3 pp.

_____. 1997c. The Daily—1996 Census: Mother Tongue, Home Language and Knowledge of Languages, 2 December 1997 [online database], Ottawa. Searched 14 July 1998; <URL: http://www.statcan.ca/Daily/English/>:15 pp.

_____. 1998a. *Canadian Economic Observer.* Catalogue no. 11-010-XPB. Ottawa: Statistics Canada.

_____. 1998b. The Daily—1996 Census: Aboriginal Data, 13 January 1998 [online database], Ottawa. Searched 14 July 1998; <URL:http:// www.statcan.ca/Daily/English/>:10 pp.

_____. 1998c. The Daily—1996 Census: Ethnic Origin, Visible Minorities, 17 February 1998 [online database], Ottawa. Searched 16 July 1998; <URL:http://www.statcan.Daily/ English/>:21 pp.

_____. 1998d. Census of Agriculture—Canada Highlights [online database], Ottawa. Searched 2 September 1998; <URL:http://www.statcan. ca/english/censusag/can.htm>:2 pp.

_____. 1998e. Table 2: Revised Statistics of the Mineral Production of Canada, by Province, 1996 [online database], Ottawa. Searched 3 September 1998; <URL:http://www.nrcan.gc.ca/ mms/efab/mmsd/production>:11 pp.

_____. 1998f. Canadian Statistics—Export of Goods on a Balance-of-Payments Basis [online database], Ottawa. Searched 3 September 1998; <URL:http://www.statcan.ca:80/english/ Pgdb/Economy/International/>:1 p.

_____. 1998g. *Canadian Forestry Statistics 1995.* Catalogue no. 25-202-XPB. Ottawa: Ministry of Industry, Science and Technology.

Further Reading

Found, William C. 1996. 'Agriculture in a World of Subsidies'. In *Canada and the Global Economy: The Geography of Structural and Technological Change*, edited by John N.H. Britton, Chapter 9. Montréal–Kingston: McGill-Queen's University Press.

Rice, Murray D. 1996. 'Functional Dynamics and a Peripheral Quaternary Place: The Case of Calgary'. *Canadian Journal of Regional Science* XIX, no. 1:65–82.

Semple, R.K. 1994. 'The Western Canadian Quaternary Place System'. *Prairie Forum* 12:81–100.

Stabler, Jack C., and M.R. Olfert. 1992. *Restructuring Rural Saskatchewan: The Challenge of the 1990s*. Regina: Canadian Plains Institute.

Chapter 9

Overview

Atlantic Canada, located on the eastern margins of the country, is far from the centres of power. As an 'old' resource hinterland, Atlantic Canada's economic growth lags far behind Canada's other geographic regions. As well, Atlantic Canada suffers from high out-migration, which slows its population growth. While the North American Free Trade Agreement has provided access to the larger New England market, Atlantic Canada has limited resources suitable for export. The demise of its major resource, the northern cod stocks, has seriously hurt the region's economy, especially in Newfoundland, which has the highest unemployment rate in Canada. While the discovery of nickel at Voisey's Bay and offshore oil and gas developments hold promise for the future, Atlantic Canada has so far been unable to quicken its economic pace and break out of its resource-based economy.

Objectives

- Describe Atlantic Canada's physical and historical geography.
- Present the basic characteristics of Atlantic Canada's population and economy from the perspective of an 'old' resource hinterland.
- Examine these two elements within the context of Atlantic Canada's physical setting, especially its fragmented geography.
- Explore Atlantic Canada's role as a resource hinterland within Canada and North America.
- Focus on the cod fishery as an example of the Tragedy of the Commons.

Atlantic Canada

Introduction

Stretching along the country's eastern coast, the region of Atlantic Canada consists of two parts: the Maritimes (Nova Scotia, Prince Edward Island, and New Brunswick) and Newfoundland/Labrador (Figure 1.1). Despite the differences this physical separation has created between the Maritimes and Newfoundland/Labrador, Atlantic Canada is united by a rich sense of place that has grown out of the region's history and geographic location. The Atlantic Ocean has dominated the history of this region of Canada from the days of the early

Cape Breton's rugged topography culminates in the highlands at its northern cape, North Harbour. *(Corel Photos)*

British and French settlers, who depended on the sea for their livelihood, to the 'Golden Age of Wooden Ships and Iron Men'. Today, Atlantic Canadians continue to rely heavily on the sea, fishing on the continental shelf and conducting energy exploration below it. Most important of all, the sea provides a link between Atlantic Canada and foreign countries. As in the region's early history, the North Atlantic Ocean asserts itself in the affairs of Atlantic Canada. In this chapter, the *Key Topic* is the Fishing Industry.

Atlantic Canada Within Canada

As Canada's oldest hinterland, Atlantic Canada has experienced both growth and decline over the years. Today, Atlantic Canada is, in Friedmann's regional version of the core/periphery model, a downward transitional region. It is a region troubled by past exploitation of its renewable resources and the exhaustion of its most accessible and richest non-renewable resources.

As a resource hinterland Atlantic Canada has not been faring well. As the region enters the twenty-first century, the people of Atlantic Canada are searching for better ways to manage the region's resources in a more sustainable fashion; to achieve a higher degree of processing of its non-renewable resources; and, at the same time, to expand its base of high-technology industries. Some changes are already taking place. Offshore oil and gas developments have stimulated the growth of highly specialized manufacturing firms in both Newfoundland and Nova Scotia, while high-technology firms are locating in the major cities of Atlantic Canada. Information technology is also growing in the Maritimes. For instance,

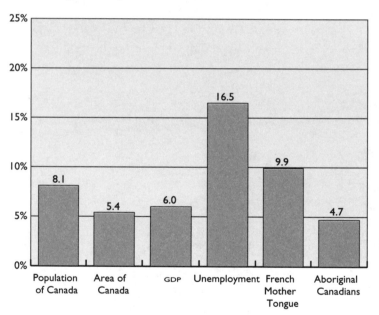

Figure 9.1 Atlantic Canada, 1996. Atlantic Canada has a weak economy. The region has 8.1 per cent of Canada's population but produces only 6.0 per cent of the country's GDP. The high rate of unemployment is another indication of its poor economic performance—it is the highest in the country. Acadians account for the high number of French-speaking Canadians in Atlantic Canada, which has the second highest percentage by region in Canada. Aboriginal peoples, however, form a small percentage of Atlantic Canada's population.

Sources: Statistics Canada, 1997a:Table 1; 1998b:Table 30; 1997b:1 to 3 of 3; 1997c:4 and 5 of 15; 1998c:1 and 2 of 10; 1998d:15 to 20 of 21; Stanford 1998:185.

Table 9.1
Basic Statistics for Atlantic Canada by Province, 1996

Province	Population (000)	Population Density (per km²)	GDP (per cent)	GDP/ Capita ($)	Unemployment (per cent)
Prince Edward Island	134.6	23.8	0.3	21,222	13.8
Newfoundland	551.8	1.4	2.4	19,330	25.1
New Brunswick	738.1	10.3	2.0	21,377	15.5
Nova Scotia	909.2	17.2	1.3	21,622	13.3
Atlantic Canada	2,333.7	4.4	6.0	21,377	16.5
Canada	28,846.8	2.9	100.0	28,431	10.1

Sources: Statistics Canada, 1997a:Table 1; 1998b:Table 39; 1998f:1 of 4.

Bathurst, Fredericton, Halifax, Moncton, Saint John, and St Stephen are telephone call centres for firms operating in Canada, such as Air Canada, AT&T Canada, Canadian Pacific, IBM, Northern Telecom, Purolator Courier, Royal Bank of Canada, and UPS. (Call centres enable customers to phone a 1-800 number to obtain the services that companies provide.) However, these information service jobs are at the low end of the labour market in terms of pay and skills, and, unlike the innovative high-technology industries, provide few spin-offs.

Despite a few signs of change, Atlantic Canada's economy is struggling. The economic growth that is taking place is concentrated in a few urban centres and in select industrial sectors. The economy in rural Atlantic Canada continues its steep decline. Except for the tourist industry, the economy of the rural countryside is stagnant. Jobs in the primary sector, such as logging and fishing, are disap-

pearing as firms replace workers with more advanced machinery. Economic decline has created another problem—out-migration. With few prospects in sight, many in Atlantic Canada have been leaving the region to seek opportunities in other parts of the country. This makes the task of rebuilding the region's economy even more difficult.

One measure of Atlantic Canada's overall economic performance is reflected in the region's per capita gross domestic production figures and its level of unemployment (Figure 9.1). In 1996, Atlantic Canada's GDP per capita was the lowest in Canada, while the region's unemployment rate was the highest (Table 9.1). The primary reasons for Atlantic Canada's weak economic performance are: (1) the geographic division of Atlantic Canada into subregions adds to companies' transportation costs and discourages the emergence of an integrated economy in which economies of scale might occur; (2) the region has a limited resource base; (3) the region has small internal markets; and (4) the region is far from Canada's major markets. All these factors have made it extremely difficult for economic development to flourish in Atlantic Canada. Furthermore, Atlantic Canada has, over time, become heavily dependent on Ottawa for economic support through equalization payments and social programs. This cycle of dependence will only end once the region is able to revitalize its stagnant economy.

Atlantic Canada's Physical Geography

Atlantic Canada is in two of Canada's physiographic regions: the Maritimes and New-

foundland are part of the Appalachian Uplands, and Labrador is part of the Canadian Shield. Within the Appalachian Uplands, the region's landforms are dominated by a rugged coastline and an upland interior consisting of hilly to mountainous terrain. In the Canadian Shield, where arable land is even more scarce than in the Appalachian Uplands, the geography consists of a coastline with fjords, a rugged interior, and the lofty Torngat Mountains.

Atlantic Canada's physiography lends itself to a resource hinterland. The region has little arable land, except on Prince Edward Island, in the Saint John River Valley, and in the Annapolis Valley (Vignette 9.1). Beyond these few favoured lowlands, the land is too rough and rugged for settlement. As a consequence, most settlements in Atlantic Canada are clus-

Atlantic Canada's landscape is characterized by a rocky coastline and a hilly terrain. These features have made agriculture in the region difficult. *(Corel Photos)*

Vignette 9.1

The Annapolis Valley

The Annapolis Valley is a low-lying area in Nova Scotia. At its western and eastern edges the land is at sea level, but it rises to about 35 m in the centre. The area is surrounded by a rugged, rocky upland that reaches heights of 200 m and more. Beneath the valley's fertile soils are soft sedimentary rocks. The sandy soils of the Annapolis Lowlands originate from marine deposits that settled there about 13,000 years ago. After glacial ice retreated from the area, sea waters flooded the land, depositing marine sediments that consisted of minute sand and clay particles. Isostatic rebound then caused the land to lift and slowly these lowlands emerged from the sea. In the seventeenth century, the favourable soil of the Annapolis Valley attracted early French settlers, who later became known as Acadians. Today, the Annapolis Valley's stone-free, well-drained soils and its gently rolling landforms provide the best agricultural lands in Nova Scotia.

tered along the coastline, where for years fishers have made their living harvesting the riches of the ocean. Further inland, settlements evolved where mineral and forest resources were profitable. For instance, in the Appalachian Uplands, a lush forest bolstered the economies of the Maritime provinces, especially New Brunswick. Similarly, mining has long been a way of life for some Atlantic Canadians—from the famous coal mines of Cape Breton Island, Nova Scotia, to the recent discovery of nickel-copper-cobalt in Voisey's Bay, Newfoundland. Furthermore, the region has long benefited from its extensive continental shelf, where fishers pursued the cod fish, and recent discoveries of petroleum and natural gas have led to the development of major oilfield sites like Hibernia on the Grand Banks.

Atlantic Canada is not a homogenous region but 'a region of geographical fragmentation' (Macpherson 1972:xi). It consists of three subregions: the Maritimes, Newfoundland, and Labrador. Macpherson argued that this physical division has impacted on all aspects of life in Atlantic Canada, hindering the emergence of a common political will, a common regional consciousness, and an economic union among the four Atlantic provinces. Whether or not an economic union is a desirable political arrangement for Atlantic Canada is uncertain, though the topic of union has long been debated.

The natural division of Atlantic Canada has led to different historical developments, which have affected the region's political evolution and nurtured a diverse cultural geography. The settlement of Atlantic Canada began the process of cultural diversity. For instance, Nova Scotia was settled first by the Acadians, who were later expelled and replaced by the New England Planters; German Protestants arrived at Lunenberg; the Loyalists, including about 3,000 Black refugees, settled in Halifax and other parts of Nova Scotia; and Gaelic-speaking Highlands Scots, both Catholic and Protestant, made Cape Breton and the Northumberland shore their home. In addition to these different settlement patterns, the physical obstacles between the subregions and the separation of the region from the rest of Canada fostered and maintained cultural differences:

- The Maritimes, though a relatively cohesive economic unit, has closer ties to New England than the rest of Canada.
- The island of Newfoundland stands alone, separated from the rest of Atlantic Canada by the Atlantic Ocean. This subregion does not have close trade links with the Maritimes.
- Labrador, while part of the province of Newfoundland, is drawn closer into Québec's economic orbit. Examples include: (1) the transmission of hydroelectric energy to markets across southern Québec; (2) the shipment of Labrador's iron ore to the port of Sept-Îles, Québec; and (3) the Labrador–Québec highway that runs from Labrador's largest town, Happy Valley–Goose Bay, to Baie-Comeau and Québec's provincial highway system.

Despite the differences between Atlantic Canada's provinces, many similarities exist. In addition to a common geographic location—the waters of the Atlantic Ocean wash up on the shores of all four Atlantic provinces, which are all physically separated from the rest of Canada—the Atlantic provinces have all undergone a similar economic process of growth, decline, and dependence. As the region where people first settled in the country, Atlantic Canada has undergone a very long period of resource exploitation. Its role as a resource hinterland brought early prosperity, which was followed by decline as its non-renewable resources were exhausted and some of its renewable resources were overexploited. This economic malaise drew Ottawa into the affairs of Atlantic Canada, often at the request of the Atlantic provinces, which turned to Ottawa for financial assistance in times of crisis. Ottawa continues to have a strong presence in the region, and, some say, an unduly strong influence. Ottawa's powerful role in the management of the Atlantic fisheries, for example, is particularly resented. Nevertheless, Atlantic Canada relies on the support provided by Ottawa during periods of economic hardship. The challenge of the future will be to break the cycle of dependence by revitalizing Atlantic Canada's economy.

Atlantic Canada's Historical Geography

Atlantic Canada was the first part of North America to be discovered by Europeans. While the Vikings arrived earlier, the first documented discovery of land in North America was made by John Cabot. The Italian navigator, employed by the king of England, reached the rocky shores of Atlantic Canada (the exact location is disputed to be either Cape Bonavista, Newfoundland, or Cape Breton Island, Nova Scotia) on 24 June 1497. Like Columbus, Cabot was searching for a sea route to Asia. In England, Cabot's report of the abundance of groundfish—such as cod, grey sole, flounder, redfish, and turbot—in the waters off Newfoundland lured European fishers to make the perilous voyage across the Atlantic to these rich fishing grounds. Canada's Atlantic coast quickly became a popular area for European fishers and though landings on shore took place—for drying the fish and establishing temporary habitation during the fishing season—permanent settlements were slow to take hold in this part of North America.

The first permanent settlement in North America north of Florida was a cluster of French settlers at Port-Royal on the Bay of Fundy in 1605. French settlement then spread into the Annapolis Valley and other lowlands in the Maritimes. The expanse of French-settled lands which evolved during the seventeenth and part of the eighteenth centuries became known as Acadia. Acadia consisted of a collection of French settlements united by culture, language, and a common economy.

While French settlers were populating the Maritimes, British settlers began arriving in 1610 at Conception Bay, Newfoundland. Over the next 100 years, British settlements developed because fishers preferred to remain in Newfoundland at the close of the fishing season rather than make the long journey back to Britain. By the 1750s, there were over 7,000 permanent residents living in hundreds of small fishing communities along the shores of the island of Newfoundland.

During the first half of the eighteenth century, struggle between the two European colonial powers in North America—England and France—was almost continuous. During that time, the French forged an alliance with the Micmac and Malecite, drawing them into the conflict with the English and their Iroquois al-

lies. Under the terms of the 1713 Treaty of Utrecht, France surrendered Acadia to the British. However, many French-speaking settlers (known as Acadians) remained in this newly won British territory, which was renamed Nova Scotia. During the previous century, the Acadians had established a strong presence in the Maritimes with settlements and forts. Most Acadians lived along the Bay of Fundy coast, tilling the soil. Until the mid-1700s, Britain made little effort to colonize these lands and the Acadians were the vast majority of settlers in this British-held territory. By 1750, Acadians numbered over 12,000.

British settlement began to spread to the Maritimes in the mid-1700s, with the first serious attempt occurring in 1749 when the British government began to recruit settlers for its newly acquired lands in Nova Scotia. Halifax, founded in 1749, would become the centre of British influence and military power in Nova Scotia and would provide a counterbalance to the well fortified French military base of Louisbourg on Cape Breton Island. Through a series of wars culminating with the Seven Years' War (1756–63), the British would gradually gain control of Atlantic Canada.

In 1756, the final struggle for North America, known as the Seven Years' War, began. Britain and France engaged in an all-out battle for supremacy around the world. In North America, the British commanders began preparing for war in 1755. When the Acadians refused to swear an oath of unconditional loyalty to the British Crown, the British deported 6,000 Acadians to more secure parts of the British colonies in North America. After the

British defeated the French, all French territories in North America were under British control. The dream of a French North America was dead. Under the terms of the Treaty of Paris in 1763, France ceded all its territories to Britain, except for the islands of Saint-Pierre and Miquelon near the southern shore of Newfoundland. Today, most Acadians live in northern New Brunswick, where they constitute the majority of the province's population (just under 40 per cent).

The next event to influence the evolution of Atlantic Canada was the American Revolution. Following the outcome of the revolution, approximately 40,000 Loyalists made their way to Nova Scotia and New Brunswick. (See 'The Loyalists' in Chapter 3 for more details.) Most settled along the shores of the Bay of Fundy, in the Annapolis Valley, on Cape Breton Island, and around the British stronghold of Halifax. Halifax became known as the 'Warden of the North' because it provided a superb port for the ships of the British Navy. Over the next fifty years, the arable land was cleared and occupied. During that time, more British settlers came to the Maritimes. Nova Scotia alone received 55,000 Scots, Irish, English, and Welsh. Most Scots went to Cape Breton and the Northumberland shore. While many newcomers tried their hand at farming, the fishery still dominated life in the Maritimes and Newfoundland. Through the years, until the end of the nineteenth century, Atlantic Canada changed rapidly: the population increased, towns grew, and the economy evolved from one based on subsistence production to one emphasizing commercial production.

In the early-nineteenth century, the harvesting of Atlantic Canada's natural wealth increased. This frontier hinterland of the British Empire exploited its rich natural resources—the cod fish in Newfoundland and the virgin forests in the Maritimes—and became heavily dependent on transatlantic trade of these resources. Overseas trade with other British colonies in the Caribbean, as well as with Britain, formed the cornerstone of Atlantic Canada's transatlantic trade and prosperous economy. In the first half of the nineteenth century, trade with New England was relatively limited because both the Maritimes and New England had almost identical resource-oriented economies. However, after the American Civil War, New England would industrialize, leading to greater trade between the two regions. In addition, Britain's move to free trade in 1849 would mean the loss of Atlantic Canada's protected markets for its primary products, resulting in even greater interest in the American market, especially in the Maritimes.

Historic Head Start

With the fishery and the timber trade well underway in the early-nineteenth century, Atlantic Canada had a head start in developing a commercial economy. Furthermore, the availability of timber and the region's favourable, seaside location provided the ideal conditions for shipbuilding in Atlantic Canada. By 1840, Atlantic Canada entered the 'Golden Age of Sail', with Nova Scotia and New Brunswick becoming the leading shipbuilding centres in the British Empire. Exports from Atlantic Canada were primarily cod and timber, while imports were manufactured goods from England and sugar and rum from the British West Indies. Several world events in the mid-nineteenth century added strength to this economic boom in the Maritimes. Demand for its exports rose, especially for timber, which was used to build British merchant ships and warships. The Crimean War (1853–6), the Reciprocity Treaty with the United States (1854–66), and the American Civil War (1861–4) opened new markets for fish, minerals, and lumber from the Maritimes and cod from Newfoundland. International trade was increasing and Atlantic Canada was looking outward to the rest of the world, while the Province of Canada was more consumed with internal developmental issues at that time.

Just before Confederation, however, external events dampened Atlantic Canada's resource-based economy. Iron was replacing wood in shipbuilding; the three-way trade with Britain and its Caribbean possessions collapsed, largely because a world glut of sugar caused prices to drop; and the end of the Reciprocity Treaty (1866) cut off access to the Maritimes' natural trading partner, New England. These events resulted in the deterioration of Atlantic Canada's economic position within the Atlantic trading nations and colonies.

Confederation

The provinces of Atlantic Canada joined Canada at different times. Nova Scotia and New Brunswick joined at the time of Confederation (1867); Prince Edward Island quickly followed in 1873; while Newfoundland did not join Canada until 1949. At the time of Confederation, Newfoundlanders, being so isolated, saw little advantage to joining Canada. In addition to the geographic reasons, the fear that Newfoundland would be dominated by the larger provinces also discouraged union. Besides, Newfoundland's trade relations at the time were tied to Britain, not Canada. For the Maritimes, the decision to join Canada was not an easy one. In spite of the Maritime provinces' historic head-start and their excellent access to sea and world markets, they were on the margins of the new country and their once-booming economy was collapsing.

Confederation favoured economic growth in Central Canada. The federal government's national policy reinforced Central Canada's advantage, placing the Maritimes' fledgling manufacturing base in danger. In order to stimulate Canadian manufacturing, Canada imposed duties on imported goods. In retaliation, Americans increased their trade barriers on Canadian manufactured goods. Canada even placed a tariff on foreign-produced coal to protect the high-cost coal mines of Nova Scotia. In effect, Canada had created a 'closed' economy that facilitated the industrialization of Central Canada by delegating the rest of Canada to the dual roles of domestic market for Canadian-manufactured goods and resource hinterland for Central Canada, the United States, and the rest of the world (see Chapters 1, 3, and 4).

At Confederation, the Maritimes did have a fledgling manufacturing base, but local markets were too small and the Maritimes' natural market, New England, was now less accessible because of high US tariffs. What Atlantic Canada needed was access to the market of

Central Canada. Ottawa's answer was the Intercolonial Railway (completed in 1876). Built to fulfil the terms of Confederation, the Intercolonial Railway provided rail access to Central Canada, but the railway was never a commercial success, though it did stimulate economic growth in the Maritimes. The railway was operated by the federal government; freight rates were kept low in order to promote trade; and substantial annual deficits were paid by Ottawa. With low-cost transportation to the national market and the possibility of firms achieving economies of scale, a general economic surge occurred in the Maritimes. Cotton mills, sugar refineries, rope works, and iron and steel manufacturing plants were established or expanded to serve the much larger national market and to grab a share of the expanding western market. The railway boom in the Canadian Prairies had a positive impact on Nova Scotia in the form of heavy investment in the province's steel mills. With the construction of new railway lines, steel production was focused on manufacturing steel rails and locomotives. By taking advantage of Cape Breton's coalfields and iron ore from Bell Island, the steel industry in Sydney prospered, accelerating Nova Scotia's economic growth well above the national average.

The Maritime economy, however, suffered a deadly blow when, in 1919, freight rates were increased to levels common in Central Canada. Immediate access to the national market became more difficult and, with the loss of sales, many firms had to lay off workers, while others were forced to shut down their operations. Even before these troubled times, the Maritimes' economy was unable to absorb its entire workforce, leading many to migrate to the industrial towns of New England and southern Ontario. After the First World War, the out-migration increased.

After the Second World War, the Maritimes grew but at a slower pace than other parts of Canada. Newfoundland joined Canada in 1949, mainly for economic reasons. Newfoundland stood to gain from the social programs available to Canadians—for instance, unemployment insurance for seasonal workers like fishers. Above all, Newfoundlanders hoped that Canada's booming economy would pull Newfoundland from its poverty and its dependence on the fishery. Even during periods of national affluence—for instance the 1950s and 1960s—Atlantic Canada experienced limited economic growth; even numerous federal loans and grants to firms in Atlantic Canada made little difference. No doubt this region languished because of its scattered and relatively small population, its narrow resource base, and the high cost of transporting goods to the national market. In 1968, Professor David Erskine at the University of Ottawa drew a rather dismal picture of Atlantic Canada as a hinterland:

> The region is, in the Canadian context, one of 'effort' rather than of 'increment'. Small scale resources once encouraged small scale development, but only large scale resources encourage modernization. The small scale and lack of concentration of its resources makes the region one in which government investment is easily dispersed without bringing about growth. Low levels of professional services and

low levels of education result from the high taxes from low incomes; thus, still further retardation of economic growth occurs (Erskine 1968:233).

Atlantic Canada Today

Speaking of Atlantic Canada as a whole belies the geographic differences among its four provinces: New Brunswick, Nova Scotia, Prince Edward Island, and Newfoundland. One difference lies in history. For example, Newfoundland joined Confederation over eighty years later than the Maritime provinces.

Another difference is related to physical geography. Newfoundland and Labrador, for instance, were not well endowed by nature. Both have climates less conducive to agriculture than the Maritimes, partly because both Newfoundland and Labrador are located much farther north and because of the chilling effect of the Labrador Current. Added to this climatic disadvantage, neither Newfoundland nor Labrador have much soil. Instead, they consist mainly of rocky landforms. Also, Newfoundland is an irregularly shaped island jutting into the North Atlantic, which makes access to and from Canada more difficult, while Labrador is part of the mainland, alongside Québec. With little fertile land and limited timber stands, Newfoundland and Labrador's inhabitants were forced to turn to the sea for their survival.

In contrast, the Maritime provinces are a comparatively more compact unit with rail and road links to Montréal and the markets of Central Canada. The Maritimes also rely less on the sea than Newfoundland because they

have a milder climate that allows for agriculture and a more varied resource base. Prince Edward Island, blessed with gently rolling farmland and red, fertile soils, has an agricultural base as well as good fishing grounds. In Nova Scotia and New Brunswick, a combination of agriculture, forestry, and tourism play a strong role in their economies. The Maritimes' proximity to New England also attracts American tourists more so than the more complicated automobile route and ferry crossing to Newfoundland. In fact, the importance of land links for tourism is evident after the completion of Confederation Bridge from New Brunswick to Prince Edward Island, which resulted in a sharp increase in the number of out-of-province vehicles and visitors to this island.

Economic variations also exist in Atlantic Canada, and many of these are related to the provinces' different geographies and resource bases. Nova Scotia and New Brunswick, with their more diversified resources, have the strongest economies, while Newfoundland, with its heavy reliance on the seasonal and declining fisheries, has by far the weakest economy. Prince Edward Island trails behind New Brunswick. These economic differences are illustrated in the region's unemployment rates, with Newfoundland having the highest rate at 25.1 per cent in 1996 and Nova Scotia having the lowest rate at 13.3 per cent (Table 9.1).

Population

Since Confederation, Atlantic Canada's population has increased very slowly, well below the national average. In recent years, its population growth has virtually stalled. From 1991 to 1996, Atlantic Canada experienced a population increase of only 0.5 per cent compared to the national average of 5.7 per cent. The population increase that did occur in Atlantic Canada varied among the four provinces, largely reflecting differences in economic conditions. From 1991 to 1996, Prince Edward Island had the highest rate of increase at 3.7 per cent, while Newfoundland actually suffered a population decline (Table 9.2). The slow rate of increase in this region is a measure of the large number of people leaving Atlantic Canada to search for jobs in more prosperous regions of Canada, especially southern Ontario, Alberta, and British Columbia.

Since the 1920s, the economic circumstances in Atlantic Canada have been on a downward cycle, driving hundreds of thousands to seek their fortunes elsewhere. Over the years, Atlantic Canada has lost more people through out-migration than it has gained through in-migration. In fact, few immigrants to Canada choose to settle in Atlantic

Canada because they see few opportunities. The resulting loss of people, while boosting economies elsewhere, has hindered economic growth in Atlantic Canada. An indication of the serious nature of this net outflow is shown in the proportion of Canadians living in Atlantic Canada. In 1949, when Newfoundland joined Canada, Atlantic Canada constituted 11.6 per cent of the nation's population. By 1996, it had fallen to 8 per cent.

Until the United States placed restrictions on the movement of Canadians into the United States in the 1930s, New England was the favourite destination of migrants from the Maritimes. Many went to Boston. After the 1930s, most Maritime migrants settled in Ontario. When Newfoundland joined Confederation in 1949, its migrants went mainly to Ontario. From 1986 to 1996, about 175,000 left Atlantic Canada for Ontario, Alberta, and British Columbia. During that ten-year period, Ontario was the destination of just over half of

Table 9.2
Population Change and Density in Atlantic Canada, 1991–1996

Province	Population 1996	Change 1991–6 (per cent)	Population Density 1996 (per km²)
Prince Edward Island	134,557	3.7	23.8
Newfoundland	551,792	(2.9)	1.4
New Brunswick	738,133	2.0	10.3
Nova Scotia	909,282	1.0	17.2
Atlantic Canada	2,333,764	0.5	4.4

Source: Adapted from Statistics Canada, 1997a:1.

Vignette 9.2

The Newfoundland Resettlement Program

Since the first British and French families arrived in Newfoundland, the province's settlement pattern has consisted of isolated coastal fishing outports (often with only three or four families each). By 1945, nearly 1,400 tiny outports with populations under 200 had been established along the coast of Newfoundland (Staveley 1987:257). But by 1998, there were only about 400 outports with populations under 200. This decline was the result of two forms of relocation—voluntary and government-assisted.

Between 1945 and 1975, fishing families living in approximately 300 tiny outports with populations under 200 abandoned their homes and moved to either larger outports or to one of the cities, such as St John's. Relocation was prompted by unfavourable fishing conditions—first overcrowding then declining stocks. At first the abandonment of outports was voluntary. Between 1946 and 1954, forty-nine small communities were abandoned without government assistance. Then government became involved. First, the Newfoundland government made financial help available under the Centralization Program. Under the Newfoundland scheme, all households in the community had to agree to relocate before public assistance would be made available. In total, 110 communities were abandoned and the residents resettled in larger centres of their choice. In 1967, the federal government joined forces with Newfoundland. The federal-provincial resettlement scheme, the Newfoundland Resettlement Program, added new conditions to relocation. Migrants had to relocate to one of seventy-seven designated 'growth centres'—selected communities that planners thought would offer more social and economic opportunities. (Most of these larger centres had fewer than 2,000 people.) Another stipulation was that at least 75 per cent of the households in the outport community must agree to move in order for the migrants to receive public relocation funds. Under the joint program, another 150 communities were abandoned from 1967 to 1975. After this program ended, the process of relocation continued, but at a slower pace.

Because of deteriorating fishing conditions, small Newfoundland fishing villages, like this one, were the focus of a government-sponsored resettlement program between 1945 and 1975. (Courtesy Aninda Chakravarti)

the migrants from Newfoundland, nearly half from Nova Scotia, about 40 per cent from New Brunswick, and one-third from Prince Edward Island. Within Ontario, Toronto attracts most of the migrants from Atlantic Canada.

Dispersed Population

Atlantic Canada's population of over 2 million is highly dispersed along its coastline in both small and large centres. Geography explains why this dispersed settlement pattern developed. The original selection of sites for fishing villages, particularly in Newfoundland and Nova Scotia, was largely determined by small, sheltered harbours suitable for mooring boats, and quick access to good fishing grounds. While many of these coastal villages still remain, they are anachronisms in the late-twentieth century. Delivering basic public services, such as education and health care, in these coastal places is expensive because populations are small and are located in isolated, hard-to-reach areas.

From an economic perspective, this dispersed pattern of population distribution is a major weakness for three reasons:

- The highways and rail lines required to connect the urban centres in Atlantic Canada are more extensive than in regions with a more compact population geography, and are therefore more expensive to build and maintain.
- High construction and maintenance costs for transportation networks result in higher transportation costs for firms operating in Atlantic Canada, thus discour-

aging companies from investing and locating in the region.
- Because of the absence of a concentrated population and concentrated economic activity, firms are hampered from achieving economies of scale and **economies of association** (having parts and service firms close to the manufacturing plant).

The rural-to-urban migration, common to other regions of Canada, is occurring in Atlantic Canada, but may have been slowed because many fishers prefer living in coastal outports and, with limited education and skills, see few opportunities in urban centres. Furthermore, the cost of living in an outport is much lower than that in a regional centre, which further discourages rural-to-urban migration. 'Planned' relocation is fraught with two dangers: social disruption and economic transfer of rural poverty to urban poverty (Vignette 9.2). However, migration of younger members of outport families is occurring, while older members may prefer to live in the outport, despite the fishery problems.

Access to External Markets

Atlantic Canada's limited access to external markets has hindered economic development in the region. Manufacturing firms in Atlantic Canada have had to overcome high transportation costs to deliver their products to Canada's major markets in Central Canada. Competition from firms in Central Canada often drove Atlantic-based firms from the marketplace for two reasons:

- Manufacturing firms in Ontario and

Québec could achieve economies of scale, making their production costs lower and resulting in lower prices for consumers.

- Manufacturing firms in Ontario and Québec had much lower transportation costs in reaching their customers than similar firms in Atlantic Canada.

Unable to compete within the national marketplace, most manufacturers in Atlantic Canada had to limit their production to the local market and keep wages low in order to compete with imported products from Central Canada.

Before FTA and GATT, trade barriers made it difficult for manufacturing businesses in Atlantic Canada to access foreign markets. Even agricultural and resource products had to overcome American tariffs to reach markets in New England. For example, prior to FTA, sales of fish and potatoes to consumers in New England were hindered by tariffs and quotas. Since FTA and GATT, trade barriers to the United States and other foreign countries have diminished and will eventually disappear. In addition, a low Canadian dollar and an expanding American economy during the 1990s have created ideal conditions for an increase in exports to the United States and in the number of US tourists visiting Atlantic Canada.

Provincial Barriers to Trade

In addition to the obstacles that Atlantic Canada has faced in trading with external markets—Canadian and foreign—the region has also grappled with trade barriers between its provinces. Provincial barriers divide the region's population into four provincial markets, often impeding interprovincial trade and labour mobility. The purchasing policies of provincial governments and their various agencies frequently favour local producers and act as non-tariff barriers to producers from other provinces. Other barriers take the form of provincial marketing boards, provincially controlled liquor sales, and provincial-preference hiring practices in the construction industry. Such internal barriers are a major impediment to stimulating economic growth within Atlantic Canada. The very division of the region into four provinces results in higher administrative costs and thus higher tax rates, which exert an equally negative impact on economic development. Simply having four provincial governments for a relatively small population costs taxpayers in Atlantic Canada about $400 million annually, making them bear a higher tax burden than residents of other provinces.

From time to time, there has been talk of an Atlantic economic union or the creation of a single province to overcome this regional fragmentation. Such an amalgamation would not overcome the differences within the region's physical geography, but it might consolidate Atlantic Canada's economic and political power. It might even reduce the cost of government. As early as 1949, Joey Smallwood, Premier of Newfoundland, proposed a union of the four eastern provinces to enable them to increase their bargaining power in Ottawa. In the 1970s, the Deutsch Commission on Maritime Union recommended a 'full political union as a definite goal'. However, there was no support within Atlantic Canada, and before the 1970s had ended, Nova Scotia Premier

Table 9.3
Employment by Industrial Sector in Atlantic Canada by Province, 1995

Industrial Sector	NFLD (000)	PEI (000)	NS (000)	NB (000)	Atlantic Canada (000)	Atlantic Canada (per cent)
Primary	17	4	23	19	63	6.6
Agriculture	—	4	8	6	18	
Fishing	10	—	7	5	22	
Logging	—	—	4	5	9	
Mining	6	—	4	4	14	
Secondary	24	9	62	57	152	15.9
Tertiary	156	46	299	238	739	77.5
Total	197	59	384	314	954	100.0

Source: Adapted from Statistics Canada, 1996:Table 40.

Gerald Regan pronounced the proposal for a Maritime Union 'as dead as a door nail'. For the residents of Atlantic Canada, the advantages of a union are not readily apparent, and Atlantic premiers have not shown much enthusiasm for this proposal.

How much the taxpayers of Atlantic Canada would save with a single government is difficult to estimate, but political union should achieve considerable savings by eliminating overlapping public agencies and provincial trade barriers. Unfortunately, an Atlantic union would likely lead to massive layoffs in the public sector. For a region with high unemployment, such a prospect makes a union much less attractive.

Industrial Structure

Atlantic Canada's three primary resources are its fish, forests, and minerals, but the region's economic future lies in the development of its tertiary sector, especially high-technology industries. Employment in primary activities, as well as secondary and tertiary activities, provides an overall picture of the basic economic structure of Atlantic Canada and of each of its provinces. In Table 9.3 employment in primary activities accounts for 6.6 per cent of the labour force in Atlantic Canada; secondary employment constitutes 15.9 per cent; and tertiary employment 77.5 per cent.

Significant variations among the provinces' employment figures are also illustrated in Table 9.3. Newfoundland, as expected, ranks first in employment in the fisheries, but it also ranks first in mining, primarily because of the large iron-mining operations in Labrador. Employment in manufacturing is much stronger in Nova Scotia and New Brunswick than in Newfoundland and Prince

Edward Island, mostly because there is greater processing of local raw materials in these two provinces—from the manufacturing of agricultural, fish, and wood products to the processing of mineral products. However, manufacturing has also suffered its setbacks. For instance, heavy industry in Sydney used to employ many workers, but its steady decline has hurt the economy of Nova Scotia and made Cape Breton's unemployment figures among the highest in Atlantic Canada.

Atlantic Canada's Economy

As a result of its long-term economic problems, Atlantic Canada has become heavily dependent on Ottawa, which sends money to the region through cash subsidies. The most important subsidies are equalization payments and unemployment insurance payments. Unemployment insurance payments to workers in Atlantic Canada are extremely important, particularly to those engaged in seasonal industries, such as self-employed fishers and those employed in fish plants and logging operations. Ottawa first extended unemployment insurance to self-employed fishers in 1947. Before the cod moratorium in 1992, inshore fishers caught fish during the summer and drew unemployment insurance payments in the winter.

Ottawa has tried to overcome the economic malaise in Atlantic Canada by providing funds for new businesses and for improving the region's transportation system. Such initiatives were meant to create jobs in areas of high unemployment. Since the 1960s, federal grants for businesses have totalled over $6 billion. Most federal funds have been administered by several key federal agencies. In 1962,

the Atlantic Development Board was established to stimulate businesses and strengthen the region's public infrastructure. In 1969, the federal Department of Regional Economic Expansion (DREE) began to offer incentives to encourage companies to locate in Atlantic Canada and other less-favoured areas of the country. DREE also spent money on highway construction, schools, and municipal services. In the late 1980s, the Atlantic Canada Opportunities Agency was formed to develop the economy through business and job opportunities. But in the mid-1990s, Ottawa began to reduce its regional subsidies in order to combat the federal debt, thereby threatening the financial aid upon which Atlantic Canada relied.

Ottawa's ongoing influence on Atlantic Canada's economy was evident after the closure of the cod fisheries in 1992. After it announced a moratorium on cod fishing because of the dismal state of the stocks, the federal government promised additional financial assistance for the short term. The Atlantic Groundfish Strategy (TAGS) began in May 1994. TAGS provided nearly $2 billion, distributed in monthly payments over five years, to about 40,000 fishers and employees displaced by the moratorium. While cod fishers are found in all parts of Atlantic Canada, the greatest impact of the cod moratorium occurs in the isolated fishing communities that are totally dependent on the fishing industry. TAGS was intended to shift fishers into other types of work through retraining and education programs, but most fishers considered TAGS to be another form of unemployment insurance. In their hearts, the fishers hoped that the cod would return before the payments from TAGS ran out. The success of TAGS was hindered by the following factors: alternative employment opportunities in remote fishing communities are virtually non-existent; few fishers choose to leave their homes and village for an uncertain future in larger towns or cities; and given the age and education level of most fishers, retraining becomes an almost impossible task.

By early 1998, TAGS was coming to a close but neither had the cod returned to their former numbers nor had the number of cod fishers declined. Ottawa was faced with a dilemma. Pressure from Atlantic Canada, particularly Newfoundland, was intense. In August 1998, Ottawa announced that an additional $730 million would be available to cod fishers under the retraining program initiated by TAGS (MacAfee 1998:A3). Such funding would extend TAGS payments to cod fishers for another year but Ottawa made it clear that this would be 'the end of the line' (Greenspon 1998:A4). If the cod fishery is not reopened for the summer of 1999, the economic and social impact on cod fishers and their communities will be devastating.

As an older resource hinterland, Atlantic Canada is suffering from a problem that is all too common to such regions. Over the years, Atlantic Canada has depended heavily on the exploitation of its natural wealth. Except for the iron and steel industry based in Cape Breton Island, Atlantic Canada's natural wealth was exported to the United States, Britain, and other European countries where the resources underwent further processing. Therefore, a strong manufacturing and industrial economy was never established and much of the region's natural wealth was consumed. This pattern of economic development has limited future alternative for growth in the region.

Vignette 9.3

Bull Arm Complex

In 1990–1 the Bull Arm complex, an elaborate and highly specialized construction site, was built at a cost of nearly $500 million. The complex was designed to build the Hibernia offshore oil-drilling platform, which was towed out to the Grand Banks and began production in 1997. The Newfoundland government hoped that the Bull Arm complex could continue to build offshore components for the oil and gas industry's future energy developments. In 1998, these hopes were confirmed when Petrocan chose the Bull Arm complex to build the components for its production vessel (a project valued at $100 million). Though the steel hull for this ship is under construction at shipyards in South Korea, the key components for the vessel, including an oil-processing unit and a flare stack to burn off natural gas, are being built at Bull Arm in Newfoundland. In 2000, this floating oil-production system is expected to be completed and will then be positioned above the Terra Nova oil deposit to extract the oil.

Source: Adapted from Brent Jang, 'Hibernia Halves Output', *The Globe and Mail*, 5 March 1998:B1. Reprinted with permission from the Globe and Mail.

One of the alternatives that may grant the region a second chance is the vast wealth of energy lying below its continental shelf. However, the pattern of resource exploitation and export is the same. For example, the oil from Hibernia, the $6.2-billion, offshore oil project east of St John's, is destined for refineries along the eastern seaboard of the United States, while a gas pipeline will send the natural gas from the Sable Island deposit to New England. To counter the pattern of exploitation and export, provincial governments have tried to intervene in the marketplace to obtain more benefits from large-scale resource developments, but have had only limited success. For example, the Newfoundland government has obtained some concessions from the oil industry—the key components for the Terra Nova oil-production vessel will be manufactured within the province at Bull Arm, the construction site for the Hibernia platform (Vignette 9.3).

As Atlantic Canada searches for its place in the highly competitive global economy, diversification of its resource hinterland role is essential. Economic diversification lies in two directions: greater emphasis on value-added activities (i.e., manufacturing and processing of its natural resources, especially fish stocks) in the primary sector, and an expansion of high technology in the tertiary sector.

Key Topic:
The Fishing Industry

Nature has given Atlantic Canada a vast continental shelf that provides an excellent physical environment for fishing: the warm ocean currents from the Gulf of Mexico and the cold Labrador Current create ideal conditions for fish reproduction and growth. The continental shelf extends almost 400 km offshore (Figure 9.2). In some places, where the continental shelf is raised, the water is relatively shallow. Such areas are known as banks. The largest banks are the Grand Banks of Newfoundland and Georges Bank near Nova Scotia (Vignette 9.4).

The diversity that characterizes Atlantic Canada extends to the fishery as well. Although each province relies on the fishery, there are striking differences between them (particularly Newfoundland and the Maritimes provinces) in the type of their catch and the location of their fishery. Before the ban on cod fishing, Newfoundlanders depended mostly on cod fishing, while Maritimers harvest a variety of sea life, including cod, lobster, and scallops. Fishers from the Maritimes ply the waters of Georges Bank and smaller banks just offshore of Prince Edward Island and Nova Scotia. In these waters, Maritimers find cod, grey sole, flounder, redfish, and shellfish. Newfoundlanders, on the other hand, primarily fish in the waters of the Grand Banks, around the island of Newfoundland, and along the shore of Labrador, where cod and other **groundfish** congregate in the shallow areas of the continental shelf.

The fishing industry consists of an inshore fishery, an offshore fishery, and fish-processing plants. Traditionally, Newfoundland and Maritime fishers used a flat-bottomed skiff called a dory and stayed close to the shore. Only the larger skiffs ventured to the Grand Banks and other offshore banks. Today, inshore fishers still use smaller boats to fish close to the shore.

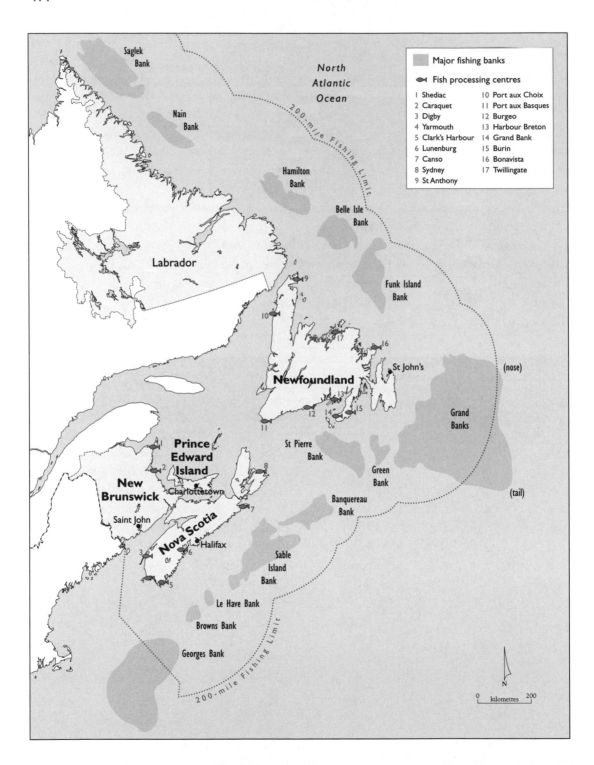

Major fishing banks

Fish processing centres

1 Shediac	10 Port aux Choix
2 Caraquet	11 Port aux Basques
3 Digby	12 Burgeo
4 Yarmouth	13 Harbour Breton
5 Clark's Harbour	14 Grand Bank
6 Lunenburg	15 Burin
7 Canso	16 Bonavista
8 Sydney	17 Twillingate
9 St Anthony	

Saglek
Bank

Nain
Bank

*North
Atlantic
Ocean*

200-mile Fishing Limit

Hamilton
Bank

Belle Isle
Bank

Labrador

Funk Island
Bank

Newfoundland

St John's

(nose)

Grand
Banks

St Pierre
Bank

Prince
Edward
Island

New
Brunswick

Charlottetown

Green
Bank

Banquereau
Bank

(tail)

Saint John

Nova Scotia

Halifax

Sable
Island
Bank

Le Have Bank

Browns Bank

Georges Bank

200-mile Fishing Limit

N

0   200
  kilometres

Vignette 9.4

Georges Bank

As part of the Atlantic continental shelf, Georges Bank is a large, shallow-water area that extends over nearly 4000 km². Water depth usually ranges from 50 to 80 m, but in some areas the water is 10 m or less. Georges Bank is one of the most biologically productive regions in the world's oceans because of the tidal mixing that occurs in its shallow waters. This brings to the surface a continuous supply of regenerated nutrients from the ocean sediments. These nutrients support vast quantities of minute sea life called plankton. In turn, large stocks of fish (such as herring and haddock), as well as scallops, feed on plankton. Cod also feed in these waters, but their numbers have dwindled due to overfishing, especially in international waters. Properly managed, Georges Bank can sustain annual fishery yields of about 400 000 tonnes.

During the winter, rough waters and ice prevent the smaller craft from leaving port. Offshore fishers travel in steel-stern trawlers to the Grand Banks and other offshore fishing banks where they fish for weeks. Upon returning to port, they unload hundreds of tonnes of iced groundfish at a fish-processing plant. Offshore vessels, often much larger than 25 tonnes, are equipped with radar, radios, sonar detectors, and other sophisticated equipment that enable them to operate in winter as well as summer.

The modernization of the Canadian fishing industry, with its larger trolling ships and more efficient nets and gear, was accompanied by huge investments by fish companies and individuals and an expansion of the fishing fleet in number and size. Atlantic Canada's fishery, particularly in Newfoundland and Nova Scotia, was able to increase its catch and its income. Now, fishing companies usually own the ships, so fishers no longer share in the value of the catch but receive a salary. However, modernization had many downsides:

- Fewer fishers were required by this more efficient fishing system.
- The new fishing system created the circumstances for overfishing.
- Overfishing and the collapse of the northern cod stocks forced inshore fishers to turn to government assistance to survive.

Management of Fish Stocks

The expansion and modernization of the world fishing fleets has threatened world fish stocks. With no management of fish stocks in

◄ **Figure 9.2 Major fishing banks in Atlantic Canada.** The Atlantic coast fishery operates within a vast continental shelf that extends some 400 km eastward into the Atlantic Ocean, southward to Georges Bank, and northward to Saglek Bank. Within these waters, there are at least a dozen areas of shallow water known as 'banks'. The Grand Banks of Newfoundland is the most famous fishing ground while Georges Bank contains the widest variety of fish stocks.

Fishers unload a catch. Fishing as a way of life is epitomized by Atlantic Canada's inshore fishers, who use small boats and remain close to the shore to fish. (Courtesy Aninda Chakravarti)

international waters, pressure on selected fish stocks grew. By the 1970s, the world fishing fleets had greatly increased their take of groundfish in the Grand Banks. As is the case in British Columbia with the salmon, the issue of fishing limits in Atlantic Canada was crucial in trying to manage the fish stocks in a sustainable way. Management of the international fish stocks in Atlantic Canada was so important that in 1977, Ottawa claimed the right to manage the fisheries within a 200-nautical mile zone off the East Coast of Atlantic Canada. Washington also extended its fishing zone to the 200-nautical mile limit, making Georges Bank an area of dispute between the two countries.

Georges Bank is one of the richest fishing banks, where fish and shellfish, such as scallops and mussels, are found. Fishers from both the Maritimes and New England harvest the fish stocks on this bank, which is located at the edge of the Atlantic continental shelf between Cape Cod and Nova Scotia. In the 1970s, fish stocks, especially herring, dropped to dangerous levels due to overfishing. A boundary decision made by the International Court at The Hague in October 1984 allocated five-sixths of Georges Bank to the United States. However, the easternmost one-sixth that was awarded to Canada is rich in groundfish and scallops. The Nova Scotia fishers have greatly benefited from this decision. With the scallops now under Canadian management, sustainable harvesting of the scallops and groundfish has brought a measure of stability to this sector of the fishing industry.

However, overexploitation of international fish stocks remains a worldwide problem. The root of the problem is that public resources are decimated by the selfish actions of individuals who have no regard for the well-being of the resource. Such mismanagement is known as the Tragedy of the Commons. Such ecological disasters occur when no one is responsible for ensuring the proper management of resources. Northern cod stocks fell victim to this tragedy twice: in the 1970s when there was no management of fishing in the Grand Banks, and again in the 1990s when the resource was mismanaged (Vignette 9.5). Ottawa's mismanagement was based on two factors: (1) Ottawa issued quotas for cod fishing that were too high because federal fisheries officials' estimate of the size of the cod stock was too high; and (2) Ottawa had no control over foreign fishing along the nose and

tail of the Grand Banks. The circumstances and consequences of the demise of cod stocks in Atlantic Canada can lead to a greater appreciation of the need for sustainable harvesting in the fishing industry.

Overfishing and the Cod Stocks

The Atlantic cod (*Gadus morhua*) is the major natural resource in Atlantic Canada. While the Atlantic cod (sometimes called the northern cod) is found as far north as the southern tip of Baffin Island and as far south as Cape Hatteras, cod stocks are concentrated off Newfoundland and along the Labrador Coast. The largest single concentration of northern cod is at the Grand Banks.

Historically, the Atlantic cod has been one of the world's leading food fishes. Cod is sold fresh, frozen, salted, and smoked. Salted cod was very popular in the past, but now most of the cod sold is frozen. Cod has been fished on the Grand Banks for centuries. During the sixteenth century, the catch was small, probably averaging less than 25 000 metric tonnes a

---

Vignette 9.5

An Ecological Crisis

The decline in groundfish stocks in Atlantic Canada is an ecological crisis. The Canadian Atlantic Fisheries Scientific Advisory Committee revealed in mid-1992 that stocks had declined sharply since the 1960s, when the **biomass** (the volume of fish aged three years or older) was over 3 million tonnes. By 1991, the biomass had dropped to between 530 000 and 700 000 tonnes, possibly the lowest level ever observed. **Spawning biomass** (generally seven years or older) represents the reproductive parts of the fish stock. In 1991, it amounted to between 50 000 and 110 000 tonnes. Thirty years before, the spawning biomass had been as high as 1.6 million tonnes, and even four years earlier, it was 400 000 tonnes. Faced with such a serious crisis, the Canadian government announced in 1992 a moratorium on cod fishing in the waters of Atlantic Canada.

year. By the end of the seventeenth century, however, the annual catch may have reached 100 000 metric tonnes. Annual cod landings during the nineteenth century were about 150 000 to 400 000 metric tonnes. As fishing technology advanced, catches of cod rose to nearly 1 million metric tonnes in the 1950s. By the 1960s, the annual catch reached a peak of almost 2 million metric tonnes. European trawlers accounted for most of this catch.

With the emergence of a modern fishing fleet, fish stocks around the world were over-fished. Several fish stocks, including the California sardine, Peruvian anchovy, and Namibian pilchard, collapsed in the 1980s from overfishing—not so much from local fishing operations but from the highly modernized fishing fleets that scan the world's oceans for fish. Overfishing of the northern cod has been attributed to foreign and Canadian fishing fleets that employ sophisticated fishing equipment and huge nets that indiscriminately trap all groundfish regardless of age, species, and value. Waste is enormous because 'non-commercial' fish are simply discarded.

In the case of cod, Ottawa's ability to manage the fish stocks was complicated by the fact that parts of the Grand Banks, known as the nose and tail, extend into international waters. This made it difficult for Ottawa to monitor and enforce catches in these areas. An agreement was reached with the countries belonging to the European Common Market, but monitoring the actual catches proved difficult. Some European fishing vessels, particularly those from Spanish ports, were exceeding their fish quotas and were catching undersized cod by using nets with a small mesh, which is illegal. During the 1980s, groundfish stocks, especially the northern cod stocks, dropped significantly—perhaps as much as 60 per cent from a high of 775 000 tonnes (Cashin 1993:19). This collapse translated into a decline in fish landings and processing. Cod landings were reduced by as much as 40 per cent from 1991 to 1992. Ottawa was forced to impose a moratorium on the cod fishery. The fear was that the northern cod stocks had collapsed and further fishing would make their recovery more difficult. Scientists are unsure as to how long it will take for the northern cod stocks to recover. Since cod must reach the age of seven before the fish can reproduce, the recovery may take a decade, perhaps longer.

The impact of the cod moratorium on Atlantic Canada, particularly Newfoundland, was severe. The cod has been the mainstay of Newfoundland's economy. Quite justifiably, the Atlantic cod was called 'Newfoundland currency'. In the past, the total catch by Newfoundland fishers was close to four-fifths groundfish. By comparison, the groundfish would constitute only about half of the catch in Nova Scotia and about one-third of the catch in Prince Edward Island and New Brunswick.

Before the moratorium, most jobs were in the fish-processing plants. Often these jobs lasted for only a short period, sometimes less than a month. However, employment at fish-processing plants provided two highly valued items—much-needed cash and enough weeks of work to qualify for unemployment insurance. Since May 1984, fish-processing employees have been eligible for financial support from TAGS and the extension to TAGS announced in 1998. In Newfoundland, the fish-processing sector was dominated by two firms, Fisheries Products International and National Sea Products. Prior to the moratorium on cod fishing, these two firms accounted for about half of the fish-processing jobs or nearly 8 per cent of the total number of employed people in Newfoundland.

The fishing industry in Atlantic Canada has suffered under the July 1992 moratorium on cod fishing. Since 1995, however, Newfoundland's fishing industry has recorded higher and higher production values because of high fish prices for crab, shrimp, lobster, clams, turbot, and other stocks. In this highly managed industry, only a few fishers have licences for these special kinds of catches. For example, only about 1,200 licences to catch crab are issued to Newfoundland fishers each year. On the other hand, before the cod moratorium, approximately 7,000 licences were issued for groundfish. Without cod, wealth from the fishery is concentrated in fewer and fewer hands. By 1997, small catches of cod by inshore fishers were permitted, but, in spite of strong pressure from fishers, Ottawa is still not prepared to allow the cod fishery to reopen. In Atlantic Canada, fishing is not only an important economic industry, it is a way of life. If the cod stocks are unable to recover, and if the overexploitation of other stocks occurs, the results would be devastating for all fishers in Atlantic Canada.

Atlantic Canada's Land Wealth

Though Atlantic Canada's early settlement and subsequent development revolved very much around the harvesting of the sea, the land mass of the four provinces has also provided the region's economy with important and profitable natural resources (Figure 9.3). The following section discusses the forest, mining, and agricultural industries within Atlantic Canada.

Forest Industry

Beyond the settled areas of Atlantic Canada, the land is covered by forest. Atlantic Canada has 22.9 million ha of forest land. Within the region, the forest industry plays an important role in the economy, particularly in the economy of New Brunswick. In the 1990s, the average annual value of forest products was approximately $3 billion. Almost half of this production originated in New Brunswick. Most trees harvested, including logs from private woodlots and especially spruce, are sold to local pulp mills. There are ten pulp plants in New Brunswick, four in Nova Scotia, and three in Newfoundland. These mills account for most of the exports. With the exception of Saint John, where there are two pulp mills, mill towns have no other industrial activities and are therefore single-industry communities.

Employment in forestry is declining due to the introduction of efficient logging equipment and the popularity of clear-cutting operations. A single-grip timber harvester with one operator can cut as many trees in a day as twenty men using chain-saws.[1] Capital investment is also replacing high-priced labour in wood processing. The forestry companies' goal is to maximize profits by keeping their products competitive in the global marketplace. Unfortunately, this process adversely affects

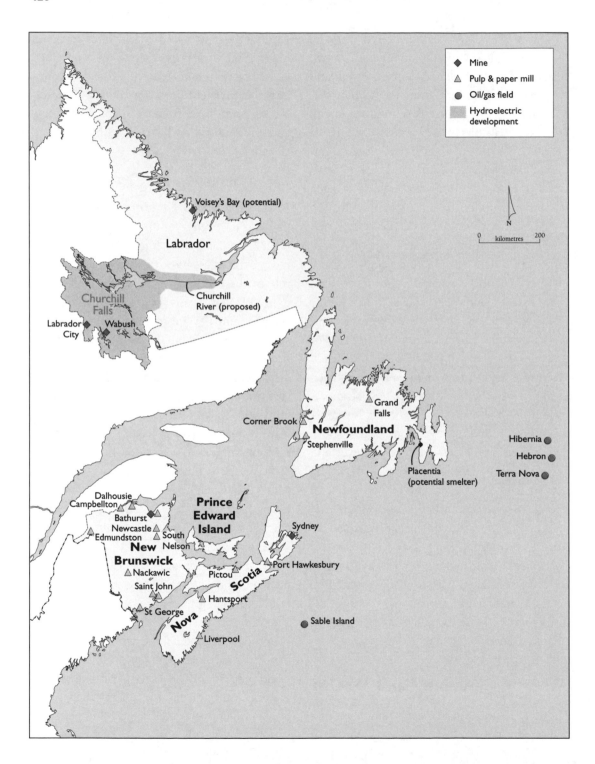

Mine

Pulp & paper mill

Oil/gas field

Hydroelectric development

N

0   kilometres   200

Voisey's Bay (potential)

Labrador

Churchill Falls

Churchill River (proposed)

Labrador City   Wabush

Grand Falls

Corner Brook

Newfoundland

Stephenville

Hibernia

Hebron

Terra Nova

Placentia (potential smelter)

Dalhousie

Campbellton

Prince Edward Island

Bathurst

Newcastle

Edmundston   South Nelson

New Brunswick

Sydney

Nackawic

Pictou

Port Hawkesbury

Saint John

Scotia

Hantsport

St George

Nova

Sable Island

Liverpool

resource hinterlands by reducing the size of their labour force, leaving many unemployed. Mechanization has also made logging difficult for the small logging operator, who cannot afford the high cost of current machinery and who often owns a small woodlot where such machinery is unsuitable. Small operators, including those with woodlots, are finding it difficult to survive in the face of such stiff competition.

Because Atlantic Canada was settled first, its hardwood and softwood forests have been harvested for a very long time—over 300 years. In Nova Scotia and New Brunswick, some areas have been logged three or four times. Logging is an important occupation in Atlantic Canada. Unlike in the rest of Canada, where forest land is usually Crown land (public), the proportion of private timber lands to Crown lands in the Maritimes is extremely high. Private timber lands comprise 92 per cent of the commercial forest in Prince Edward Island, 70 per cent in Nova Scotia, and 50 per cent in New Brunswick. In Newfoundland, like the rest of Canada, private ownership makes up only 2 per cent of the forested area. On average, the rate of logging on private lands is very high, sometimes exceeding the annual allowable cut estimated by the province. Exceeding the allowable cut impedes sustainable harvesting practices in the region. The high rate of logging often takes place on farms where timber sales are an important source of income.

Another problem being experienced in the forest industry in Atlantic Canada relates to Aboriginal peoples' right to timber lands. Aboriginal Canadians are demanding greater access to timber. Most Crown land in Atlantic Canada has been leased to large companies, mainly to the Irving Forest Corporation. In 1998, a dispute erupted over cutting rights on Crown land. Micmac Indians in New Brunswick and the provincial government are trying to negotiate a compromise whereby some Crown land that is currently leased to large companies would be reallocated to Indians. At this time, the matter remains unresolved.[2]

Mining Industry

Atlantic Canada is endowed with a number of minerals with commercial value. In 1996, mineral production in this region included metallics (such as iron), non-metallics (such as potash), and fuels (such as coal). There are vast iron ore and nickel deposits in the Canadian Shield, and coal, lead, potash, and zinc deposits in Appalachia. Historically, coal mining on Cape Breton Island was extremely important to the local economy and, at its zenith, the coal industry was the major employer in Cape Breton.[3]

In 1996, the annual value of mineral production in Atlantic Canada was $2.5 billion (Statistics Canada 1998e:1). In 1998, even with the close of the potash mine in New Brunswick, the value of oil produced at Hibernia will substantially increase this figure—

◀ **Figure 9.3 Natural resources in Atlantic Canada.** Besides fish, Atlantic Canada contains important natural resources. Iron deposits are located in Labrador, huge oil and gas reserves are found below the continental shelf, and one of Canada's largest hydroelectric facilities is situated on the Churchill River. Other natural resources, such as fertile soils and large forest reserves, are more limited.

perhaps by more than five times. The leading mining operations in Atlantic Canada in the 1990s were based on the lead-zinc and potash deposits in New Brunswick, the iron ore in Labrador, and coal in Nova Scotia.

Mining is an important industry in Atlantic Canada. For instance, the largest single employer in Labrador, the Iron Ore Company of Canada, employs nearly 400 workers. In 1993, these workers mined, processed, and shipped 13.4 million tonnes of iron in the form of pellets and concentrates to steel mills in Canada and the United States. Another 400 workers are employed at the lead-zinc mine operated by Brunswick Mining and Smelting Corp. near Bathurst, New Brunswick. This deposit, however, is a high-cost operation, so both the mine and the smelter shut down during periods of low prices for lead and zinc. Such a shut-down took place in 1993. At that time, the company announced the elimination of 112 jobs as a cost-cutting measure (Canada, Natural Resources 1994:27.2).

A Dangerous Occupation

Mining can be a very dangerous occupation—nowhere has this been more evident than in coal mining, where miners work underground in often volatile conditions. Coal has been very much a part of Nova Scotia's history. Although a coal miner's job may pay well, it is a dirty, physically debilitating, and dangerous occupation. As there are few other jobs available in Cape Breton and Pictou County, mining companies there have had no trouble recruiting miners. Danger goes with the job and, for some, the cost is their lives. In 1956, and again in 1958, Springhill Mine was the site of two underground explosions that took the lives of nearly 100 miners. In 1992, the Westray Mine in Nova Scotia was the site of another mine disaster (Vignette 9.6). A build-up of methane gas caused an underground explosion. Twenty-six men lost their lives. The mine had failed safety inspections and yet continued to operate. No doubt, the marginal nature of this coal-mining operation contributed to the substandard conditions of the mine.

Vignette 9.6

Westray: The Inevitable Disaster

At 5:18 on the morning of May 9, 1992, an explosion ripped through the Westray Mine at Plymouth in Pictou Country, Nova Scotia. Twenty-six men were trapped inside and all died underground. It was a tragic mining accident that was only one in a long legacy of catastrophes in which Nova Scotian miners paid with their lives for digging out the coal in this province. Despite significant improvements in mine safety, Westray drove home the realization that coal mining had begun as a dangerous enterprise here and, with modern safety precautions, the death toll in the coalfields might diminish but would not disappear. Once again the question would be raised: is coal worth the price of the men who would die underground?

Source: Lesley Choyce, Nova Scotia: Shaped by the Sea (Toronto: Viking, 1996):273. Copyright © 1996 by Lesley Choyce. Reprinted by permission of Penguin Books Canada Limited.

Westray was under pressure to keep its costs low to remain competitive, so work stoppages to correct the problems identified in the safety inspections would not only have been costly but might have resulted in Westray's failure to deliver sufficient coal to Nova Scotia Power. Unfortunately, as is common in 'old' resource hinterlands, the choice between closing mines where the health and safety of miners is a known risk and maintaining employment in an economically depressed region always favours the mines—until disaster and tragedy strike.

Steel, Iron, and Coal—The Rise and Fall

The iron and steel industry in Cape Breton Island near Sydney, Nova Scotia, was for a long time the heavy industrial centre of the Maritimes. In 1899, the Dominion Iron and Steel Company began to produce steel from Cape Breton coal and Newfoundland iron ore. Demand for its steel proved so great that plans were made to expand steel production. In 1901, a large, more integrated steel complex was built to supply steel rails to firms constructing new rail lines in Canada and other countries. Located at a fine harbour on the Atlantic Coast, the iron and steel mill at Sydney on Cape Breton Island had ready ocean access to raw materials and world markets. It was able to use local coal and iron ore from Bell Island, Newfoundland. It supplied markets in Canada (via the St Lawrence River and the Intercolonial Railway) and other members of the British Empire. Now called the Dominion Steel and Coal Co. (Dosco), the firm benefited from the strong demand for steel during the First World War. Under these prosperous conditions, Dosco became Canada's largest private employer.

In the 1920s, the Sydney steel mill began to fall on hard times. By 1925, Canada's railway boom had ended, weakening the demand for steel, and the demand for coal slumped as many coal customers turned to another source of energy, petroleum. A modest recovery for coal, however, occurred during the Second World War. After the war effort, it became evident that the Sydney steel mill was no longer viable without public subsidies. Sydney simply could not compete with offshore steel producers in Brazil and South Korea. The basic problems facing the Sydney steel mill were threefold: little local demand in the Maritimes, stiff competition in Canada from low-cost producers in southern Ontario and even stiffer competition in external markets, and increasing costs for its raw materials and labour.

Even with the assistance provided by government subsidies, Dosco continued to lose money. As a result, in 1965, Dosco announced plans to close its coal mines, which would eliminate 6,500 jobs. In 1967, the federal government took over the operation of the coal mines through a Crown corporation, the Cape Breton Development Corporation (Devco). The steel mill was on the brink of bankruptcy when it was sold to the Nova Scotia government in 1967. From that time on, a provincial corporation, the Sydney Steel Corporation (Sysco), not only ran the mill but, with help from the federal government, also paid for the annual shortfall. The two governments interfered because they did not wish to add to the unemployment problem in Cape Breton, an already economically depressed area with a very high unemployment rate. As Blackbourn and Putnam observed:

> The continued existence of an uneconomic and obsolete steel mill in an unsatisfactory location is a classic illustration of the importance of inertia in industrial location. The closure of the steel mill would create unemployment in a depressed province. Not only would jobs be lost in steel but the coal mines which supply the mill would probably close leading to further unemployment (Blackbourn and Putnam 1984:112).

In the late 1960s, neither the federal government nor the Nova Scotia government could accept closure of the Sydney steel mill for social and political reasons. At that time, the two governments' treasuries could absorb the annual losses, even advance more funds for plant improvements. However, by the early 1990s, governments had to deal with their enormous debts, and cutbacks were the order of the day. Even after the Nova Scotia and federal governments stepped forward with financial support, the industry continued to limp along. In 1996, the Crown corporation owned by Sydney Steel Corporation, which had received more than $2 billion in federal and provincial subsidies over the last twenty years, was sold to the Chinese government for $30 million. At the time of the sale, only 500 workers were employed at Sydney Steel Corporation, a far cry from the peak of 4,000 in the late 1960s. The Chinese government plans to relocate the mill in China, where labour costs are much lower and the demand for steel

is strong. Sydney's steel mill has now been reduced to an electric-arc furnace using scrap metal to manufacture steel rails. Without the political will to subsidize the Sydney steel mill, its future is dim.

Before the sale of the Sydney steel mill to the Chinese, demand for Cape Breton coal was already declining. With the switch at Sysco from a coal-burning steel furnace to an electric-arc furnace in 1987, a major customer for Cape Breton coal was lost. Cape Breton bituminous coal mines are operated by the federal agency, Cape Breton Development Corporation (Devco). Most coal is sold under contract to Nova Scotia Power. At first glance, coal from the Cape Breton coalfield is ideally suited for the nearby thermoelectrical stations operated by Nova Scotia Power. While the cost of transporting coal to these generating stations is low, the cost of mining Cape Breton coal is high due to its underground mining operations, which extend several kilometres. As a result, coal from deep tunnel mining at Glace Bay, Cape Breton, cost more to deliver to Nova Scotia Power than coal from Pennsylvania. With NAFTA, American coal is no longer barred from Canada by high tariffs. For that reason, Nova Scotia Power, now a privatized Crown corporation, wants to purchase lower-priced American coal to fire its thermoelectrical-generating plants in order to keep its electricity rates at their current level. But Devco has a long-term contract (to 2012) with Nova Scotia Power (at the higher price for coal). In February 1999, the government announced it was ending subsidization of Devco, which employed about 1,700 people at its two Cape Breton mines. One mine will be closed and the

Vignette 9.7

Sydney and the Sydney Steel Mill

Sydney is the principal city of Cape Breton Island and the second-largest city in Nova Scotia. Sydney has a deep ocean harbour with coal deposits nearby. Since 1900, the heart of this industrial city has been its huge steel mill. Sydney's fate is closely tied to its major industrial firm, the Sydney steel mill, and local coal mines. As the mill and mines prospered in the early part of the twentieth century, the town of Sydney expanded, but since the end of the Second World War, the mill has suffered financial losses and the size of its labour force has been sharply reduced. In turn, this has reduced the demand for coal, triggering lay-offs in the coal mines. This process of deindustrialization, while delayed by federal and provincial subsidies, has taken its toll on the people of Sydney and Cape Breton. The town's economy and population have declined. In 1961, Sydney had a population of 33,617, but by 1996, it had dropped to about 25,000. The history of Sydney illustrates its reliance upon coal and steel. The future of Sydney, without the steel and coal industry to fuel its growth, is uncertain.

other one sold to a private company, thus jeopardizing Devco's long-term contract with Nova Scotia Power. If Devco loses its last major customer, the coal mines at Glace Bay would likely close, bringing to an end the dream of a heavy industrial centre in the Maritimes (Vignette 9.7).

Megaprojects

Though some mining industries that have played an important role in the economy of Atlantic Canada are on the decline, other prospects for future developments are bright. Offshore oil and gas developments, led by the Hibernia Oil Project and the Sable Island Gas Project, the Voisey's Bay Nickel Project, and the Lower Churchill River Hydroelectric Development could make a fundamental difference in the economy of the region.

The Hibernia Oil Project, is located 315 km east of St John's, Newfoundland, on the Grand Banks above the site of a huge deposit of oil and natural gas. To tap the estimated 615 million barrels of oil from the Hibernia deposit, an innovative offshore stationary platform was needed (Vignette 9.8). About 4,000 workers built a specially designed offshore oil platform that can withstand the pounding storms of the North Atlantic and crushing blows from huge icebergs. The huge concrete and steel construction sits on the ocean floor, with sixteen 'teeth' in its exterior wall designed to absorb the impact of icebergs.

The Hibernia drilling site has an expected annual output of about 30 million barrels, and will add greatly to Newfoundland's mineral output. Based on an average price of $20 per barrel over the next twenty years, the average annual value of production is estimated at $600 million. Oil production began in 1998. By 2000, Hibernia is expected to account for 12 per cent of Canadian oil production. Oil is exported to American and other foreign refineries. According to the owners, the Hiber-

Vignette 9.8

The Hibernia Platform

The Hibernia platform is a gravity-based structure positioned on the ocean floor on the southeast corner of the Grand Banks. The 111-m-high oil platform and oil-storage units weigh over 650 000 tonnes. In the summer of 1997, the platform was placed on the ocean floor just above the oil deposits. The depth of the water at this point is about 80 m, leaving the oil platform approximately 30 m above the surface of the ocean. This structure is designed to be a platform for the oil derricks, house pumping equipment, and living quarters for the workers, as well as provide storage for the crude oil. The rig will extract huge amounts of oil from the Avalon reservoir (2.4 km under the seabed) and from the Hibernia reservoir (3.7 km deep). The crude oil is then pumped from the Hibernia storage tanks to an underwater pumping sta-

The Hibernia platform has a massive concrete base upon which sits the drilling and production facilities and workers' accommodations. *(Courtesy Keith Storey)*

tion and then through loading hoses to three 900,000-barrel supertankers for shipment to foreign refineries.

Source: Adapted from Kevin Cox, 'How Hibernia Will Cast Off', *The Globe and Mail*, 12 November 1994:D8. Reprinted with permission from The Globe and Mail.

nia Consortium, production should continue for approximately twenty years.

Another megaproject underway is at Voisey's Bay. It is the site of a recently discovered huge nickel deposit, the Ovoid deposit, which lies close to the surface. It consists of 32 million tonnes of relatively rich ore bodies—2.8 per cent nickel and 1.7 per cent copper. Another deposit, Eastern Deeps, though larger

(about 70 million tonnes), has a lower grade of nickel and copper (similar to that found at the Sudbury mines).

The Voisey's Bay deposit has three attractive features: high-grade nickel, a surface deposit that has the potential for open-pit mining, and proximity to ocean shipping. These characteristics may make it the lowest-cost nickel mine in the world. In fact, there is con-

Discovery Hill, Newfoundland, is the site of a large nickel deposit that many hope will contribute to a revitalization of the region's weak economy. (Courtesy Voisey's Bay Nickel Company Limited)

cern that production from this mine might force higher-cost producers, such as the nickel mines in Sudbury, Ontario, to close their operations. For Newfoundland, the Voisey's Bay mining and smelting operation could employ over 1,000 workers. In late 1996, Inco Ltd announced that it would build a $1-billion nickel smelter and refinery at Argentia. Approximately 750 workers would be employed at the smelter and refinery. Developments are on hold until the Innu and Inuit land claims for this part of Labrador are settled, the price of nickel increases, and the issue of building a smelter in Newfoundland is resolved.[4]

Despite the lure of megaprojects, they present two problems for regional development. First, they are capital-intensive undertakings. During the construction phase a large labour force is required, but in the operational phase relatively few employees are required. Second, megaprojects in resource hinterlands lose much of their spin-off effects to industrial areas. As a consequence, economic benefits related to the manufacture of the essential parts of a megaproject go outside the hinterland. Efforts by hinterland governments to address this classic problem of economic leakage from resource hinterlands to industrial cores have had mixed results (see Vignette 9.3 and Note 4).

Agriculture

Agriculture is limited by the physical geography in Atlantic Canada. Arable land constitutes less than 5 per cent of the Maritimes. Arable land is even more rare in Newfoundland and Labrador, making up less than 0.1 per cent of its territory. Though limited in size, agricultural production significantly contributes to the economy of Atlantic Canada. In the 1990s, the average annual value of agricultural production in the region was about $1 billion. Specialty crops, especially potatoes, contributed heavily to the value of this production.

Atlantic Canada has nearly 400 000 ha in cropland and pasture. Almost all of this farmland is in three main agricultural areas—Prince Edward Island, the Saint John River Valley in New Brunswick, and the Annapolis Valley in Nova Scotia. Specialty crops, especially potatoes and tree fruit, are extremely important cash crops. In all three agricultural areas, there are dairy cattle grazing on pasture land. The dairy industry in Atlantic Canada has benefited from the orderly marketing of fluid-milk products through marketing boards.

Prince Edward Island is the leading agricultural area in Atlantic Canada. It has almost half of the arable land in the region. Most of Prince Edward Island's 155 000 ha of farmland is devoted to potatoes, hay, and pasture, with the principal cash crop being potatoes. Since the 1980s, many potato growers have had

contracts with the island's major potato-processing plants—Irving's processing plant near Summerside and McCain's plant at Borden–Carleton now dominate the potato industry on the island. The second major agricultural area, the Saint John River Valley, is in New Brunswick. Its 120 000 ha of arable land comprise about one-third of Atlantic Canada's farmland. The Saint John River Valley has the best farmland in New Brunswick. Nova Scotia has nearly one-quarter of Atlantic Canada's farmland, with 105 000 ha. Nova Scotia's famous Annapolis Valley, the region's third agricultural area, is the site of fruit orchards and market gardens. The valley's close proximity to Halifax, the major urban market in Atlantic Canada, has encouraged vegetable gardening. In both New Brunswick and Nova Scotia, potatoes are a major cash crop. Almost all potato farmers seed their potatoes under contract to McCain Foods, which is a multinational food-processing corporation based in New Brunswick. The company has benefited from NAFTA after the removal of tariffs on its food products, especially french fries, to the United States. Newfoundland has the least amount of farmland—just over 6000 ha.

While agriculture plays a secondary role in much of Atlantic Canada, it dominates economic activity in Prince Edward Island. In 1996, the value of agricultural products in Prince Edward Island was $349 million. Nova Scotia and New Brunswick accounted for $376 million and $323 million respectively. Newfoundland, which has so little arable land, accounted for less than $75 million (Statistics Canada 1998a:1, 3, 5, and 7 of 10).

Atlantic Canada's Urban Geography

The urban geography of Atlantic Canada is characterized by few cities and a pattern of highly dispersed settlement. These features are attributed to the region's geography and natural resources, especially its fish stocks, and its small economy. It is not surprising then that a relatively small percentage of Atlantic Canada's population lives in urban places. In 1996, only about half the people in the region resided in villages, towns, and cities, making Atlantic Canada the least urbanized region of Canada (Figure 9.4).

Atlantic Canada has a coastal settlement pattern. Major cities are situated along its coastline. Trade and fishing have played a key role in the development of this pattern. The region's interior is relatively 'empty'. The re-

---

Vignette 9.9

Halifax

Halifax, the capital of Nova Scotia and the largest city in Atlantic Canada, was founded in 1749. In 1995, Halifax, Dartmouth, Bedford, and all of Halifax County formed a single city unit. By 1996, Halifax had a population of 333,000. As in the past, its strategic location allows Halifax to play a major role on the Atlantic Coast as a naval centre and international port. The economic strength of Halifax, along with the neighbouring towns of Bedford and Dartmouth, rests on its defence and port functions, its service function for smaller cities and towns in Nova Scotia, and its provincial administrative functions. Halifax also has a small manufacturing base. However, Halifax's location for manufacturing activities is marginal, as illustrated by AB Volvo's announcement on 10 September 1998 that it will close its Halifax assembly plant (Keenan 1998:B1, B4).

Table 9.4

Census Metropolitan Areas in Atlantic Canada, 1981–1996

Census Metropolitan Area	Population 1981	Population 1996	Change (per cent)
Saint John	121,012	125,705	3.8
St John's	154,835	174,051	12.4
Halifax	277,727	332,518	19.7
Total	553,574	632,274	14.2

Sources: Statistics Canada, 1982:Table 1; 1997a:Table 1.

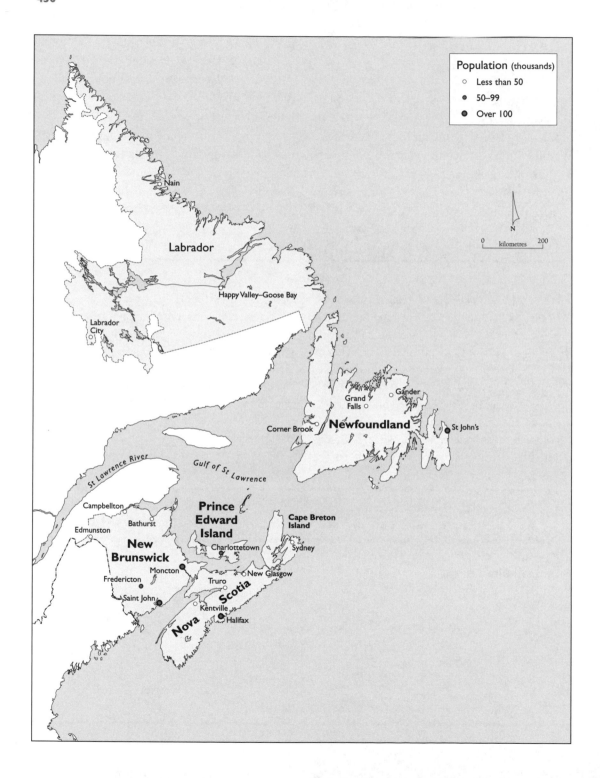

Population (thousands)

○ Less than 50

● 50–99

● Over 100

Nain

Labrador

Happy Valley–Goose Bay

Labrador
City

Gander

Grand
Falls

Corner Brook

Newfoundland

St John's

St Lawrence River

Gulf of St Lawrence

Campbellton

Bathurst

Edmunston

Prince
Edward
Island

Cape Breton
Island

New
Brunswick

Charlottetown

Sydney

Moncton

Fredericton

Truro

New Glasgow

Saint John

Nova

Scotia

Kentville

Halifax

N

0   kilometres   200

gion's physical geography has shaped this pattern because interior lands have few resources to attract settlement. The main exception is mining towns. The iron-mining town of Labrador City is an example of a single-industry community in the interior of Labrador's mining hinterland. Labrador City's population has been declining due to cost reductions and the substitution of machinery for labour.

Halifax, though the largest city in Atlantic Canada, ranks only thirteenth among Canada's major cities (Vignette 9.9). Halifax's relatively small size is due to three factors: (1) a less well-developed urban system and regional economy compared to other regions of Canada; (2) the fragmented nature of Atlantic Canada's population and resulting small markets; and (3) Atlantic Canada's relatively small economy. However, Halifax has the potential to become a 'Boston' of Atlantic Canada. The first step in this transformation would see Halifax become a super port for the eastern seaboard of North America. The need for such a port has arisen from the construction of container ships that are too large to pass through the Panama Canal. Of the three contenders—Halifax, Baltimore, and New York—Halifax has the best natural conditions, an extremely deep and sheltered harbour, but it is farther from the final destinations of the cargo. Container companies decide in 1999 which harbour will act as a super port for their ships.

Halifax and the next two largest cities in Atlantic Canada, Saint John and St John's, have experienced little population increase since 1981 (Table 9.4). Economic conditions have attracted few new immigrants to the cities of Atlantic Canada, and most of the region's residents have left for cities in other provinces. From 1991 to 1996, Halifax grew by 3.7 per cent, Saint John declined slightly by 0.1 per cent, while St John's increased by 1.3 per cent (Table 9.5). Moncton, located in northern New Brunswick, fared better. Its population increased from 107,000 to 113,000, giving it the highest rate (5.6 per cent) of the four largest cities in Atlantic Canada (Table 9.5).

Beyond these major urban cities are about a dozen smaller cities and towns that are at a lower level in the urban hierarchy of Atlantic Canada and have populations between 10,000 and 80,000. Fredericton, the capital of New Brunswick, is the largest of this group of urban places. With a population of nearly 80,000, Fredericton has been experiencing the highest rate of increase in this group of urban centres. The smaller centres did not fare so well. Over half of them lost part of their population. The iron-mining town of Labrador City suffered the largest percentage decline of 8.1 per cent, followed by three other resource towns, Sydney, Grand Falls–Windsor, and Corner Brook.

Atlantic Canada's urban geography remains on the margins—highly dispersed, broken into provincial units, and lacking a dominant metropolitan centre. This has led to an absence of cohesion, which is symptomatic of the region's economy.

◀ **Figure 9.4 Major urban centres in Atlantic Canada.** Atlantic Canada has few large cities and most urban centres are scattered across the region. Only four cities have populations exceeding 100,000. They are Halifax, St John's, Saint John, and Moncton.

Table 9.5
Urban Centres in Atlantic Canada, 1991–1996

Urban Centre	Population 1991	Population 1996	Change (per cent)
Sydney	26,063	25,025	(4.0)
Gander	12,037	12,021	0.1
Labrador City	11,392	10,473	(8.1)
Campbellton	17,183	16,867	(1.8)
Grand Falls–Windsor	21,053	20,378	(3.2)
Edmundston	22,205	22,624	1.9
Kentville	24,080	25,090	4.2
Bathhurst	25,734	25,414	(1.2)
Corner Brook	28,559	27,945	(2.1)
New Glasgow	38,676	38,065	(1.6)
Truro	42,697	44,102	3.3
Charlottetown	54,798	57,244	4.4
Fredericton	74,718	78,950	5.7
Moncton	107,436	113,491	5.6
Saint John	125,838	125,705	(0.1)
St John's	171,848	174,051	1.3
Halifax	320,501	332,518	3.7

Sources: Adapted from Statistics Canada, 1997a: Table 4, and calculations requested and provided by Statistics Canada for Sydney population 1996.

Atlantic Canada's Future

Atlantic Canada has a diverse geography and a rich history that combine to make the region a rewarding place to live, yet geography and history have also made it a difficult place in which to survive. The outflow of its people over the last century is an indication of the region's economic malaise. Even regional development funded by the federal government has not stemmed the outflow of Maritimers and Newfoundlanders. Atlantic Canada must now find its place in the integrated North American economy and the new world economy. The key to finding that niche lies in a new direction, one that more fully employs Atlantic Canada's human resources.

As a whole, Atlantic Canada's economy seems destined to lag behind that of other Canadian regions because its economy is so dispersed. Without a large population and a leading metropolitan centre, Atlantic Canada has been unable to build a strong industrial core—it therefore remains on the edge of the global economic system, making it doubly difficult to engage in new economic endeavours. Success has occurred in certain sectors—telecommunications in New Brunswick, software development in Nova Scotia, marine industries in Newfoundland, and specialty manufacturing on Prince Edward Island (DRI Canada 1994:1-40 to 1-41). Overall, however, Atlantic Canada's economy remains weak.

One problem facing Atlantic Canada is physical fragmentation. Geography has divided Atlantic Canada into three subregions: the strongest economy and largest market is in the Maritimes, followed by the island of Newfoundland, and then Labrador. Within Atlantic Canada, future prospects seem brightest for the Maritimes, which has the beginnings of a metropolitan centre in Halifax, particularly if this city is chosen as a super port for the new breed of container ships. Within other parts of the Maritimes, and in Newfoundland and Labrador, the opportunity to break out of a resource hinterland economy seems less assured.

Another challenge facing Atlantic Canada relates to the process of modernizing the primary sector of Atlantic Canada's hinterland economy. Primary activities are undergoing a technological revolution where capital is substituted for labour. Agriculture, aquaculture, fishing, forestry, and mining no longer require large workforces. As the cost of labour continues to rise, jobs will continue to be lost to machines. Consider the staple industry of Atlantic Canada, for example: even if the cod fishery regains its past productivity, the technological transformation of the fishing industry over the past twenty years means that large fishing vessels with smaller crews can catch most of the fish required by fish-processing plants. Furthermore, resource-led projects, like the iron mines in Labrador and the pulp mill at Corner Brook, have resulted in single-industry towns, which, like the hydroelectric development at Churchill Falls, have a limited impact on the provincial economy and the province's unemployment rate.

Atlantic Canada's primary sector is also facing the problem of dwindling resources. Atlantic Canada is an aging resource hinterland and many of its prime resources are gone. Those resources that remain are subject to a number of forces. For instance, in 1997, New Brunswick's potash mine closed because the mine flooded, and in 1998, the Hope Brook Gold Mine closed because of low gold prices. In fact, the price of primary products will be a determining factor in the future of the primary sector. If prices rise, Atlantic Canada may be able to diversify its economy. But if commodity prices are low, much of Atlantic Canada will remain trapped in a downward economic spiral. For example, the economic boost anticipated from Hibernia may not be realized if oil prices remain low (as they did in the late 1990s, dipping below US $15 a barrel). The future of Atlantic Canada's mining industry will also depend on the stability of commodity prices.

Ottawa's diminished capacity to subsidize Atlantic Canada's economy under international trade agreements (NAFTA and GATT) adds to the region's economic dilemma. Ottawa is not only less able to provide regional subsidies but is also less enthusiastic about them. Cape Breton Development Corporation's efforts to provide work for redundant coal miners by offering grants and subsidies to attract new industries to the region, as well as The Atlantic Groundfish Strategy's goal of assisting displaced fishers and fish-plant workers, have helped to relieve the local economic problem. But these measures have done little to resolve the fundamental weaknesses facing these two subregions and their people. Under these circumstances, the future of Cape Breton's coal and steel industry and Newfoundland's inshore cod fishery (as well as the fishing outports that rely on it) are not bright.

The global economy forces firms to reduce costs in order to survive. In Atlantic Canada, replacing workers with machines increases the number of unemployed workers. In turn, these high unemployment rates trigger an out-migration from Atlantic Canada, thereby shrinking its population base. Such a decline affects market size, the age of the labour force, and the psychological outlook of the people of Atlantic Canada. In the core/periphery model, this 'old' resource hinterland is trapped in a downward spiral. Is this the fate of Atlantic Canada?

Atlantic Canada could break out of its role as a dependent hinterland, but for this to hap-

pen, a combination of favourable circumstances, none of which is assured, would have to unfold. One of the most important would be the recovery of the fish stocks. This, coupled with a harvesting system that ensures the well-being of the fish stocks and a sharing of the catch between fishing fleets and individual fishers, would achieve two goals—a supply of fish for the processing plants and a maximization of the number of fishers.

A second economic condition would be the expansion of the tourism industry. Atlantic Canada offers tourists from New England, Ontario, and Québec a 'down home' vacation experience. The low Canadian dollar continues to attract American tourists, especially to the Maritimes. Atlantic Canada has a number of unique places that can draw selected tourists. For example, many Japanese tourists visit Cavendish in Prince Edward Island, the setting for Lucy Maud Montgomery's *Anne of Green Gables*. Others flock to Newfoundland for whale-watching.

A third economic necessity would be the development of high technology. Other regions of Canada are enjoying an expansion in high-technology industries due to Canada's highly skilled labour force. Often these innovative firms locate near universities. Halifax is particularly well-suited for such development; the success of the neighbouring American city of Boston provides an example. Such a development in Atlantic Canada would contribute greatly to the diversification of the region's economy.

Resource development is the last economic condition. Offshore oil and gas exploitation is well underway at Terra Nova and

Sable Island. In the next century, three major resource projects are possible: a hydroelectric dam on the lower Churchill Falls; an oil well at the Hebron oilfield near Hibernia; and a nickel mine at Voisey's Bay with an associated smelter near Argentia. These developments, if world prices for these resources remain high, could propel Atlantic Canada's economy into a more prosperous state—one in which unemployment rates decline, business opportunities increase, and out-migration ceases. Such an economic state may be that elusive balance between economic growth and a 'down East' way of life.

Summary

Atlantic Canada lies on the eastern edge of Canada. The region is characterized by physical fragmentation, cultural diversity, and a slow-growing economy. Atlantic Canadians rely heavily on natural resources—especially fish and energy. Canada's oldest hinterland has been unable to generate enough jobs to support its people. As a result, each year thousands leave Atlantic Canada, especially from Newfoundland.

As Atlantic Canada enters the twenty-first century, the region's future is uncertain. Some-

The Argentia Peninsula is the planned site of a smelter/refinery complex that would process the nickel mined at Voisey's Bay. (*Courtesy Voisey's Bay Nickel Company Limited*)

how Atlantic Canada must break out of its role as a resource hinterland. It must diversify its economy by expanding its tourism and by attracting high-technology industries to the region. Also, some resources remain that may be able to give the economy a boost: the vast energy resources below the sea, huge nickel deposits, and sites for hydroelectric power generation in Labrador. As well, a recovery of the cod stocks, coupled with sustainable harvesting, could bring a welcome return to a way of life for many coastal settlements in Atlantic Canada. All these possibilities hold the key to self-sufficiency and future prosperity in Atlantic Canada.

Notes

1. The FMG Timberjack 990 requires only one person—an operator—to fell a tree in a very short time. Priced at $600,000, this giant machine is one example of capital substitution for labour. The Timberjack grips the base of a tree, slices its steel blade straight across the trunk, and strips the tree of its branches. The operator then uses the Timberjack to cut the tree into measured lengths before utilizing a boom arm and clamp grip to load the cut logs onto a truck.

2. In 1997, New Brunswick's Court of Queen's Bench ruled that New Brunswick's Aboriginal peoples have a 'first' right to harvest trees on Crown land. In April 1998, the New Brunswick Court of Appeal overturned the earlier ruling. The province is negotiating with the Union of New Brunswick Indians.

3. Coal, a combustible sedimentary rock formed from the remains of plant life during the Carboniferous Age (a geological period in the Palaeozoic era), is classified into four types: anthracite, bituminous, subbituminous, and lignite. Anthracite is the highest grade of coal, while bituminous is used in the iron and steel industry and for generating thermoelectrical power. Most coal mined in Nova Scotia is bituminous coal.

4. The Labrador Inuit are close to a land-claims agreement with the Newfoundland government. This tentative agreement-in-principle contains the provision that the Labrador Inuit will receive twenty-five per cent of the revenue from mining and petroleum production in Labrador. The Labrador Innu continue their negotiations with the Newfoundland government.

The federal environmental review panel examining the possible environmental and social impacts of the Voisey's Bay Nickel Company's mine proposal are concerned about the disposal of the 15,000 tonnes of mine-tailing to be produced each day. The mining company proposes to deposit the toxic tailings in a pond and prevent the tailings from draining into surrounding streams and rivers by building two dams. While federal and provincial environmental officials are satisfied with the company's solution to the tailings problem, local people, especially the Inuit and Innu, are skeptical and worried about the effects of these toxins on the wildlife they depend on for food.

At the time of Inco's purchase of the Voisey's Bay deposit for $4.3 billion, Inco promised to build a smelter/refinery complex in Newfoundland or Labrador. In November 1996, Inco announced that Argentia was chosen as the proposed site for this complex. However, in 1998, Inco announced that the company would ship the ore to Sudbury, Ontario, rather than build a smelter/refinery at Argentia. In response, the Newfoundland government threatened to withhold approval for the mine/mill complex at Voisey's Bay. By early 1999, the issue of a smelter/refinery remained unresolved.

Key Terms

biomass
> The total quantity or weight of an organism (cod fish) in a given area.

economies of association
> Manufacturing plants that are located near their suppliers obtain lower prices because of the transportation savings associated with proximity to suppliers; localization economies.

groundfish
> Fish that live on or near the bottom of the sea. The most valuable groundfish are cod, halibut, and sole.

spawning biomass
> The total quantity or weight of a sexually mature organism that can reproduce. Cod, for example, reach sexual maturity around the age of seven.

References

Bibliography

Blackbourn, Anthony, and Robert G. Putnam. 1984. *The Industrial Geography of Canada*. London: Croom Helm.

Bradfield, Michael. 1991. *Maritime Economic Union: Sounding Brass and Tinkling Symbolism*. Halifax: Canadian Centre for Policy Alternatives.

Canada. 1993. *Canada Year Book 1994*. Ottawa: Ministry of Industry, Science and Technology.

Canada, Natural Resources. 1994. *1993 Canadian Minerals Yearbook: Review and Outlook*. Ottawa: Minister of Supply and Services.

_____. 1995. *The State of Canada's Forests: A Balancing Act*. Ottawa: Natural Resources.

Cashin, Richard. 1993. *Charting a New Course: Towards the Fishery of the Future*. Ottawa: Department of Fisheries and Oceans.

Choyce, Lesley. 1996. *Nova Scotia: Shaped by the Sea*. Toronto: Viking.

Cox, Kevin. 1994. 'How Hibernia Will Cast Off'. *The Globe and Mail* (12 November):D8.

Department of Fisheries and Oceans. 1994. *Canadian Fisheries Statistical Highlights 1992*. Ottawa: Department of Fisheries and Oceans.

DRI Canada. 1994. *Atlantic Canada: Facing the Challenge of Change*. Moncton: Atlantic Canada Opportunities Agency.

Forbes, E.R., and D.A. Muise, eds. 1993. *The Atlantic Provinces in Confederation*. Toronto: University of Toronto Press.

Greenspon, Edward. 1998. 'Ottawa Approves New Aid for Fishery'. *The Globe and Mail* (12 June):A1, A4.

Hardin, Garrett. 1968. 'The Tragedy of the Commons'. *Science* 162:1243–8.

Hutchings, Jeffrey A., and Ransom A. Myers. 1994. 'What Can Be Learned from the Collapse of a Renewable Resource: Atlantic Cod, *Gadus morhua*, of Newfoundland and Labrador'. *Canadian Journal of Fisheries and Aquatic Science* 51:2126–46.

Jang, Brent. 1998. 'Hibernia Halves Output'. *The Globe and Mail* (5 March):B1.

Keenan, Greg. 1998. 'AB Volvo to Close Halifax Plant'. *The Globe and Mail* (10 September):B1, B4.

Kirby, J.L. 1982. *Navigating Troubled Waters: A New Policy for the Atlantic Fisheries: Report of the Task Force on Atlantic Fisheries*. Ottawa: Department of Fisheries and Oceans.

MacAfee, Michelle. 1998. 'Diversity in Fisheries Paying Off: Minister'. *The Globe and Mail* (14 August):A3.

McCalla, Robert J. 1991. *The Maritime Provinces Atlas*. Halifax: Maritext.

McManus, Gary E., and Clifford H. Wood. 1991. *Atlas of Newfoundland and Labrador*. St John's: Breakwater.

Marsh, James H., ed. 1988. *The Canadian Encyclopedia*, 2nd edn. Edmonton: Hurtig Publishers.

Matthews, Ralph. 1983. *The Creation of Regional Dependency*. Toronto: University of Toronto Press.

_____. 1993. *Controlling Common Property: Regulating Canada's East Fishery*. Toronto: University of Toronto Press.

Stanford, Quentin H., ed. 1998. *Canadian Oxford World Atlas*, 4th edn. Toronto: Oxford University Press.

Statistics Canada. 1982. *Census Metropolitan Areas and Census Agglomerations with Components*. Catalogue no. 95-903. Ottawa: Minister of Supply and Services Canada.

_____. 1994. *Agricultural Economic Statistics*. Catalogue 21-603E. Ottawa: Statistics Canada.

_____. 1995. *Annual Demographic Statistics, 1994*. Catalogue no. 91-213. Ottawa: Statistics Canada.

_____. 1996. *Labour Force Annual Averages 1995*. Catalogue no. 71-220-XPB. Ottawa: Statistics Canada.

_____. 1997a. *A National Overview: Population and Dwelling Counts*. Catalogue no. 93-357-XPB. Ottawa: Industry Canada.

_____. 1997b. 1996 Census: Nation Tables—Population by Mother Tongue, Showing Age Groups, for Canada, Provinces and Territories, 1996 Census—20% Sample Data, 2 December 1997 [online database], Ottawa. Searched 15 July 1998; <URL:http://www.statcan.ca/english/census96/>:3 pp.

_____. 1997c. The Daily—1996 Census: Mother Tongue, Home Language and Knowledge of Languages, 2 December 1997 [online database], Ottawa. Searched 14 July 1998; <URL:http://www.statcan.ca/Daily/English/>:10 pp.

_____. 1998a. Canada Highlights: Census of Agriculture [online database], Ottawa. Searched 2 September 1998; <URL:http://www.statcan.ca/english/censuag>:10 pp.

_____. 1998b. *Canadian Economic Observer*. Catalogue no. 11-010-XPB. Ottawa: Statistics Canada.

_____. 1998c. The Daily—1996 Census: Aboriginal Data, 13 January 1998 [online database], Ottawa. Searched 14 July 1998; <URL:http://www.statcan.ca/Daily/English/>:10 pp.

_____. 1998d. The Daily—1996 Census: Ethnic Origin, Visible Minorities, 17 February 1998 [online database], Ottawa. Searched 16 July 1998; <URL:http://www.statcan.ca/Daily/English/>:21 pp.

_____. 1998e. Table 2: Revised Statistics of the Mineral Production of Canada, by Province, 1996 [online database], Ottawa. Searched 3

September 1998; <URL:http://www.nrcan.gc. ca/mms/efab/mmsd/production>:1 p.

_____. 1998f. 1981-1996 Census: Labour Force Activity [online database], Ottawa. Searched 2 September 1998; <URL:http://www.statcan.ca/ english/census96/mar17/labour/table6/>:1 of 4.

Stavely, Michael. 1987. 'Newfoundland: Economy and Society at the Margin'. In *Heartland and Hinterland: A Geography of Canada*, 2nd edn, edited by L.D. McCann, 247–85. Scarborough: Prentice-Hall.

Further Reading

Erskine, David. 1968. 'The Atlantic Region'. In *Canada: A Geographical Interpretation*, edited by John Warkentin, 231–80. Toronto: Methuen.

Macpherson, Alan G., ed. 1972. *The Atlantic Provinces: Studies in Canadian Geography*. Toronto: University of Toronto Press.

Savoie, Donald J., and Ralph Winter. 1993. *The Maritime Provinces: Looking to the Future*. The Canadian Institute for Research on Regional Development. Sackville: Tribune Press.

Wynn, Graeme. 1987. 'The Maritimes: The Geography of Fragmentation and Underdevelopment'. In *Heartland and Hinterland: A Geography of Canada*, 2nd edn, edited by L.D. McCann, 175–245. Scarborough: Prentice-Hall.

Chapter 10

Overview

While the Territorial North is the largest geographic region in Canada, it has by far the smallest economy and population. Its cold environment greatly limits economic development and settlement. The region is a resource frontier for both Canada and the world, with energy and mineral products accounting for most of its output. While hunting remains an important source of food for Aboriginal peoples, trapping for commercial purposes has declined sharply. With the settlement of land claims and the new territory of Nunavut, more and more Aboriginal peoples are involved in wage employment. Because Aboriginal peoples constitute nearly half of the region's population, their participation in economic development is crucial. Economic development in the Territorial North involves huge capital investments, so large-scale projects are common. The *Key Topic* in this chapter is megaprojects.

Objectives

- Describe the Territorial North's physical and historical geography, including the birth of Nunavut and modern land claims.
- Present the dualistic nature of its population and economy.
- Examine the population and economy within the context of the Territorial North's physical setting and resource potential.
- Explore the region's changing economic and political position within Canada, North America, and the emerging global economy.
- Focus on the role of megaprojects in the Territorial North's development.

The
Territorial
North

Introduction

Consisting of the Yukon, the Northwest Territories, and Nunavut, the Territorial North is Canada's largest region (Figure 1.1). In Friedmann's regional scheme of the core/ periphery model, the Territorial North would be described as a resource frontier (Table 1.5). Like other resource frontiers, the Territorial North has four principal characteristics: (1) it is far from world markets; (2) resource development is dependent on external demand; (3) the potential for resource development is limited by physical geography; and (4) its economy is sensitive to fluctuations in world prices for its resources.

The Territorial North is also a homeland for Aboriginal peoples. It is the only geographic region in Canada in which Aboriginal peoples constitute a sub-

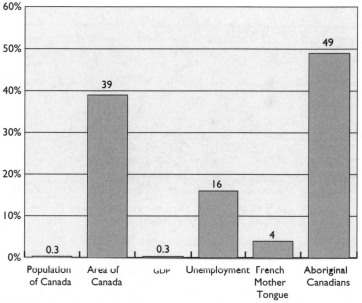

Figure 10.1 The Territorial North, 1996. Though the region is the largest in Canada, its population and economy are the smallest. Some French Canadians reside in the North but the majority of the population is Aboriginal. The Territorial North suffers from high unemployment.

Notes: The percentage of Aboriginal Canadians varies across the territories: Yukon (20.1 per cent), Northwest Territories (48.4 per cent), and Nunavut (85 per cent).

The unemployment figure is based on territorial data. Aboriginal unemployment is much higher, perhaps as high as 30 per cent.

Sources: Statistics Canada, 1997a: Table 1; 1997b:1 to 3 of 3; 1997c:4 and 5 of 15; 1998a: Table 39; 1998b:1 and 2 of 10; 1998c:15 to 20 of 21; and Stanford, 1998:185.

stantial proportion of the total population (Figure 10.1). For instance, in 1996, the Inuit constituted nearly 85 per cent of the popu-

South Nahanni River, NWT. Though the Territorial North is Canada's largest region, it is the most sparsely populated. (Corel Photos)

lation of Nunavut. Nunavut, the newly created territory, is a political expression of the homeland concept for the Inuit. (Nunavut is discussed later in this chapter.) The Territorial North then, as both a frontier and a homeland, has a dual character. This duality is a theme that runs through this chapter. The *Key Topic* explored is megaprojects in the Territorial North.

The Territorial North Within Canada

The Territorial North has the smallest economic output of the six geographic regions in Canada (Figure 10.1). As a resource frontier, the Territorial North has an export-based economy. Exploitation of non-renewable resources has spurred economic growth in the region, but this type of economic development lacks stability because it is subject to a boom-and-bust cycle. This cycle is caused by the finite nature of non-renewable resources and by fluctuations in world prices; downward price movements have caused temporary shutdowns of mining operations. Two recent examples are the Faro lead-zinc mine in the Yukon and the Colomac gold mine in the Northwest Territories.

Of the six geographic regions, the Territorial North has the smallest population. In 1998, it had fewer than 100,000 residents spread over nearly 4 million km². With a population density of only two people per 100 km², this region is one of the world's most sparsely populated areas. This demographic feature stems from three factors that inhibit the region's economic development and therefore also its population growth. These factors are: the region's cold environment, the nature of its resources (mainly non-renewable), and its distance from world markets. The first limiting factor, a cold environment, results in a low rate of biological productivity for renewable resources. Fish, for example, take much longer to reach maturity in the North than in warmer aquatic environments. In turn, the harvesting of such natural resources, particularly wildlife, has a limited capacity to support commercial enterprises. The second limiting factor, the nature of the region's resources, refers to the fact that economic development is heavily dependent on non-renewable resources. The exploitation of non-renewable resources inevitably leads to the depletion of those resources. Lacking the long-term economic stability that a more diversified economy brings, the resource-dependent region can only attract and support a limited population (or suffer high unemployment). The third factor, distance from world markets, translates into high transportation costs that inhibit resource development and population growth. Furthermore, distance from agricultural and manufacturing areas makes the Territorial North a high-cost region for imported foods and goods.

The Territorial North has two powerful and seemingly contradictory images—one is as a northern frontier, while the other is as a homeland. The traditional image of the northern **frontier** is one of great wealth that is just waiting to be discovered. For example, during the Klondike gold rush (1897–8), prospectors flooded the Yukon to pan for gold along the Klondike River. A more contemporary version of this image is one of large multinational corporations with their vast capital and advanced technology undertaking megaprojects—mining for gold, diamonds, lead, and zinc, and drilling for oil and gas. **Megaprojects** are large-scale resource developments that are financed and managed by multinational corporations designed to meet global needs for primary products. Such projects create an economic boom during the construction period, but in the operational phase of megaprojects, fewer employment opportunities are available and economic spin-offs for local businesses are limited. Because of the risks associated with developing resources in a frontier—from overcoming physical barriers unique to the Territorial North to coping with downturns in world prices for resources such projects are usually undertaken by large corporations. In return, these corporations reap large profits and supply the industrial cores of the world with raw materials and energy.

Northerners, however, particularly Aboriginal peoples, see the North as a **homeland**. This perception is based on a special, deep commitment to the North, which geographers often describe as regional consciousness. Local people have a strong appreciation for natural features, cultural traits, and the political and economic issues affecting their homeland. Regional consciousness, also known as a sense of

Offshore drilling in the Beaufort Sea. Megaprojects are common among resource frontiers, especially in the North, where adverse conditions require that developers have the investment power and experience to manage the risks. (Courtesy Bob Mahnic)

place (discussed in Chapter 1), evokes a sense of belonging to a particular place.

The dual roles of homeland and frontier are not always compatible. For instance, the interests of multinational companies relate to the profitability of the northern frontier's untapped natural wealth. The interests of northerners, however, especially Aboriginal northerners, are best served when the long-term environmental and social well-being of their communities is considered. 'Frontier' economic activities often bring jobs and investment, but these activities, by focusing on short-term profitability, have little impact on Aboriginal communities.

In the Territorial North, the concept of homeland and regional consciousness have resulted in the devolution of political power from the federal government to the territorial governments. Elected governments exist in the three territories, five land-claim agreements with First Nations have been reached, and the new Territory of Nunavut became a reality in 1999. Unlike Yukon and the Northwest Territories, Nunavut is an expression of 'ethnic' regional consciousness. The next challenge facing the Territorial North, but especially Nunavut, is to generate sufficient economic growth to break its economic dependency on Ottawa, while still preserving the homeland to which many northerners are committed.

The Territorial North's Physical Geography

The Territorial North includes the Yukon Territory, the Northwest Territories, and the new territory of Nunavut. The region extends over four of Canada's physiographic regions: the Canadian Shield, the Interior Plains, the Cordillera, and Arctic Lands (including the Arctic Archipelago) (Figure 2.1). As a result, the region encompasses a varied topography, from mountainous terrain and forest stands in the west, to barren plains, ice-covered islands, and Arctic seas further east and north. The vegetation in this region is also quite varied and includes a small portion of the boreal forest in the southwest, the tundra with its mosses and lichens further north, and a polar desert in the high reaches of the Territorial North (Figure 2.7). The region is also known for the several rivers that wind through it, the

The Yukon River Valley near Whitehorse. From Aboriginal peoples to fur traders and prospectors, this long and winding river has been an important transportation route in the history of the Territorial North. *(Courtesy R.M. Bone)*

many lakes that dot its landscape, and the Arctic Ocean that supports a range of aquatic wildlife.

However, the physical geography of the Territorial North is governed not so much by physiography as by a cold environment. Cold persists throughout most of the year and, in many ways, it affects human activities in the Territorial North. The Territorial North's cold environment includes permafrost (Figure 2.9) and long winters with subzero temperatures. The region's main climate zones, the Arctic and the Subarctic (Figure 2.6), are characterized by very short summers.[1] In the case of the Arctic climate, summer is limited to a few warm days that are interspersed by more autumn-like weather, including freezing temperatures and snow flurries. The Subarctic climate has a longer summer that lasts at least one month. During the short but warm summer, the daily maximum temperature often exceeds 20° C and sometimes reaches 30° C.

Arctic air masses dominate the weather patterns in the Territorial North. They are characterized by dry, cold weather and originate over the ice-covered Arctic Ocean, moving southward in the winter. In the Subarctic, the boreal or northern coniferous forest grows. In the Arctic, lower summer temperatures and continuous permafrost hinder the growth of trees so tundra vegetation results. In the very cold Arctic Archipelago, the ground is often bare, exposing the surface material. As there is little precipitation in the Arctic Archipelago (often less than 20 cm per year), this area's climate is sometimes described as a 'polar desert'.

The Arctic Ocean was called the 'Frozen Sea' by early explorers. This extensive ice cover, known as polar pack ice, drifts in a clockwise motion in the Beaufort Sea. Because of the extent and thickness of polar pack ice, few ships can navigate these waters without the assistance of ice-breakers. The two mining operations in the high Arctic—the Nanisivik mine on northern Baffin Island and the Polaris mine on Little Cornwallis Island—must store their ore until the summer navigation season, when ships that are reinforced against ice, sometimes aided by ice-breakers, transport the ore to European and other world markets.

The Territorial North's Historical Geography

At the time of contact with Europeans, seven Inuit tribes and seven Indian tribes belonging to the Athapaskan language family (also known as Dene) occupied the Territorial North. The Inuit tribes stretched across the Arctic: the Mackenzie Delta Inuit lived in the west; further east were the Copper Inuit, Netsilik Inuit, Iglulik Inuit, Baffinland Inuit, the Caribou Inuit, and the Sadlermiut Inuit. Inuit tribes also lived in northern Québec and Labrador. By the early-twentieth century, two tribes (most of the Mackenzie Delta Inuit and all of the Sadlermiut Inuit) would succumb to diseases that European whalers brought to the Arctic. The Indian tribes that resided in the Subarctic were the Kutchin, Hare, Tutchone, Dogrib, Tahltan, Slavey, and Chipewyan.

These Inuit and Indian tribes had developed hunting techniques that were well adapted to these two cold but different environments. The Inuit employed the kayak and harpoon to hunt seals, whales, and other marine mammals, which enabled them to occupy the Arctic Coast from the Yukon to Labrador. As a result, the Inuit lived almost totally on marine mammals and fish. The Indians hunted and fished in the northern coniferous forest, where the birch-bark canoe, the bow and arrow, and snowshoes enabled them to hunt in summer and winter. The Dene tribes relied heavily on big game like the caribou, the Chipewyans often following the caribou to their calving grounds in the northern barrens of the Arctic. Both the Inuit and the Dene moved across the land in a seasonal rhythm, following the migratory patterns of animals. Operating in small but highly mobile groups, these hunting societies depended on game for their survival. Cultural traits, such as the ethic of sharing, developed from this dependency on the land and sea for food.[2]

Though the Vikings were the first to make contact with northern Aboriginal peoples around 1000, little is known of those encounters.[3] About five centuries later in 1576, Martin Frobisher, in searching for a northwest passage to the Far East sailed to the northern Arctic. He reached Baffin Island where he encountered a group of Inuit, some on land, others in their kayaks. A skirmish between Frobisher's men and the Inuit took place in which Frobisher was wounded by an arrow. In the exchange, five of his men were lost and three of the Inuit were captured. The Inuit and one kayak were taken back to England as proof of Frobisher's discovery. All three of the captives soon succumbed to illness. Following Frobisher, various European explorers ventured into Arctic waters in search of the Northwest Passage, including John Franklin, whose famous last expedition ended in disaster (Vi-

---

Vignette 10.1

The Franklin Search

In 1845, Sir John Franklin headed a British naval expedition to search for the elusive Northwest Passage through the Arctic waters of North America. He and his crew never returned. Their disappearance in the Canadian Arctic set off one of the world's greatest rescue operations, which was conducted both on land and by sea and stretched over a decade. The British Admiralty organized the first search party in 1848. Lady Franklin sent the last expedition to look for her husband in 1857. These expeditions accomplished three things: (1) they found evidence confirming the loss of Franklin's ships (the *Erebus* and *Terror*) and the death of his crew; (2) one rescue ship under the command of Robert McClure almost completed the Northwest Passage; and (3) the massive rescue effort resulted in a greater knowledge of the numerous islands and various routes in this part of the Arctic Ocean. The exact sequence of events that led to the Franklin disaster is not known. However, archaeological work, conducted in the early 1980s on the remains of members of the expedition, revealed that lead poisoning, caused by the tin cans in the ships' food supply, probably contributed to the tragic demise of the Franklin expedition.

gnette 10.1). However, cultural exchange between Europeans and the original inhabitants of these lands remained limited until the nineteenth century, when the trade in fur pelts and whaling peaked in North America.

Whaling and the Fur Trade

Whaling, which was the first commercial venture in the Arctic, began in the late-sixteenth century in the waters off Baffin Island. During those early years of whaling, whalers had little opportunity or desire to make contact with the Inuit living along the Arctic Coast. The Inuit probably felt the same, particularly those who could recall the nasty encounter with Frobisher's men. During early summer, whaling ships set sail from British, Dutch, and German ports to Baffin Bay, where they hunted whales for several months. By September, all ships would return home. In the early-nineteenth century, the expeditions of John Ross (1817) and William Parry (1819) sailed farther north and west into Lancaster Sound. Their search for the Northwest Passage had limited success, but opened virgin whaling grounds for whalers. These new grounds were of great interest as improved whaling technology had reduced the whale population in the eastern Arctic. In fact, the period from 1820 to 1840 is regarded as the peak of whaling activity in this area. For instance, at that time, up to 100 vessels were whaling in Davis Strait and Baffin Bay.

As whaling ships went further afield to find better whaling grounds, it became impossible to return to their home ports within one season. By the 1850s, the practice of 'wintering over' (that is, allowing ships to freeze in sea ice along the coast) was adopted by English, Scottish, and American whalers. This practice allowed whalers to get an early start in the spring, providing for a long whaling season before the return trip home at the onset of the next winter. Wintering over took place along the indented coastline of Baffin Island, Hudson Bay, and the northern shores of Québec and the Yukon. Permanent shore stations were established at Kekerton and Blacklead Island in Cumberland Sound, at Cape Fullerton in Hudson Bay, and at Herschel Island on the Beaufort Sea. Life aboard whaling ships was dirty, rough, and dangerous, and many sailors died when their ships were caught in the ice and crushed.

The Inuit welcomed the whaling ships because of the opportunity for trade. The Inuit were attracted to shore stations and often worked for the whalers by securing game, sewing clothes, and piloting the whaling ship through difficult waters to promising sites for whale hunting. Some Inuit men signed on as boat crew and harpooners. In exchange for this work, the Inuit obtained useful goods, including knives, needles, and rifles, which made domestic life and hunting easier. While this relationship brought many advantages for the Inuit, there were also negative social and health aspects, including the rise in alcoholism and the spread of European diseases among the Inuit (Vignette 10.2). Perhaps the most devastating result of this trade relationship for the Inuit was the unexpected end of commercial whaling, which, for the Inuit, represented the loss of access to highly valued trade goods. Just as the twentieth century began, demand for products made from whales decreased sharply, halting the flow of

Vignette 10.2

European Diseases

Whalers, fur traders, and missionaries introduced new diseases to the Arctic. As the Inuit had little immunity to measles, small pox, and other communicable diseases like tuberculosis, many of them died. In the late-nineteenth century, the Sadlermiut and the Mackenzie Delta Inuit were exposed to these diseases. As a result, all the Sadlermiut died. The Mackenzie Delta Inuit, whose numbers were as high as 2,000, almost suffered the same fate, but managed to survive.

The Mackenzie Delta Inuit occupied the northwestern Arctic coast, in present-day Yukon, the Northwest Territories, and part of Alaska. Herschel Island, lying just off the coast of the Yukon, was an important wintering station for American whaling ships. Whalers often traded their manufactured goods with the local Mackenzie Delta Inuit, who became involved with the commercial whaling operations. Through contact with the whalers, the Inuit were infected by European diseases. By 1910, only about 100 Mackenzie Delta Inuit were left. Gradually, Inupiat Inuit from nearby Alaska and White trappers who settled in the Mackenzie Delta area intermarried with the local Mackenzie Delta Inuit, which secured the survival of these people. Today, their descendants are called Inuvialuit.

whalers, and thus trade goods, that were sailing into Arctic waters. By now, the Inuit were dependent on trade goods for their hunting activities. Somehow, they had to find another means of obtaining these useful trade goods.

Fortunately for the Inuit, the fur trade had been expanding northwards into the Arctic, thereby providing a replacement for whaling. The fur trade had already been successfully operating in the Subarctic for some time—a relationship between European traders and the Indian tribes in this part of the Territorial North was established through the trade of fur pelts, especially beaver. However, by the beginning of the twentieth century, the fashion world in Europe had discovered the attractive features of the Arctic fox pelt. Demand for Arctic fox pelts rose, which led the Hudson's Bay Company to establish trading posts in the

Arctic. Soon the Inuit were deeply involved in the fur trade. The working relationship between the Hudson's Bay Company and the Inuit was based on barter trade: white fox pelts could be traded for goods.

Until the 1950s, the fur trade dominated the Aboriginal land-based economy. It lasted for less than 100 years in the Arctic and for over three centuries in the Subarctic. Did the fur trade, as well as Arctic whaling, create a form of dependency whereby Indians and Inuit could not survive without trade goods? The answer is a qualified yes. At first, Aboriginal peoples had a form of partnership with European traders and whalers. Each side had power—for instance, the European traders needed the Indians to trap beaver and the Indians needed the traders to obtain European goods and technology. Gradually, however, the power relationship shifted in favour of the traders. By the nineteenth century, the fur companies controlled the fur economy. Fur-trading posts dotted the northern landscape. Indians, who had long ago integrated trade goods into their traditional way of life—including their hunting techniques and their migration patterns—were therefore heavily dependent on trade. In fact, when game was scarce, tribes relied on the fur trader for food. Ironically, by securing game for the traders, Indians reduced the number of animals that would be available for their own sustenance. In the Territorial North, game became scarce around fur-trading posts not from natural causes but from overexploitation.

The problems of a growing dependency on European goods and a changing way of life for northern Aboriginal peoples were compounded after the arrival of Anglican and Catholic missionaries in the 1860s and the Royal Canadian Mounted Police (RCMP) in the 1890s. The Indians and Métis were confronted with the full force of Western culture in the late-nineteenth century and the Inuit in the early-twentieth century. The Western ideas and rules introduced by the missionaries and police who now lived near the trading posts had a profound impact on Aboriginal culture. The Royal Canadian Mounted Police imposed Canada's system of law and order on Aboriginal peoples, while the missionaries challenged their spiritual values and encouraged the Inuit, Indians, and Métis to remain in the settlements. Also, both Anglican and Catholic missionaries placed young Aboriginal children in church-run residential schools, where they were taught in either English or French. In this attempted assimilation, most children learned to read and write in English or French, but were inadequately prepared for northern life. As they lost the opportunity to learn from their parents about how to live on the land, they were ill-prepared for such a life. Under these circumstances, many lost their indigenous language, animistic beliefs, and cultural customs. Fur traders opposed many of these induced Western cultural adaptations because the fur traders needed the Aboriginal peoples on the land to trap. Nevertheless, the influence of the church, the power of the state, and the number of non-Aboriginal residents in the North increased in the twentieth century, placing Aboriginal cultures under siege and crippling their land-based economy. However, political changes were occurring at this time that would first lead to territorial governments and then to land-claim agreements.

Political and Territorial Evolution

When Canada was formed in 1867, the Territorial North remained a British possession. Britain had claimed British North America on the basis of settlement, trade, and exploration. In the Territorial North, the British declared their ownership of the Arctic islands, basing their claim on the British Navy's efforts to find the Northwest Passage, including the search for Franklin's missing ship. In the rest of the Territorial North, the Hudson's Bay Company had established a number of fur-trading posts in the forested lands of the Mackenzie Basin and the Yukon. By extending its fur-trading economy over this area, the British government claimed these Subarctic lands. Canada came into possession of the Territorial North with the purchase of Rupert's Land in 1870 and the transfer of the Arctic islands by Britain to Canada in 1880.

At first, these northern territories were governed from Ottawa. The first territorial government was established in the Yukon after the population soared to 30,000. The demand for self-government came not from the Aboriginal population but from those who had been lured north by the Klondike gold rush. In 1898, the Yukon became a separate territory—a territorial government was formed that consisted of a federally appointed commissioner and council located in Dawson City.[4] All appointed officials were residents of the newly formed territory. In 1899, the process towards electing members of council began. At that time, two were elected and by 1908, all ten members were elected. People began to leave the territory when the production of gold declined. This population loss led Ottawa to withdraw some of the territory's powers, including reducing the size of its elected council. After the Second World War, Yukon's population again increased, reaching 9,096 by 1951, so the number of elected council members increased from three to five. By 1996, Yukon's population was 30,766, permitting seventeen elected members.

The evolution of the Northwest Territories was much different. In 1905, much of the Northwest Territories south of 60° N was assigned to two new provinces, Saskatchewan and Alberta. Another adjustment to the territories' borders took place in 1912, when Manitoba's boundary was extended to 60° N (see Figures 3.4 to 3.7). The boundaries of the Northwest Territories underwent another

change with the establishment of Nunavut in 1999. From 1905 until after the Second World War, the NWT was governed by the appointed commissioner and council, which was composed entirely of senior civil servants based in Ottawa. When its population reached 16,004 in 1951, elected members were gradually added to the previously all-appointed council until it became a fully elected body in 1975. Until 1963, the commissioner was a deputy minister in the federal department in charge of the administration of the Northwest Territories. In 1964, the first full-time commissioner was appointed to a separate territorial office. In 1967, the seat of territorial government was moved to Yellowknife and the commissioner was relocated there with the nucleus of what has since become a territorial public service. By 1996, the population has reached 64,402 and the council of the Northwest Territories consisted of twenty-four elected members.

The territories, including Nunavut (discussed later in the chapter), do not have all the powers given to the provinces through the Canadian Constitution. In fact, territorial powers do not stem from the Constitution but are assigned to the territories by the federal government. Territorial powers include education, social services, tax collection, highways, and community services. Given the importance of wildlife to Aboriginal peoples, the territorial governments are also responsible for wildlife. But the territories do not control tax revenues from their natural resources, which go to Ottawa. To offset this tax loss, the territories receive most of their revenue as transfer payments from the federal government. Without these transfer payments, the territorial governments would not be able to afford their existing programs.

Forgotten Frontier, Confederation to 1939

Since the Territorial North was transfered to Canada in 1880 to the Second World War, the region was a forgotten part of Canada. The Territorial North was forgotten because the region had little agricultural land and few commercial resources. For the federal government, the Territorial North was not a 'priority' region and thus received little attention from Ottawa. Aboriginal peoples, the vast majority of the population, continued to engage in hunting and trapping and, in times of need, turned to the fur trader for assistance. Ottawa adopted a *laissez-faire* policy towards the Territorial North to minimize federal expenditures. At the same time it left the fur traders and missionaries to deal with the needs of a hunting society.

The federal presence in the Territorial North was through the Royal Canadian Mounted Police, first assigned to Dawson City in Yukon during the Klondike rush. At numerous fur-trading posts across the North, the RCMP established detachments that not only enforced Canadian law but also provided a variety of other administrative services. During the winter, the local constable and his Native assistant would visit the various hunting camps by dog team to ascertain the well-being of the people at these camps and, if necessary, deal with law-and-order matters and administer basic health care.

Before 1939, the only commercial interest expressed in this region of Canada came in the form of gold seekers. Individual prospectors of the Klondike gold rush in the late 1890s were followed by mining companies that invested in the region and established gold mining operations. As a remote hinterland, the Territorial North had only a few mining operations, most of which were gold mines. Gold and other precious metals have such a high value per unit of weight that they can overcome the 'barrier of distance'; that is, the cost of transporting them to external markets. The cost of shipping other minerals that could be mined in the region—such as low-value minerals like copper, lead, nickel, and zinc—was simply too costly at this time. Coal, however, was mined in the Yukon because it supplied local heating needs.

Strategic Frontier, 1939–1990

A new role for the Territorial North began to emerge during and after the Second World War. This new role took two forms—one was as a strategic military zone, the other as a resource frontier. When the Second World War broke out, the United States military quickly recognized the strategic importance of Canada and its northern territories, a strategic importance that would last until the collapse of the Soviet Union in 1991. For the Americans, Canada's North provided a secure transportation link to the European theatre of war and, in 1942, to Alaska where the Japanese threatened to attack the United States. The transportation links were named the Northwest Staging Route and Project Crimson. Each consisted of a series of northern landing strips that would enable American and Canadian warplanes to refuel and then continue their journey to either Europe or Alaska. In the north-east, Project Crimson involved constructing landing fields at strategic intervals to allow Canadian and American airplanes to fly from Montréal to Frobisher Bay (now Iqaluit) and then to Greenland, Iceland, and, finally, England. In Canada's northwest, American aircraft came to Edmonton and then flew along the Northwest Staging Route to Fairbanks, Alaska, where their major military base was located. The Alaska Highway, which was built at the same time, provided road access to the various landing fields and a truck convoy route to Alaska. The US Army command had decided that the oil needed by the American armed forces in Alaska must be made secure by increasing the oil production at Norman Wells in the NWT and sending the oil by pipeline across several mountain ranges to Whitehorse and then northward to the military facilities at Fairbanks. Known as the Canol project, the oil pipeline was completed in 1944 but, with the disappearance of the Japanese threat, it was closed within the year.

After the Second World War, the geopolitical importance of northern Canada changed. The Territorial North was now a buffer zone between the United States and the Soviet Union. The North's new strategic role was to warn of a surprise Soviet air attack. The defence against such an attack was a series of radar stations that would detect Soviet bombers and allow sufficient response time for American fighter planes and (later) American missiles to destroy the Soviet bombers. In the 1950s, twenty-two radar stations, called the Distant Early Warning line, were constructed in the Territorial North along 70° N.

During this time, world demand for resources increased, transforming the Territorial North's fur economy into a resource frontier (Figure 10.5). By the 1960s, multinational corporations had turned their attention to the region's mineral wealth, spending millions of dollars exploring to find profitable mineral deposits. By the early 1970s, three types of major resource development were underway: (1) major oil companies, such as Gulf Canada and Dome Petroleum, were drilling for oil and gas in the Mackenzie Delta and the Beaufort Sea; (2) Cominco, a large mining corporation, began extracting lead and zinc from a mine at Pine Point in the Northwest Territories; (3) Cyprus Anvil, now a defunct mining company, began mining lead and zinc at Faro, Yukon.

Arctic and Subarctic Settlements

In addition to geopolitical and resource developments, the Territorial North underwent many social changes in the second half of the twentieth century—particularly among its Aboriginal peoples. The biggest change was the increase in the size of Arctic and Subarctic settlements.

At first, trading posts had few people except for the trader, the police officer, and the missionary. After the 1950s, however, trading posts evolved into tiny settlements due to the influx of Aboriginal families. Relocation is a controversial subject in northern history. For some, relocation was a badly planned attempt at social engineering, while Ottawa saw it as a necessary step in the 'modernization' of northern Aboriginal peoples. The death by starvation of about sixty Caribou Inuit was the event that triggered relocation. In the early 1950s, the Caribou Inuit living in the interior of the barren lands had difficulty finding game, particularly caribou. Reports of deprivation and even cases of death by starvation among the Caribou Inuit had reached Ottawa, but no action was taken. By 1958, Ottawa made the decision to relocate the Caribou Inuit to settlements, such as Baker Lake and Eskimo Point, but by then, starvation had taken its toll on the Caribou Inuit—their population had dropped from 'about one hundred and twenty in 1950 to about sixty in 1959' (Williamson 1974:90). At the same time, Ottawa extended this relocation program to coastal Inuit and Indians in the Subarctic who also lived off the land. However, Ottawa was unprepared for the economic, psychological, and social consequences of settlement life for hunting peoples.

Most Aboriginal peoples moved to trading posts. These settlements had certain attractions: ready access to food and other supplies at the Hudson's Bay store, medical services from the nursing station, and employment opportunities. Few jobs existed, however, because these settlements had no commercial purpose except fur trading. Most cash income came from transfer payments, such as child allowances and old-age pensions. A few traditionally oriented families remained on the land, but eventually they too came to live in settlements. Most adults spent some time hunting and trapping to escape from the stresses of settlement life and to enjoy once more the 'old ways' with family and friends. Within a generation of settlement living, however, many young, settlement-born Aboriginals were no longer interested in trapping.

Life in settlements had a number of social consequences. For example, access to food and medical services helped reduce the infant mortality rate, which caused the birth rate to rise. Since the 1940s, the Aboriginal birth rate has remained well above the national average, often more than twice as high. As a consequence, most northern settlements have tripled their population since 1951.[5] Even so, most are small by southern standards (under 1,000 people). For example, the largest centre in Nunavut is Iqaluit, with 4,220 residents in 1996. However, Iqaluit is expected to grow rapidly in the twenty-first century.

Life in settlements—affected as it was by western ideas and goods—also meant that the pressure to obtain cash income rose sharply. As well, the cost of hunting and trapping increased because transportation to the hunting and trapping grounds was now by snowmobiles rather than dog teams.[6] The two principal sources of income became wages and various forms of government payments, including social assistance. However, employment opportunities in these former trading posts remained limited, while the Aboriginal population continued to grow rapidly. This dilemma is revealed in statistics for the Northwest Territories: from 1986 to 1996, the number of Aboriginal employees increased by 27 per cent, while the number of social assistance cases has more than doubled (Government of the Northwest Territories 1997:11, 23).

Settlement life for northern Aboriginal peoples has been an established fact for nearly fifty years. During that time, Aboriginal society has become part of urban Canada and, in the process, its people have had to accept many of the cultural values and ways of the

The settlement of Pangnirtung on Baffin Island has undergone many of the stages of northern development: first the site of European whaling, then fur trading, today the community is known for the art its local Inuit inhabitants create. (Courtesy Alec Aitken)

dominant society. Aboriginal children have learned English through the school system. Knowledge of English is necessary for employment in both private companies and public institutions. While some have succeeded in finding suitable employment in the community, many have not as northern settlements have few job opportunities. The major employer is the government. Ironically, senior administrative positions and professional jobs are often held by southerners, who have the necessary education and job experience. Efforts to correct this situation through equity programs in the Northwest Territories have met with limited success. The Nunavut government will try to break this pattern by hiring people who speak both Inuktitut and English. However, basic problems remain. For instance, most Aboriginal students are not completing their high-school education. Also, most jobs that are available are found only in capital cities (Whitehorse, Yellowknife, Iqaluit). Still, the idea of moving to a capital city where there are more job opportunities may be appealing, particularly for young Aboriginals who have graduated from high school and are attracted by urban amenities.

The Territorial North Today

The Territorial North consists of three territorial governments: the Yukon, the Northwest Territories, and (as of 1 April 1999) Nunavut. The capital cities of these three territories are Whitehorse, Yellowknife, and Iqaluit. Territorial governments have fewer powers than provincial governments and, in this sense, they are political hinterlands. For example, the federal government retains power over natural resources and collects a substantial amount of tax revenue from companies using natural resources, so this important source of revenue is not available to territorial governments. Instead, they depend on Ottawa for transfer payments. Without these transfer payments, the territorial governments could not offer the basic services that are available in southern Canada. For example, Nunavut receives 90 per cent of its revenue from Ottawa. The level of fiscal dependency is somewhat lower for the Yukon and Northwest Territories.

The Territorial North is changing, especially politically. One recent political change was the division of the Northwest Territories to create a new territory, Nunavut. The second political change is being triggered by the resolution of outstanding land claims between Aboriginal peoples and Ottawa through comprehensive land-claim agreements that transfer land, cash, and administrative powers to northern Aboriginal organizations.

The Territory of Nunavut

The new territory of Nunavut was made possible through a land settlement agreement between Canada and the Inuit of the Eastern Arctic in 1993. The terms of the Nunavut Land Settlement Agreement included: the use of Crown lands for the Inuit to hunt, fish, and trap; and the transfer of part of the land to the Inuit, with a portion of this area involving rights to subsurface minerals. The same year the land agreement was reached, the federal government made the commitment to create the new territory of Nunavut by passing the Nunavut Act. The Nunavut Act, which provided the legal basis for the creation of a distinct territory and territorial government, also allowed for a six-year transition period, giving the Inuit time to form their government, recruit their civil servants, and select a capital city. By means of a plebiscite, Iqaluit was selected as the capital of the new territory. Following an election in February 1999, the 19 members of the Nunavut Assembly were selected. On 1 April 1999, the newly elected government of Nunavut began the process of governing the people of Nunavut.

The word 'Nunavut' means 'our land' in Inuktitut. The vast majority of the approximately 25,000 people who reside in Nunavut are Inuit. A primary goal of the Nunavut government is to have the government reflect more accurately the people's aims and aspirations. In addition to expanding economic development and increasing the number of jobs in the new public service for residents of the very decentralized government of Nunavut, the new government will strive to promote Inuit culture and the Inuktitut language. For many Inuit, Nunavut is a dream—one that ensures their cultural survival within Canadian society. The dream also brings with it the challenge of building a northern economy that can provide jobs for its growing labour force and thereby reduce its financial dependency on Ottawa.

Land Claims

Aboriginal land claims are based on the peoples' long-time use of the land for hunting and trapping. In Canada, Aboriginal land claims are settled by treaty: the Aboriginal tribe surrenders its claim to all the land in exchange for title to a smaller amount of land and a cash settlement. Other issues involved in land-claim negotiations are self-govern-

ment, management of the wildlife resources, and preservation of language and culture. In simple terms, land claims are an attempt by Aboriginal peoples and the federal government to resolve the issue of Aboriginal rights. (See Chapter 3 for more information on these rights and the types of land claims.)

During the latter part of the twentieth century, Indian, Inuit, and Métis organizations in the Territorial North made land claims on behalf of their members. In the 1970s, there were three major land claims made in the region: Yukon First Nations (Yukon), Dene/

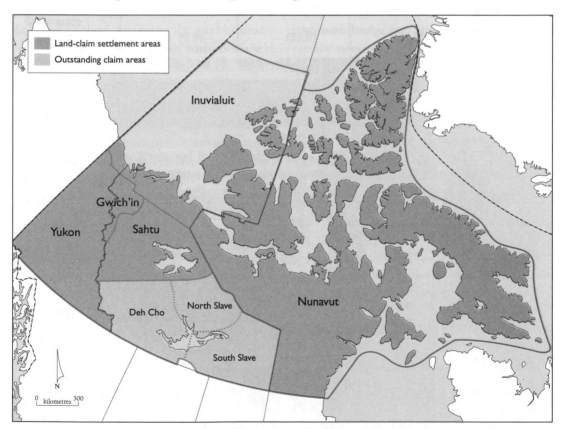

Figure 10.2 Land-claim agreements in the Territorial North, 1999. Five comprehensive land-claim agreements have been finalized while three land claims, the Deh Cho, North Slave, and South Slave, remain outstanding.

Métis (Denendeh), and Inuit (Nunavut). Over the next two decades, the Inuit land claim was split into the Western Arctic (Inuvialuit) and the Eastern Arctic (Inuit/Nunavut), and the Dene/Métis land claim for Denendeh was divided into five separate land claims (Gwich'in, Sahtu/Métis, North Slavey, South Slavey, and Deh Cho).[7] By 1984, the Inuvialuit reached an agreement with Ottawa. They were followed by the Gwich'in in 1992 and the Sahtu/Métis, Inuit, and Yukon First Nations in 1993.

In 1984, the first comprehensive land-claim agreement was signed between the Inuvialuit and Canada. The Inuvialuit Final Agreement (IFA) was based on the Inuvialuit's traditional use and occupancy of lands in the Western Arctic; that is, the land that they and their ancestors used for hunting and trapping. The Inuvialuit's land claim was originally part of the Inuit land claim to the entire Arctic. However, the Inuvialuit broke away from the Inuit and, as a result, there are two **settlement areas** in the Arctic, the Inuvialuit and the Nunavut settlement areas.[8] The goals of the In-

uvialuit were to preserve their cultural identity and values within a changing northern society; to enable themselves to be equal and meaningful participants in the northern and national economy and society; and to protect and preserve the Arctic's wildlife, environment, and biological productivity (Canada 1985:1). Canada's goal, on the other hand, was to extinguish the Inuvialuit claim to the Western Arctic.

As the first comprehensive land-claim agreement in Canada, the IFA served as a model for subsequent ones. Four additional comprehensive land-claim agreements were reached: Gwich'in (1992), Inuit (Nunavut) (1993), Sahtu/Métis (1993), and Yukon (1993). Each agreement is very similar to the IFA. One common feature is that each agreement has created economic and environmental administrative sectors. In the case of the IFA, the economic sector is the Inuvialuit Regional Corporation, which manages and invests the cash settlement received as part of the agreement.[9] The second sector, the Inu-

Table 10.1
Comprehensive Land-Claim Agreements in the Territorial North

Aboriginal Peoples	Date of Agreement	Cash Value	Land (km²)
Inuvialuit	1984	$ 45 million (1977)	90 650
Gwich'in	1992	$ 75 million (1990)	22 378
Sahtu/Métis	1993	$ 75 million (1990)	41 000
Inuit (Nunavut)	1993	$ 580 million (1989)	350 000
Yukon First Nations	1993	$ 243 million (1989)	41 440

Sources: Canada, 1985:6 and 31; Canada, 1991:3; 'Mackenzie Land Claim Settled', 1993:A1; Canada, 1993a:3; Canada, 1993b:81 and 215.

vialuit Game Council, is responsible for environmental issues that affect their hunting economy. The creation of these two sectors was the Inuvialuit's attempt to straddle two worlds; their old world was based on harvesting game from the land, while their new world is part of the global industrial economy.

For the Inuvialuit, the agreement was an important step towards defining their place within Canadian society. In exchange for cash and land, the Inuvialuit gave up their Aboriginal rights, including their claim to the vast lands of the Western Arctic. In exchange, they received legal title to about 90 600 km² of land, which is slightly less than 20 per cent of the settlement area (Canada 1985:6). The land to which the Inuvialuit gained title lies within the Inuvialuit settlement area, which extends from the Arctic mainland into the Arctic Ocean (Figure 10.2). A number of islands, including Banks Island, Prince Patrick Island, and the western parts of Melville and Victoria islands, are situated here. Of the total settlement land, 12 950 km² include surface and subsurface mining rights for the Inuvialuit. The Inuvialuit enjoy exclusive rights to hunting, trapping, and fishing over the remaining 77 700 km².

Population

The population of the Territorial North is small and concentrated into settlements. In 1996, the Territorial North had nearly 100,000 people. At that time, the Northwest Territories had 65,000 residents and Yukon 31,000. Overall, Aboriginal peoples make up close to 49 per cent of the northern population. However, the percentage of Aboriginal peoples varies widely between the three territories. In 1998, Nunavut had the highest percentage of Aboriginal peoples at 85 per cent, followed by the Northwest Territories at 48 per cent and the Yukon at 20 per cent.

Because of the nature of the environment, almost everyone in the Territorial North lives in a settlement, town, or city. Most urban centres are very small (Figure 10.3). In 1996, for example, approximately three-quarters of these urban centres had populations under 1,000, and more than 40 per cent of the Territorial North's population was in three cities: Whitehorse (20,075), Yellowknife (17,275), and Iqaluit (4,220). By function, urban centres in this region fall into three categories: Native settlements, resource towns, and regional service centres. Most Aboriginal peoples reside in Native settlements, where they form more than half the population. A growing number, however, now live in regional centres, particularly the capital cities. This population shift has occurred because there are more job opportunities, particularly in the public sector, in these larger cities (especially the capitals). Also, a small number of Aboriginal families have relocated to cities in southern Canada for similar reasons.

In 1996, the rate of natural increase in the Territorial North was approximately 2 per cent. In comparison, the national figure was only 0.5 per cent. The reason for this significant difference is the much higher birth rate in the Territorial North (23 births per 1,000 people compared to the national figure of 12.5 births per 1,000 people) and a lower death rate in the Territorial North (3.5 deaths per 1,000 people compared to the national

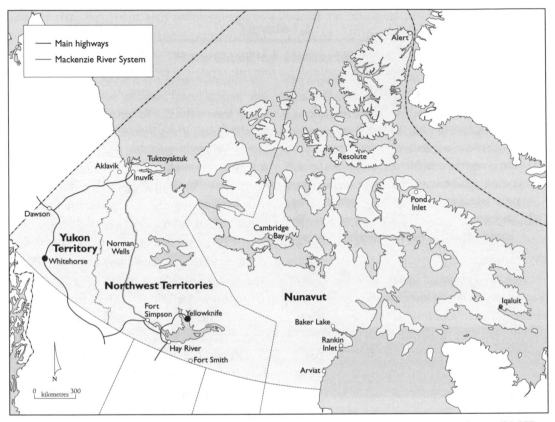

Figure 10.3 Major urban centres in the Territorial North. The major cities are Whitehorse (20,075 in 1996) and Yellowknife (17,275 in 1996). The next tier of urban centres by size have populations under 5,000. The largest city in this tier is Iqaluit with about 4,220. The smallest, but more common settlements form a third tier and have populations under 1,000.

figure of 7.2 deaths per 1,000 people). Within the Territorial North, the highest birth rates and lowest death rates are among Aboriginal peoples, especially among the Inuit in Nunavut. (See 'Aboriginal Population' in Chapter 4 for details on Aboriginal population trends.)

The population of the Territorial North is also affected by in-migration. Migration to the North normally occurs when economic expansion creates jobs. Over the past forty years, many southerners have moved north to take

jobs in the mining industry, the public service, and the business sector. However, many newcomers only remain for a few years, creating a rapid turnover in the non-Aboriginal population. During an economic downturn, the flow of non-Aboriginal peoples returning to southern Canada increases. There are two basic reasons for this out-migration. First, these migrants have an education and job skills that allow them to find employment in many areas of Canada, so they are economically mobile. Second, with friends and relatives living in southern Canada, they consider southern Canada their 'homeland'.

Two Societies

A distinctive feature of the Territorial North's population is that it can be divided into two societies. These two societies have evolved from two different cultures. The first society is a product of Western culture and early French and English presence in North America. The second society has its roots in Aboriginal cultures and the fur trade.

At the time of Confederation, many officials believed that Aboriginal peoples would assimilate into Canadian society, while others feared that diseases would destroy Aboriginal peoples. Neither has happened. In the Territorial North, there are obvious signs of Indian, Inuit, and Métis culture, particularly in Native settlements and hunting camps. Aboriginal culture, which is founded on a land-based economy, is prominent in this region of Canada because of the high percentage of Aboriginal peoples in the Territorial North— they comprise about 49 per cent of the total population. Aboriginal presence is most prominent in the smaller communities, where residents have a very close relationship with the land. Here they often form 90 per cent of the population.

In sharp contrast, cities like Whitehorse and Yellowknife not only house the majority of the non-Aboriginal population but are similar in function and structure to cities in southern Canada. There are visible signs of a western industrialized society, such as resource development, regional centres, and capital cities. The transportation and communications network spread across the North not only illustrates the importance of establishing connections between many small centres across an extensive land mass, they also show the influence that western society has had on this part of Canada.

Over time, the cultural differences between the two societies of the Territorial North have been narrowing. Three factors that reduce the differences are: (1) English is the common language used by all people in the Territorial North; (2) Aboriginal and non-Aboriginal children attend the same schools and are exposed to similar knowledge and ideas; and (3) Aboriginal men and women are entering the wage economy in greater numbers. Nevertheless, each society retains certain distinct characteristics. Three factors that illustrate a continuing difference between the two societies are:

- Aboriginal peoples are beneficiaries of comprehensive land-claim agreements, which have given them more control over the environment and wildlife in their settlement areas—control that is critical to their land-based economy.
- Aboriginal peoples remain culturally attached to the land. For instance, they harvest and eat game in much greater quantities than non-Aboriginal peoples.
- Aboriginal peoples, by taking greater pride in their culture, are reclaiming their identity through education and the renaming of geographic places in the Territorial North. For example, in 1996 Fort Norman was changed to Tulita (meaning, where the waters meet).

The Aboriginal Economy

The Aboriginal economy has been, and continues to be based on the harvesting of land and sea resources, but this economy is not static. In fact, the Aboriginal economy is a dynamic one that has constantly incorporated new technologies and ideas to make use of natural resources. The Aboriginal economy has passed through two stages—the hunting stage and the trapping stage—and it is now entering the commercial stage. In all three stages, the Aboriginal economy has focused on the utilization of land and sea resources, but whereas the first two stages were based on subsistence and trade, the current, commercial stage involves a greater interest in deriving 'cash' for resources.

The commercialization of the Aboriginal economy was a natural outcome of the integration of the Territorial North into the global economy and the relocation of Aboriginal peoples into settlements, where they became more reliant on consumer goods. Aboriginal commercial enterprises include the sale of fish and meat, such as the marketing of muskox meat, earnings generated from eco-tourism, such as polar bear sports hunting, and revenue resulting from the sale of carvings and prints (these carving and prints being an artistic representation of the land and its people).

Potentially, the Aboriginal economy could provide a sustainable use of northern resources primarily involving Aboriginal workers and owners. Yet, there remains a fundamental weakness, namely its marginal economic nature: the income generated by these activities is limited, resulting in low wages for Aboriginal workers and low profits for Aboriginal entrepreneurs. For example, the value of Aboriginal commercial enterprises is insignificant compared to the value of mineral production. These commercial enterprises do not generate the necessary cash to support Aboriginal families who live in settlements, so these families must supplement their income by either wages (in the non-Aboriginal economy) or welfare payments.

Wages can provide a partial answer but the number of Aboriginal workers far outstrips the number of jobs. Two other problems exist: geographic location and education. Most jobs are found in regional centres and capital cities, while most Aboriginal workers live in small communities, and educational levels among Aboriginal workers are much lower than those in the national labour force. Given this situation, the strengthening of the commercial land-based economy seems the only logical solution (another solution would be the relocation of Aboriginal peoples to larger centres—like the Newfoundland relocation program of the 1950s). Unfortunately, the path leading to a more secure and prosperous land-based economy has so far eluded both government planners and Aboriginal leaders.

Some land-based activities have experienced great instability over the years, due to changes in the commercial economy. For instance, since the 1950s, declining prices for furs and efforts by animal rights groups to ban the sale of seal pelts in Europe made the situation for Inuit seal hunters very difficult. While there is still a small domestic market, the number of seal pelts has declined from nearly 15,000 in 1982/3 to just under 5,000 in 1996/7. As shown in Figure 10.4, the value of fur production in the Northwest Territories dropped drastically from the mid-1980s to the late 1990s. In 1992/3, for example, the value of fur sales dropped below $1 million for the first time. In the subsequent four years, trapping recovered somewhat by exceeding $1 million in sales. Nevertheless, the revenue generated by trapping has generally declined over the last thirty years while the cost of trapping has risen sharply.

Despite the transition of the Aboriginal economy into the commercial stage, subsistence activities, like trapping, hunting, fishing, and gathering, persist among Aboriginal peoples in the North, though to a lesser degree. These activities remain not because of their commercial value but largely because of culture. For instance, country food remains a core cultural feature among northern Aboriginal families.

Country Food

Though cultural change has occurred among Aboriginal peoples, such as the growth of the wage economy, some core cultural elements have remained. For Aboriginal peoples, the core elements are a strong attachment to the land, to country food, and to the ethic of sharing. **Country food** is food obtained from the land, a preferred source of meat and fish. As equivalent store-bought foods are expensive, most Aboriginal northerners keep their food costs low by consuming country food. While it is true that there are substantial costs expended in harvesting country food, to some

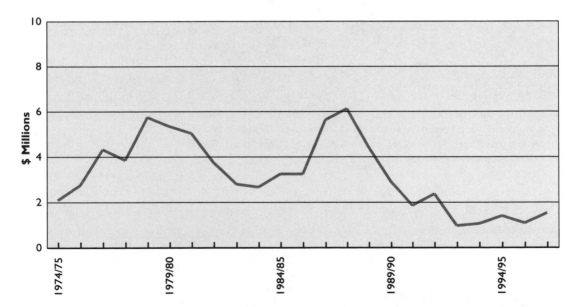

Figure 10.4 Value of fur production in the Northwest Territories. Since peaking in the late 1980s, the value of fur production has been steadily declining, signalling the shrinking importance of the commercial trapping economy in the North.

Source: Government of the Northwest Territories, 1997.

Country food among Aboriginal peoples in the Territorial North is not only an alternative to costly store-bought food but is also an intrinsic feature of their culture. (*Courtesy Northwest Territories Information Bureau*)

vesting country food are not available. However, estimates suggest that the 'substitute value' of country food (the value of equivalent store-bought food) was about $40–50 million per year in the mid-1980s (Usher and Wenzel 1989). More recent estimates are not available, but a visit to an Aboriginal community quickly reveals the importance of wildlife harvesting to the daily diet. While some store-bought foods can be found on the tables of Aboriginal families, most of the meat and fish they consume is from the land.

Country food is so important to Aboriginal northerners that they have made wildlife a key topic in land-claim negotiations. For instance, in northern Québec, both the Cree and Inuit were able to obtain financial support for their hunters and trappers in the James Bay and Northern Québec Agreement. Though such financial support was not included in the comprehensive land-claim agreements of the Territorial North, they did assign harvesting of wildlife exclusively to the Aboriginal claimants in their respective settlement areas. The agreements also establish co-management committees that administer the environment and

degree this cost is offset by the pleasure and spiritual rewards of being on the land and participating in hunting and fishing. Sharing also remains an important component in the harvesting and distribution of country food among family members, relatives, and close friends. However, this ethic does not apply to the wage economy—wages, for instance, are not shared.

Figures on the cost, size, and value of har-

wildlife. For example, they approve (or reject) proposals for new industrial projects and determine the total allowable harvest of wildlife based on biological principles and **traditional ecological knowledge**. In the case of the Inuvialuit Final Agreement, the members of the co-management committees are named by the Inuvialuit Game Council and the three governments (Canada, Northwest Territories, and Yukon). The members of the Inuvialuit Game Council are selected by the Hunters and Trappers Associations found in each of the six Inuvialuit communities. Under the direction of these committees, professional scientists record and monitor the state of the environment and wildlife. As a result of comprehensive land-claim agreements, power to control the environmental and wildlife now rests, in part, with most Aboriginal peoples in the Territorial North.

Industrial Structure

Since the 1970s, when oil prices rose sharply and petroleum companies rushed into the North, there has been a growing interest in the resource wealth, especially mineral and petroleum, of the Territorial North (Figure 10.5). Since then the farthest reaches of Canada's North have fallen under the sway of the global economy. At the same time, the federal government increased its financial support for the public sector in the North. As a result, the contemporary economic structure of the Territorial North mirrors these two powerful external forces, making the tertiary and primary sectors all-powerful. The secondary sector, manufacturing, is almost non-existent in the Territorial North.

The resource industries, comprised of energy and mineral developments, export their production to world markets. The northern transportation system is geared for that role. The value of production from the primary sector accounts for nearly 90 per cent of all production in the Territorial North. While the value of mineral output varies from year to year, it hovered around $1.5 billion for most of the 1990s. However, this highly efficient and capital-intensive industry employs relatively few people and generates few indirect benefits for the Territorial North because companies purchase their equipment and supplies from firms in southern Canada or in foreign countries.

Another drawback to the resource industry is that its labour force often redirects money to other parts of the country. For instance, mineral deposits are often found in remote places, where the mining industry has sometimes built resource towns to house its labour force. But commuting by air is a common alternative to living in an isolated resource town. Several mining companies fly their workers to the mine site and then transport them back to their home community, often in more southern parts of the country. The Lupin gold mine just east of Great Bear Lake transports workers by air from Edmonton and Yellowknife to the mine site once every three weeks. For the Territorial North, such air commuting systems are a drawback for the region's economy because: (1) workers spend their wages in their home community in southern Canada, thereby stimulating provincial, not territorial, economies; and (2) workers who reside in a province but work in

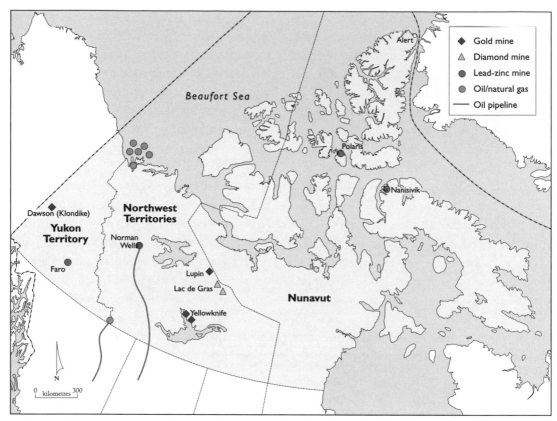

Figure 10.5 Resource development in the Territorial North. The mineral wealth of the Territorial North lies mainly in the 'western' Northwest Territories. Diamonds, gold, and oil drive this territory's resource economy. In sharp contrast, the resource economies of the Yukon and Nunavut are much smaller and are based on lead and zinc mines.

the territories pay personal income tax to provincial, not territorial governments, thereby depriving territorial governments of valuable personal income tax.

In the 1990s, the Territorial North employed about 40,000 workers per year, but only 15 per cent of those employees worked in the primary sector. Just over 80 per cent of the labour force was employed in the tertiary or service sector, by far the largest in the Territorial North. Most workers in the tertiary sector are employed by one of three gov-

ernments: federal, territorial, or local. Local governments are settlement councils, band councils, and other organizations funded by a higher level of government. The reason the tertiary sector is so large in the Territorial North is that geography demands such an investment of people and capital to ensure the delivery of public services. Territorial governments must spend more money per resident than do provincial governments to provide basic services. Much of the cost differential is attributed to overcoming distance in the Territorial North and hiring staff for small communities. The drawback is that economies of scale are difficult to achieve in small communities where teachers and nurses often have a relatively small number of students and patients compared to those working in larger centres.

However, the social importance of the public service sector goes beyond the number of employees. The wide geographic distribution of public jobs across the North is a major social benefit to those in small communities. As a result, employment opportunities, while concentrated in the three capital cities, exist in every community. In small, remote communities, where employment rates are often as high as 30 per cent, virtually all jobs are associated with one or more levels of government. A second social benefit is the governments' ability to implement social policies in their hiring practices. For example, Nunavut's government seeks to employ mainly Inuit in its civil service.

Resource Development in the Territorial North

Since the sixteenth century, explorers—now replaced by multinational companies—sought to exploit the North's natural wealth, particularly its minerals. The search for precious metals began soon after explorers reached North America. In Mexico and South America, the Spanish and Portuguese were particularly successful in obtaining vast quantities of gold and silver. In North America, English and French explorers searched for similar wealth. Northern Indians trading at the Hudson's Bay posts often wore copper ornaments and had

copper tools. Richard Norton, the factor or chief trader at Fort Prince of Wales, ordered one of his servants, Samuel Hearne, to find the source of this copper and assess its commercial worth to the company. In one of the great overland journeys of all time, Samuel Hearne travelled with Chipewyan leader Matonabbee's tribe from Fort Prince of Wales to the mouth of the Coppermine River. In 1771, Hearne found the source of copper, but this natural copper only occurred at the surface in small amounts. Others seeking mineral wealth in the North were more fortunate. In 1896, George Carmacks, an American prospector and his Indian brothers-in-law, Skookum Jim and Tagish Charley, found gold nuggets in a small tributary (later named Bonanza Creek) of the Yukon River. When word of their find reached the outside world, the Klondike gold rush began.

The Klondike Gold Rush

The discovery of gold in the Yukon sparked the greatest gold rush of all time. It caught the imagination of the world and gave credence to the myth of a northern Eldorado. While the discovery of gold took place in 1896, the Klondike gold rush did not begin until the following year. Dawson City was the centre of this frenzy. Thousands of prospectors flooded the area to scour the sandbars of the rivers and pan for gold. Within two summers, the more accessible placer gold was gone. When word of a new gold find in Alaska reached Dawson City, most prospectors moved down the Yukon River into Alaska.

By 1900, gold mining in the Klondike had changed from an individual pursuit to a large-scale capital-intensive operation. Now that the accessible placer gold was gone, the only gold that remained was deep in the frozen **terraces** found along the sides of the Klondike River. Permafrost is widespread at latitudes well above 60° N. Mining in permafrost was beyond the abilities of a prospector, so a second phase of mining began, organized by companies with both capital and technology. Mining companies soon purchased the leases to large sections of the land surrounding the rivers. Gold mining became highly organized, capital intensive, and technically advanced. To expose the gold locked in the frozen terraces, hydraulic and steam-mining techniques were used to thaw the ground and force the sand, gravel, and fine gold particles into separating devices commonly known as sluice boxes (Vignette 10.3). The emergence of commercial, large-scale gold mining marked the birth of the mining industry in the Territorial North.

Environmental and Social Impacts

In the late-nineteenth century, society regarded the North as a resource frontier and ignored negative impacts on the environment and Aboriginal peoples. Development was the priority. The Klondike gold rush exemplifies this attitude. The North, popularized by the writings of Jack London and Robert Service, and by Charlie Chaplin's portrayal of a prospector in the motion picture *Gold Rush*, was seen as a frontier, a place of adventure and excitement. In those days, the Yukon's environment was at the mercy of prospectors and mining companies.

To extract the gold, hydraulic mining of the frozen sands and gravels of the valley terraces on the tributaries of the Yukon River was necessary. Gold was recovered after thawing and sorting these deposits, but in the process, trees were cut down for firewood and building materials, river banks were destabilized, and tailings (the huge piles of discarded sand and

---

Vignette 10.3

Gold Rushes

Placer gold was found in commercial quantities mainly in the western Cordillera region from California to Alaska where a series of gold rushes occurred in the mid- to late-nineteenth centuries. The discovery of placer gold in some remote places triggered a sudden influx of prospectors, traders, gamblers, merchants, bankers, prostitutes, and various hangers-on. All sought to increase their wealth either by finding gold or by 'extracting' gold from the miners.

Placer gold can be worked cheaply by amateurs. It involves the separation of gold nuggets and particles from sand and gravel. The common practice is to pan for gold; that is, to jiggle the sand, gravel, and gold in a pan, causing the heavier gold to separate from the other material. Sluicing is another more elaborate method of separating the gold from other loose material obtained from the stream bed. This requires sluice boxes—channels or troughs fitted with grooves to separate gold from gravel.

gravel) were scattered across the landscape. Environmentally, gold mining had a destructive impact on the Yukon landscape and those scars are still visible today.

The Klondike gold rush also had social impacts on the Territorial North, but these were more subtle. The influx of miners, administrators, and eventually, police officers, transformed the southern Yukon into an organized territory within Canada. The impact on Aboriginal tribes in the Yukon was overwhelming—now they were not only a minority in their own country but were also living within a Western culture with values that were often very different from their own. Aboriginal peoples found themselves on the margins in this new world.

The disregard for the environmental and social impacts of resource development in the North continued until after the Second World War. Following the Berger Report (1977) on the effects of building a gas pipeline in the North, the public began demanding that Ottawa take action to deal with the hidden costs of resource development.[10] The federal government responded by enacting legislation that requires an environmental and social assessment of development projects before they are approved by Ottawa. The federal agency responsible for enforcing these regulations is the Canadian Environmental Assessment Agency. By their very nature, industrial projects alter the natural and social environments. This federal agency's task is to review the proposals of planned developments, to assess the degree of their impact on the natural and social environments, and, if necessary, to recommend measures that will mitigate any identified impacts.

However, this process does not always manage to reconcile the interests of all the parties that are affected by developments. For instance, in 1994 the NWT Diamond Project was submitted to Ottawa for review under the Federal Environmental Assessment and Review Process. The BHP Diamonds Inc. and the Blackwater Group proposed to construct open-pit and underground diamond mines in a remote area about 300 km northeast of Yellowknife near Lac de Gras. The ensuing review process did not satisfy everyone. Aboriginal peoples complained that they did not have enough funding and time to examine and understand the project. They also complained that the public review process did not hold enough meetings in their communities. (Public hearings were held in eight Native communities but most took place in Yellowknife.) The Northern Environmental Coalition expressed concern that the panel did not pay enough attention to the cumulative effects of the project on the environment (O'Reilly 1996). Nevertheless, in 1996, the federal environmental assessment panel filed its report, recommending that 'the Government of Canada approve the NWT Diamond Project', subject to a series of recommendations (MacLachlan 1996:1).

Though resource developers often promise long-term economic development, non-renewable resource projects often have a short life span and therefore fail to generate lasting economic benefits. Because of their boom-and-bust nature, mining operations cause social dislocations when they cease to operate. The

BHP **Diamonds Inc.'s Ekati diamond mine. The value of mineral production from this mine far exceeds that of all other mines in the Territorial North.** *(Courtesy Jiri Hermann/BHP Diamonds Inc.)*

question of the long-term economic and social value of resource developments is further complicated by the North's need to attract investment and create jobs. As a result, the region is extremely vulnerable. This vulnerability is very evident when developers request public subsidies for their projects. For instance, in the case of the Izok Lake lead-zinc deposit located about 250 km south of Kugluktuk (formerly called Coppermine), a multinational company (Metall Mining Corporation) proposed to develop a lead-zinc mine at this site, but the company wanted considerable public financial support. Metall argued that if the government built a highway

to an ocean port in this remote area, Metall would open a mine and other minerals could be developed. In the *Slave River Journal*, two opinions, one in favour of development and the other opposed, capture the essence of the debate surrounding such development projects (Vignette 10.4).

Development in Remote Locations

In the Territorial North, the development of known mineral deposits in remote areas is often prevented by high transportation costs. The pattern of mining developments in the North clearly indicates the importance of transportation, especially water transportation.

Vignette 10.4

Izok Lake Mining Proposal: Pro and Con Views

Pro: The plan for the Izok Lake mining project is the centrepiece of northern Canada's economic strategy. The proposed lead-zinc development near Coppermine would secure a prosperous future for the Western Arctic and Nunavut. The minister of economic development for the Northwest Territories, John Todd, knows all too well the importance of Izok Lake and now that Metall Mining Corporation is apparently having second thoughts, he is courting the mining company with financial assistance to convince them to go ahead with the project. Financial support for needed infrastructure, such as an all-weather road north from Yellowknife and a deep sea port at Coppermine, would make the project economically viable for Metall and create thousands of good jobs in the NWT. In addition to the Izok Lake development on the Arctic Coast, many new mineral deposits along the road will become viable, both because of the road access and because the new port will provide access for large shipping vessels, providing a cost-effective method of transporting raw materials. A development of this scale will shore up the economy of the central Arctic, providing tremendous opportunities for business people and jobs for the locals, both during the project construction phase and while the ore is being extracted. The next two decades will be looked after. It will be a major boost for the sluggish Yellowknife economy, especially as it suffers from the loss of jobs due to the division of the NWT. And it would ultimately greatly reduce the NWT's financial dependence on Ottawa. Everyone will win with this one.

Con: Why invest hundreds of millions of dollars in an operation that is marginal and will likely fail? Remember the pain in shutting down Pine Point? The proposed lead-zinc development at Izok Lake near Coppermine is teetering and the NWT government wants to prop it up—supporting a multinational corporation with our tax dollars. Lead and zinc are low in value and even if the prices rise, they will likely fall again. Look at how long pundits have been predicting the resurgence of the gold price. What then, a bailout? The developers will never stop coming back once they get a taste of free government money. And the more marginal the operation, the more environmental protection laws will be overlooked. Didn't we just finish cleaning up a bunch of old industrial sites in the north at a cost of millions of dollars to Canadian taxpayers? It is admirable to want to create jobs, but why put everything we have into a megaproject with a limited life span? There are other northern roads half finished that will be forgotten and small, long term community-based economic projects that will be lost in the stampede to make quick, big bucks. Twenty years is not a long time and once the ore has been high-graded and the mine is being mothballed, what will the workers do? What will the cost of social problems in the northern communities be then? Are all these costs being considered in that equation? It seems John Todd, the minister of economic development for the NWT, has a vision that lines the pockets of the shareholders of Metall Mining Corporation and leaves (most) Northerners out in the cold.

Source: 'Opinion: Pick a Side', *Slave River Journal*, 9 March 1994:4. Reprinted from the *Slave River Journal*, Fort Smith, NWT, Canada.

In the mid-1930s, underground gold mining at Yellowknife was possible because the mine was located on the shores of Great Slave Lake and therefore had excellent access to the Mackenzie River transportation system. Mines located far from water transportation, such as the Lupin gold mine (about 250 km northeast of Yellowknife), required the construction of a winter road. The need for roads in mining is not always so much for transporting minerals to external markets but for bringing equipment and supplies to the mine site. Mines along the Arctic Coast can take advantage of ocean transportation. Three examples are the former nickel mine at Rankin Inlet on the shores of Hudson Bay, the lead-zinc mine at Nanisivik near the northern tip of Baffin Island, and the Polaris lead-zinc mine on Little Cornwallis Island in the Arctic Archipelago.

In assessing the cost of transportation, the value of the mineral is taken into account. For example, copper, lead, nickel, and zinc are all low-grade ores (much of the ore has no commercial value). Even after the first-stage separation of some of the waste material from the valuable mineral, the enriched ore still remains a bulky product, with significant waste material remaining that can only be removed through smelting. Shipping such a low-value commodity is very expensive. Ideally, such ore is transported to a smelter by ship. Railways are the second most effective transportation carrier for low-grade ore. The critical nature of transportation for such mines is clear in the example of the lead-zinc deposit at Pine Point in the Northwest Territories. Discovered in 1898 by prospectors heading overland to the Klondike gold rush, the Pine Point deposit was not developed until 1965, when a railway was extended to the mine site. With a means of transporting the ore to a smelter, the company, Cominco, could begin sending massive amounts of ore by rail to its smelter at Trail in British Columbia. Unfortunately, the mine closed in 1983, transforming Pine Point into a ghost town.

Resource development is also expected to provide employment and business opportunities to communities in the North. Unfortunately, most jobs end up in southern Canada. For instance, many companies use air commuting to transport employees from southern Canada to their remote mining sites. Most Northerners, in fact, work in the public sector. Few Aboriginal workers are employed in the mining industry. The Rankin Inlet nickel-mining operation proved to be an exception to this. The majority of the employees at this mine were from the local community. The local Inuit had a chance at these jobs because it was difficult to recruit outside workers to work in such a remote place, the mining operation was small, and mine officials were willing to hire entire Inuit crews. This meant that most underground communications were in Inuktituk.

Key Topic: Megaprojects

The Territorial North has entered a new phase of resource development characterized by megaprojects controlled by multinational companies. Megaprojects usually cost more than $1 billion and require several years to complete. This phase of development has integrated the Territorial North's economy into the global economy, thereby firmly locking the North into a resource hinterland role in the world economic system. The most recent megaproject in the Territorial North is the NWT Diamonds Project (Canada 1996). Preliminary construction began in the early 1990s, but the project was not completed until 1998.

Proponents of resource development describe megaprojects as the economic engine of northern development, though others challenge this assumption, stating they offer few benefits to the region (Vignette 10.4; Bone 1992:143-5). These large-scale ventures are designed for the export market. By injecting massive capital investment into the construction of giant engineering projects, megaprojects create a short-term economic boom. However, most construction monies are spent outside hinterlands because the manufactured equipment and supplies are produced not in hinterlands but in core industrial areas. This reduces the benefits of megaprojects on the hinterland economy and virtually eliminates any opportunity for economic diversification. As well, since all megaprojects in the Territorial North are based on non-renewable resources, these developments have a limited life span. At the end of these projects, the local economy suffers a collapse. Three examples are the closure of mines at Faro, Yukon; Pine Point, Northwest Territories; and Rankin Inlet, Nunavut.

However, megaprojects can more readily inject much-needed capital and create development in the region. Megaprojects in resource hinterlands are high-risk ventures. Multinational companies can reduce their

risks in three ways. First, they can create a consortium of companies and thereby spread the investment risk among several firms. Second, they can arrange for long-term sales of the product at a fixed price before proceeding with construction. Third, they can obtain government assistance, which often takes the form of low-interest loans, cash subsidies, and tax concessions. Two megaprojects are discussed in the following sections: the Norman Wells Oil Expansion and Pipeline Project, and the NWT Diamonds Project. Each project was approved after an environmental impact assessment. The Norman Wells Project received its approval in 1980, while the NWT Diamonds Project was approved in 1996.

The Norman Wells Project

The Norman Wells Project, which cost almost $1 billion, was hailed by industry as a model megaproject for the North. Both the federal government and Esso Resources Canada believed that the project would not harm the natural and social environments in the construction impact zone, which stretched from Norman Wells to Fort Simpson and then to the Alberta border. Aboriginal organizations, such as the Dene Nation, and environment groups, such as the Canadian Arctic Resources Committee, did not agree.

Plans for this major oil and pipeline construction effort by Esso Resources Canada began in the late 1970s, when the price of oil was rising sharply. Once the proposal was approved by the federal government, construction began, running from 1982 to 1985. The petroleum pipeline route begins at the resource town of Norman Wells in the

A drilling rig near Norman Wells. The Norman Wells Project allowed for a significant increase in oil production and supply to southern markets. *(Courtesy R.M. Bone)*

Northwest Territories and ends at Zama in northern Alberta, the northern terminus for the national oil pipeline system. Much of the terrain north of 60° N lies in the zone of discontinuous permafrost. As a result, special construction measures were necessary. Limiting construction to the winter months prevented excessive damage to the terrain and vegetation by trucks and other heavy equipment, and efforts to revegetate the pipeline route following construction were designed to minimize soil erosion and the warming of ground temperatures by solar radiation. The danger from higher ground summer temperatures is that it could accelerate the melting of ground ice, leading to the subsidence (sinking) of the ground and possible rupturing of the buried oil pipe.

The local oilfields were first discovered in 1920. By 1982, oil exploration investigations had revealed the quantity of proven reserves to be as high as 100 million m³ (Bone 1992:145). The purpose of the Norman Wells Project was to increase the oil produced and shipped to southern markets by means of a pipeline. Prior to 1985, the output from the Norman Wells oilfields served only the local Mackenzie Valley market. At that time, annual output was less than 180 000 m³. With the completion of the pipeline and the expansion of oil production by a factor of ten, the Norman Wells oilfields had the capacity to send vast quantities of oil to southern customers. Since 1985, annual output has exceeded 1 million m³ and almost reached 2 million m³ several times.

The Norman Wells Project represents a successful and profitable operation. For over 15 years, Esso has produced and transported crude oil to markets in southern Canada and the United States. During that time, while some subsidence of the ground along the pipeline route has occurred, the pipeline has not been damaged and there have been no oil spills. Equally important, the operation of the Norman Wells Project has not triggered social problems in the communities along the pipeline route. From the oil industry's perspective, the Norman Wells Project demonstrated that pipelines can be built in areas with permafrost. However, from the Dene Nation's perspective, land claims should have been settled before the Norman Wells Project was approved. Under such circumstances, the Dene would have become business partners in the construction and ensuing operation of the Norman Wells Project.

The NWT Diamonds Project

In 1996, the federal government approved the NWT Diamonds Project, a joint venture between BHP Diamonds Inc. and the Blackwater Group (BHP). This project entails the development and operation of five diamond mines in the Lac de Gras area of the Northwest Territories for about twenty-five years. Five diamond-bearing **kimberlite pipes** are being mined, four of them located within a few kilometres of each other. All five kimberlite pipes lay under lakes that had to be drained to facilitate mining operations.

During the operations of this project, between 35 and 40 million tonnes of waste rock are removed from the Kimberlite pipes each year. Since only a small proportion of that rock contains diamonds, the rest of the rock is placed in piles near the mines. Recovery of diamonds from the ore takes place in a processing plant where the ore is crushed and diamonds are separated. Final sorting is done using X-rays to separate the diamonds from the remaining waste material. BHP then sells rough-cut diamonds at international diamond markets. The mining operation employs about 800 workers with an estimated annual wage bill of $39 million (MacLachlan 1996:50). BHP uses an air commuting system to bring workers from Yellowknife to its mine site on a two-weeks-in and a two-weeks-out rotation.

Diamond mining has quickly become the backbone of the mining industry in the Territorial North. By value of production, number of employees, and spin-off effects, BHP Diamonds is the leading mining company in the North. In 1998, the company produced $500 million worth of diamonds; had, with 800 miners, the largest number of employees of all northern mines; and had beneficial effects for the northern economy that were far greater than other mining operations. One spin-off was the designation of Yellowknife as the pick-up point for miners who commute by air on a rotational scheme to the mine site. This reverses the usual trend of cash and tax flowing out of the territories. The second spin-off was the location of a diamond cutting and polishing business in Yellowknife. Furthermore, several other diamond firms plan to establish offices in Yellowknife if they can obtain rough gem diamonds from BHP Diamonds. These circumstances will inject much-needed jobs and investment into the economy of the Territorial North.

Megaprojects: Achilles Heel

Megaprojects in the Territorial North are based on non-renewable resources, which consist of petroleum deposits and mineral bodies. Forests, water, and wildlife are classified as renewable resources. The value of non-renewable resources far outweighs the economic value derived from renewable resources. In 1996, mineral production in the Territorial North was valued at $1.2 billion, which was almost 2.5 per cent of Canada's mineral output. That same year, approximately 66 per cent of the Territorial North's production came from the Northwest Territories. The leading minerals by value were gold, zinc, oil, and lead. The major mines were located at Faro (lead-zinc), Nanisivik (lead-zinc), Polaris (lead-zinc), Con (gold), Giant (gold), and Lupin (gold). Placer gold is obtained from the valleys of the Yukon River and its tributaries. Norman Wells is the only oil-producing site in the Territorial North, but there are much larger oil and gas deposits in the Sverdrup Basin and the Beaufort Sea (Vignette 10.5).

Non-renewable resource development, because of such projects' limited life span, subject the Territorial North to a boom-and-bust economic cycle. Since mineral and petroleum deposits are in finite quantities, they do not offer long-term economic stability for the Territorial North. This economic cycle is the Achilles heel of northern development. In addition, because resource development is based on world demand, production fluctuates with this demand, creating great instability in resource communities. Fluctuations in mineral production provide a measure of this instability. For instance, over the past twenty years, the value of mineral production in the Territo-

Vignette 10.5

Beaufort Sea

The Beaufort Sea is part of the Arctic Ocean. Named after the British hydrographer, Sir Francis Beaufort, this 'frozen' sea occupies an area of approximately 450 000 km². Most of its surface is permanently covered by the polar pack ice, which is often over 5 m thick. Only large ice-breakers can penetrate this formidable ice barrier. In August 1994, two ice-breakers, the US *Polar Sea* and the Canadian *Louis S. St Laurent*, ploughed through this ice on a scientific voyage to the North Pole. During the short summer, the shore ice melts, leaving a narrow stretch of open water between the shore and the polar pack ice. Small ships take advantage of this open water to bring supplies to communities located along the coast of the Beaufort Sea. Most of these supplies are transported by barge northward along the Mackenzie River.

The shallow waters of the Beaufort Sea are often less than 50 m, so offshore drilling for oil and gas is possible. Sparked by high oil prices in the 1970s, oil companies drilled hundreds of wells in the Beaufort Sea over the subsequent twenty years. Significant deposits of oil and gas were discovered, but the high cost of production and transportation to southern markets has prevented further offshore oil and gas development.

rial North reached a high of $1.7 billion in 1989 and a low of $350 million in 1976. In 1996, it was $1.2 billion but with the closure of the Faro mine it dropped to $1 billion or 17 per cent. The negative economic and social effects of such a drop are deeply felt in resource towns. Mines even suspend their operations during periods of low prices. In 1997, for example, the Faro mine was closed for the fourth time in its short history. Again, low metal prices made the zinc, lead, and silver-mining operation unprofitable. The Faro mine had been the principal mining activity for thirty years in the Yukon. Each time the mine closed, the value of mineral production in the Yukon dropped drastically. For example, in 1992 the Yukon accounted for nearly half a billion dollars worth of minerals. After the Faro mine closed in 1993, production dropped to $81 million in 1994.

Megaprojects have made important contributions to the northern economy. These contributions include: generating the bulk of the North's GDP, expanding the northern transportation infrastructure, and providing high wages for its employees. The diamond industry holds out some hope of even greater spinoffs such as the processing of diamond gems in Yellowknife. Megaprojects, however, though touted as the engine of northern development in the 1970s, have so far failed to transform the Territorial North's economy. Megaprojects, while generating profits for the corporations, have failed to diversify the northern economy, have been unable to solve the massive unemployment problem, and have increased the region's economic vulnerability to sudden changes in the world de-

mand for its primary products. Worse yet, all northern megaprojects are based on non-renewable resources which, having a fixed life-span, cannot provide the basis for a stable, long-term economy. From this perspective, megaprojects are not the engine, but rather are the Achilles heel of the northern economy.

The Territorial North's Future

Old ways and structures are changing in the Territorial North. The people of the North's two cultures are building new worlds within the region. To that end, the political landscape has changed dramatically. Power has shifted in two directions. Land-claim agreements have resulted in a devolution of economic and environmental powers from Ottawa to the organizations representing Aboriginal claimants in the North. A second shift took place in 1999, when Nunavut became a territory and political powers were transferred from Yellowknife to Iqaluit.

The economic landscape is changing too. Megaprojects will draw the Arctic, as they have the Subarctic, into a resource hinterland serving the global economy. A sign of this transformation is evident in the diamond mines along the southern edge of the Arctic. Before the turn of the century, the Territorial North will become one of the major diamond producers in the world. However, resource development, particularly non-renewable development, has not yet triggered sufficient economic growth to create a stable economy nor reduce the Territorial North's financial dependency on Ottawa.

Nevertheless, in the next century, multinational corporations will participate in more and more resource developments. For example, if the price of oil rises beyond US $30 a barrel, then exploitation of the Beaufort Sea oil and gas reserves will become attractive. Given the vast reserves of oil, the Western Arctic may become the Alberta of the North. The Inuvialuit Regional Corporation would benefit in such a situation, especially if oil is discovered on Inuvialuit lands. Most resource development will involve non-renewable resources.

More economic changes are taking place through land-claim agreements in the Yukon, much of the Northwest Territories, and Nunavut. These agreements, which so far involve over three-quarters of the Aboriginal peoples in the Territorial North, provide capital and create an Aboriginal business corporation to manage and invest this capital on behalf of its members. But there are risks in such ventures and economic success is not assured. The economic place of Aboriginal peoples in the Territorial North will become better defined in the next century. Much depends on the extent to which land-claim agreements can protect wildlife and the environment. If successful, then harvesting of country and commercial food will continue and eco-tourism may well become an important 'sustainable' industry. Business corporations owned and operated by the Inuvialuit, Gwich'in, Inuit, Sahtu/Métis, and Yukon First Nations may provide the vehicle for entering the market economy and still allow them control over decision making, including hiring Aboriginal peoples to run these businesses. Aboriginal peoples who have agreements are controlling

their future. Unfortunately, the future of those Aboriginal groups who have not yet negotiated a land-claim settlement is less clear.

The social landscape in the Territorial North is changing more slowly. While a rural-to-urban migration has taken place, most settlements have no economic purpose. A second migration, driven by economic considerations, is underway, drawing Aboriginal residents to regional centres and capital cities within the Territorial North, or to metropolitan centres in southern Canada. Until there is a better geographic match between economic opportunities and Aboriginal peoples, social problems, including high unemployment rates, will continue to plague the Territorial North. This population imbalance is fuelled even further by the exceptionally high birth rates among Aboriginal peoples, especially among the Inuit. Another obstacle to northern development is the limited schooling achieved by Aboriginal workers, which either blocks their chances to enter the labour market or relegates them to 'low-end' jobs. Can the social landscape improve? The political and economic changes in the Territorial North are an important first step in that direction.

Summary

The Territorial North is both a resource frontier and a homeland for Aboriginal peoples. It is the only geographic region in Canada where Aboriginal peoples form a substantial proportion of the total population. Nunavut is a political expression of this homeland concept for the Inuit. Comprehensive land-claim agreements in five geographic areas have equipped

Aboriginal peoples with capital, land, and control over wildlife to chart a new future. Although they are involved in the market economy, Aboriginal peoples retain a strong attachment to the land, country food, and the ethic of sharing. Such cultural traits strengthen their recent political and economic advances.

As a resource hinterland, the Territorial North has three principal characteristics: it is far from world markets; resource development is dependent on external demand; and its economy is sensitive to fluctuations in world prices for resources. The economy of the Territorial North has two serious flaws, namely, that it is based on non-renewable resources and that economic benefits from resource development accrue mainly to firms and workers in southern Canada and foreign countries. These two flaws combined lead to a state of underdevelopment known as a staple trap.

In recent years, megaprojects have drawn the Territorial North more closely into the global economy as a resource frontier. The implication of such resource development means that stable economic growth and economic diversification remain elusive goals. Without a more diversified economy, the Territorial North will continue to suffer from a boom-and-bust economic cycle and its three governments will remain financially dependent on Ottawa.

Notes

1. There is a third climatic zone in the Territorial North—the Cordillera. However, the Cordillera climate, often described as a mountain climate, is affected by elevation (as elevation increases, temperature drops). North of 60° N, the Cordillera climate is also affected by latitude so that boreal natural vegetation is found at lower elevations and tundra natural vegetation at high elevations.

2. Sharing food was essential for small hunting groups living together harmoniously. By sharing in all hunters' successes, they could adjust for the vagaries of individual luck and reduce the threat of starvation. Today, the sharing of food remains a pivotal component of Aboriginal culture and the Native economy.

3. The Vikings made contact with the ancestors of the Inuit, the Thule (c.AD 1000 to 1600). The Thule originated in Alaska where they hunted bowhead whales and other large sea mammals. They quickly spread their whaling technology across the Arctic, travelling in skin boats and dog sleds. With the onset of the Little Ice Age in the fifteenth century, climate conditions affected the distribution of animals and the Thule who were dependent on them. An increased amount of sea ice blocked the large whales from their former feeding grounds, resulting in the collapse of the Thule whale hunt. With the loss of their main source of food, the Thule had to rely more and more on locally available foods, usually some combination of seal, caribou, and fish. By the eighteenth century, the Thule culture had disappeared and was replaced by the Inuit hunting culture.

4. Territorial governments, unlike provinces, are governed under delegated powers from the federal government. In that sense, they are dependent on Ottawa for their political powers. Under the Cana-

dian Constitution, Ottawa has the power to govern its territories. In the past, responsible government occurred when a part of these territories has a sufficiently large population and tax base to warrant having an elected council. Until that time, the governing members were appointed by the federal government.

5.   Social programs may have encouraged large families. For example, the Family Allowance Program provided a payment for each child. For a family of five, the annual cash derived from a family allowance was often greater than the cash a father might earn from the sale of fur pelts.

6.   Today, capital and operating costs are still serious problems for hunters. Transportation from a settlement to hunting/trapping areas is a major expenditure. Snowmobiles, for example, may cost as much as $20,000 in a northern retail store. (Ten years ago, they cost $5,000 to $10,000.) With a continuing rise in the prices of manufactured goods, the cost of a snowmobile ten years from now may double again. Since a snowmobile used for long-distance travel has a short life span, the hunter/trapper must replace it about every three or four years.

7.   The first comprehensive land-claim submission came from the Dene/Métis in 1974. The land claimed by the Dene/Métis extended over most of the Mackenzie Basin north of 60° N. This land was called Denendeh. In 1988, an agreement-in-principle was signed by the two negotiating parties (the federal government and Dene Nation). This agreement called for a cash payment of $500 million over 15 years plus title to 181 230 km² of land. Nearly 6 per cent of this land (10 000 km²) would include subsurface rights for the Dene/Métis. As well, the Dene/Métis would obtain a share of federal resource royalties, including those generated

by the Norman Wells oil field. Chiefs and elders from the Great Slave Lake area refused to approve this agreement for two main reasons—the agreement-in-principle contained no reference to self-government and it called for the surrendering of Aboriginal rights. The issue had already caused the Gwich'in and Sahtu/Métis to break away from the umbrella group. They would subsequently negotiate their own agreements.

8.   The Inuvialuit broke away from the other Inuit who were seeking a common land-claim settlement that would have stretched from the Arctic Coast of the Yukon to Baffin Island. The Inuvialuit wanted to advance their own land claim before the huge oil and gas deposits in the Beaufort Sea were developed. The Inuvialuit, with their separate agreement, hope to obtain land with oil and gas deposits and achieve a taxation arrangement similar to that of the Arctic Slope Regional Corporation of the Inupiat near Prudhoe Bay. This municipality received substantial tax revenues from the oil companies.

9.   Under the IFA, Canada agreed to transfer $45 million in 1977 dollars. By 1984, these funds were valued at $152 million. The payments to the Inuvialuit Regional Council, the business corporation of the IFA, began on 31 December 1984 with $12 million. By the year 1997, there were three annual payments of $1 million beginning 31 December 1985; five annual payments of $5 million beginning 31 December 1988; four annual payments of $20 million beginning 31 December 1993; and a final payment on 31 December 1997 of $32 million (Canada 1985:107).

10. Until the 1970s, Canadians did not question the merit of northern industrial projects. The Mackenzie Valley Pipeline Inquiry headed by Thomas Berger made Canadians aware of the hid-

den costs of building a gas pipeline from Prudhoe Bay along the Arctic Coast to the Mackenzie River and then southwards along the Mackenzie Valley to markets in southern Canada and the United States. The hidden costs identified by Berger were: the potential threat to wildlife, especially the Porcupine caribou herd; the ground subsidence that could result from a buried pipeline; and the feared social impacts on Aboriginal peoples living in the construction zone. For more on this subject, see Bone 1992:164–6.

Key Terms

country food
> As hunters, Indians, Inuit, and Métis fished and hunted for food. Now settlement dwellers, they still fish and hunt for cultural and economic reasons. Such food is called game or country food.

frontier
> The perception of the Territorial North as a place of great mineral wealth that awaits development by outsiders.

homeland
> People who live in a region develop a strong attachment to that place; a sense of place.

kimberlite pipe
> An intrusion of igneous rocks in the earth's crust that takes a funnel-like shape. Diamonds are sometimes found in these rocks.

megaprojects
> Large-scale construction projects that exceed $1 billion and take more than two years to complete.

settlement area
> Each comprehensive land claim has a geographic extent that is known as a settlement area. While less than 25 per cent of this land is allocated to the Aboriginal beneficiaries as a collective (not individual) land holding, the entire area is subject to the environmental and wildlife regulations exercised by the settlement area's co-management boards.

terraces
> Streams and rivers form terraces when they cut downward into their flood plain and form a new and lower flood plain. The old flood plain (now a terrace) is found along the sides of the stream or river.

traditional ecological knowledge
> Aboriginal peoples are very familiar with their natural surroundings. This knowledge of the environment and wildlife, which has accumulated over the centuries, is known as traditional ecological knowledge, or TEK.

References

Bibliography

Anderson, Robert B. and Robert M. Bone. 1995. 'First Nations Economic Development: A Contingency Perspective'. *The Canadian Geographer* 39, no. 2:120–30.

Berger, Thomas R. 1977. *Northern Frontier Northern Homeland: The Report of the Mackenzie Valley Pipeline Inquiry*. Vols. 1 and 2. Ottawa: Minister of Supply and Services.

Bockstoce, John R. 1994. *Whales, Ice and Men: The*

History of Whaling in the Western Arctic. Seattle: University of Washington Press.

Bone, Robert M. 1995. 'Power Shifts in the Canadian North: A Case Study of the Inuvialuit Final Agreement'. *Proceedings of 'Canada in Transition' Symposium*, edited by Roland Vogelsang. Bochum: Universitätsverlag Brockmeyer.

_____, and Shane Long. 1995. 'Population Change in Aboriginal Settlements in the Mackenzie Basin North of 60°. *Proceedings of the 1995 Symposium of the Federation of Canadian Demographers.* Ottawa: St Paul University.

Bone, Robert M. 1998. 'Resource Towns in the Mackenzie Basin'. *Cahiers de Géographie du Québec* 42, no. 116:249–59.

Canada. 1985. *The Western Arctic Claim: The Inuvialuit Final Agreement.* Ottawa: Department of Indian Affairs and Northern Development.

_____. 1991. 'Comprehensive Land Claim Agreement Initialled with Gwich'in of the Mackenzie Delta in the Northwest Territories'. *Communique* 1-9171. Ottawa: Department of Indian Affairs and Northern Development.

_____. 1993a. 'Formal Signing of Tungavik Federation of Nunavut Final Agreement'. *Communique* 1-9324. Ottawa: Department of Indian Affairs and Northern Development.

_____. 1993b. *Umbrella Final Agreement between the Government of Canada, Council for Yukon Indians and the Government of the Yukon.* Ottawa: Department of Indian Affairs and Northern Development.

_____. 1996. *NWT Diamonds Project.* Report of the Environmental Assessment Panel. Canadian Environmental Assessment Agency. Ottawa: Minister of Supply and Services.

Coates, Ken, and Bill Morrison. 1992. *The Forgotten North: A History of Canada's Provincial Norths.* Toronto: Copp Clark.

Crowe, Keith J. 1991. *A History of the Original Peoples of Northern Canada*, 2nd edn. Montréal–Kingston: McGill-Queen's University Press.

Frideres, James. 1993. *Native Peoples in Canada.* Scarborough: Prentice-Hall.

Government of the Northwest Territories. 1997. *Statistics Quarterly*, vol. 19. Bureau of Statistics. Yellowknife: Department of Public Works & Services.

Grant, Shelagh D. 1988. *Sovereignty or Security? Government Policy in the Canadian North, 1936–1950.* Vancouver: UBC Press.

Hamelin, Louis-Edmond. 1978. *Canadian Nordicity: It's Your North, Too.* Translated by William Barr. Montréal: Harvest House.

_____. 1982. 'Originalité culturelle et régionalisation politique: Le project Nunavut des Territoires-du-Nord-Ouest (Canada)'. *Recherches Amérindiennes au Québec* 12, no. 4:251–62.

Johnston, Margaret E. 1994. *Geographic Perspectives on the Provincial Norths.* Northern and Regional Studies Series, vol. 3. Centre for Northern Studies, Lakehead University. Mississauga: Copp Clark Longman.

Keith, Robbie. 1998. 'Arctic Contaminants: An Unfinished Agenda'. *Northern Perspectives* 25, no. 2:1–3.

Kenney, Gerard. 1994. *Arctic Smoke and Mirrors.* Prescott: Voyageur Publishing.

Légaré, André. 1993. 'Le project Nunavut: Bilan des revendications des Inuit des Territoires-du-Nord-Ouest'. *Études/Inuit/Studies* 17, no. 2: 29–62.

_____. 1996. 'Le gouvernement du Territoire du Nunavut'. *Études/Inuit/Studies* 20, no. 1:7–43.

McGhee, Robert. 1996. *Ancient People of the Arctic*. Vancouver: University of British Columbia Press.

'Mackenzie Land Claim Settled'. 1993. *The Globe and Mail* (7 September):A1.

MacLachlan, Letha. 1996. *NWT Diamonds Project: Report of the Environmental Assessment Panel*. Ottawa: Canadian Environmental Assessment Agency.

Marcus, Alan R. 1995. *Relocating Eden: The Image and Politics of Inuit Exile in the Canadian Arctic*. Hanover: University Press of New England.

Marsh, James H., ed. 1988. *The Canadian Encyclopedia*, 2nd edn. Edmonton: Hurtig Publishers.

Notzke, Claudia. 1994. *Aboriginal Peoples and Natural Resources in Canada*. Toronto: Captus Press.

O'Reilly, Kevin. 1996. 'Diamond Mining and the Demise of Environmental Assessment in the North'. *Northern Perspectives* 24, no. 1-4:1–6.

Purich, Donald. 1992. *The Inuit and Their Land: The Story of Nunavut*. Toronto: Dormer.

Ray, Arthur J. 1990. *The Canadian Fur Trade in the Industrial Age*. Toronto: University of Toronto Press.

Rowley, Graham W. 1996. *Cold Comfort: My Love Affair with the Arctic*. Montréal–Kingston: McGill-Queen's University Press.

Royal Commission on Aboriginal Peoples. 1996. *Report of the Royal Commission on Aboriginal Peoples*, vols 1–5. Ottawa: Minister of Supply and Services.

Saku, James A., Robert M. Bone, and Gérard Duhaime. 1998. 'Towards an Institutional Understanding of Comprehensive Land Claim Agreements in Canada'. *Etudes/Inuit/Studies* 22, no. 1:109–21.

Slave River Journal. 1994. 'Opinion: Pick a Side'. *Slave River Journal* (9 March):4.

Stanford, Quentin H., ed. 1998. *Canadian Oxford World Atlas*, 4th edn. Toronto: Oxford University Press.

Statistics Canada. 1997a. *A National Overview: Population and Dwelling Counts*. Catalogue no. 93-357-XPB. Ottawa: Industry Canada.

_____. 1997b. 1996 Census: Nation Tables—Population by Mother Tongue, Showing Age Groups, for Canada, Provinces and Territories, 1996 Census—20% Sample Data, 2 December 1997 [online database], Ottawa. Searched 15 July 1998; <URL:http://www.statcan.ca/english/census96/>:3 pp.

_____. 1997c. The Daily—1996 Census: Mother Tongue, Home Language and Knowledge of Languages, 2 December 1997 [online database], Ottawa. Searched 14 July 1998; <URL: http://www.statcan.ca/Daily/English/>:15 pp.

_____. 1998a. *Canadian Economic Observer*. Catalogue no. 11-010-XPB. Ottawa: Statistics Canada.

_____. 1998b. The Daily—1996 Census: Aboriginal Data, 13 January 1998 [online database], Ottawa. Searched 14 July 1998; <URL:http://www.statcan.ca/Daily/English/>:10 pp.

_____. 1998c. The Daily—1996 Census: Ethnic Origin, Visible Minorities, 17 February 1998 [online database], Ottawa. Searched 16 July 1998; <URL:http://www.statcan.Daily/English/>:21 pp.

Usher, Peter J. 1982. 'The North: Metropolitan Frontier, Native Homeland?'. In *Heartland and Hinterland: A Geography of Canada*, edited by L.D. McCann, 411–56. Scarborough: Prentice-Hall.

———, and George Wenzel. 1987. 'Native Harvest Surveys and Statistics: A Critique of Their Construction and Use'. *Arctic* 40, no. 2:145–60.

———, and ———. 1989. *A Strategy for Supporting the Domestic Economy of the Northwest Territories*. Report for the Legislative Assembly's Special Committee on the Northern Economy. Ottawa: P.J. Usher Consulting Services.

———, and ———. 1989. 'Socio-Economic Aspects of Harvesting'. In *Keeping on the Land: A Study of the Feasibility of a Comprehensive Wildlife Harvest Support Program in the Northwest Territories* by Randy Ames, Don Axford, Peter Usher, Ed Weick, and George Wenzel, Chapter 1. Ottawa: Canadian Arctic Resources Committee.

Wenzel, George. 1991. *Aboriginal Rights, Animal Rights: Ecology, Economy and Ideology in the Canadian Arctic*. Toronto: University of Toronto Press.

Williamson, Robert G. 1974. *Eskimo Underground: Socio-Cultural Change in the Canadian Central Arctic*. Occasional Papers II. Uppsala, Sweden: Almqvist & Wiksell.

Further Reading

Bone, Robert M. 1992. *The Geography of the Canadian North: Issues and Challenges*. Toronto: Oxford University Press.

Dickason, Olive Patricia. 1997. *Canada's First Nations: A History of Founding Peoples from Earliest Times*, 2nd edn. Toronto: Oxford University Press.

Elias, Peter Douglas. 1995. *Northern Aboriginal Communities: Economies and Development*. North York: Captus Press.

French, Hugh M., and Olav Slaymaker. 1993. *Canada's Cold Environments*. Montréal–Kingston: McGill-Queen's University Press.

Hamilton, John David. 1994. *Arctic Revolution: Social Change in the Northwest Territories, 1935–1994*. Toronto: Dundern Press.

Chapter 11

Overview

Canada is a country composed of six major geographic regions. In this final chapter, geography's role in shaping Canada's complex regional nature is summarized. This regionalism is based on Canada's physical and historical geography, its core/periphery economic structure, and the tensions existing in the social and political structures of Canada. The principal tensions or faultlines are between centralists and decentralists, Aboriginal and non-Aboriginal Canadians, and between French-speaking and English-speaking Canadians. Canada's future, while full of uncertainty, contains one truth—Canada will remain a country where cultural and regional diversity exist. Though problems have arisen because of this diversity in the past, and they will continue to exist in the future, Canadians have recognized that these differences cannot be obliterated. Instead compromises must and can be found.

Objectives

- Demonstrate that Canada is and will remain a country of regions.
- Re-examine the core/periphery model in the light of globalization.
- Re-evaluate Canada's three faultlines not as centrifugal forces but centripetal ones.
- Stress that Canada's future lies not in division but in the 'process' of seeking compromises.

Canada:
A Country
of Regions

Introduction

'Canada is big—preposterously so', wrote geographer Kenneth Hare (1968:31). The sheer size of Canada has meant that the country spans a diverse physical geography, varying climate conditions, and a wide range of vegetation. These geographic variations, combined with regional histories, have transformed the country into a country of regions. Geography and history have woven Canada's regions into a federated nation and, in so doing, have shaped and then reshaped Canada's identity and the identities of its regions.

During the historical evolution of the country, tensions between certain groups have arisen. These tensions, known as faultlines, are manifested in three different contexts: between centralists and decentralists, between French-speaking and English-speaking Canadians, and between Aboriginal Canadians and non-Aboriginal Canadians. However, these cracks, rather than fragmenting the country, have forced Canadians to build bridges— between regions and governments and between different peoples. Canada, by incorporating cultural diversity and regional differences within its socio-political structure, provides an alternative to monolithic nation-states, where force settles differences. In this concluding chapter, five issues will be addressed: (1) Canada's regional character; (2) Canada's core/periphery spatial structure; (3) Canada's faultlines; (4) Canada's past; and (5) Canada's future.

Regional Character

The geographic complexity of Canada begins with its six geographic regions—Ontario, Québec, British Columbia, Western Canada, Atlantic Canada, and the Territorial North. These regions, though primarily created by geography, were shaped further by Canada's historical development. For instance, southern Ontario presented some of the most favourable geographic conditions for economic growth and industrialization. It had fertile soils and a moderate climate, which supported agriculture and a growing population, as well as geographic proximity to water transportation routes and to US markets. Complementing these favourable, natural features, historically Ontario was supported by economic conditions that allowed it to prosper, such as the Reciprocity Treaty and then protective tariffs—policies that ensured a stable market for the region's industries.

All of Canada's regions have been similarly shaped by a combination of geography and history. For instance, Québec has been shaped by the St Lawrence River and by its French-speaking inhabitants. British Columbia has been shaped by the Cordillera, which separates it from the rest of Canada, and by the railway that first connected it to the country. Western Canada has been shaped by its fertile soils and by a sense of alienation from the country's centre of power. Atlantic Canada has been shaped by its coastal geography and by economic downturns in its resource-oriented industries. The Territorial North has been shaped by its cold and harsh climate and by Aboriginal land-claim agreements. This is to name but a few of the geographic and historical features that have defined Canada's regions.

From each region's history and geography, a sense of place has developed. Regional geography is very much determined by a sense of place, the power of which has been elegantly expressed by Canadian writer, John Ralston Saul:

> Because, in spite of intellectual claims to the contrary, not religion, not language, not race but place is the dominant feature of civilizations. It decides what people can do and how they will live (1997:69).

Within each region of Canada, the collective experiences of people, both past and present, evoke a sense of belonging to a place. Like other geographic concepts, this psychological sentiment operates at several geographic levels—national, regional, local. In this book,

attention has been directed to regions and the common experiences within those regions that have fuelled a sense of place—for instance, grain farming in Western Canada.

Within a federation of regions, sense of place can take on political connotations as divisions between regions feed geographic 'separatism'. Though traditionally, Québec is identified with this sense of separation, almost every region experiences it to some degree. For instance, historian Margaret Ormsby (1958:257) claimed that 'British Columbia was *in*, but not *of*, Canada', meaning that union with Canada for British Columbia was not based on sentiment but on material advantage, such as the promise of a railway and the absorption of BC's debt by Ottawa. In British Columbia, history created a sense of independence while geography formulated a sense of isolation from the rest of Canada. That Canada 'ends' at the Rocky Mountains is part of this region's sense of identity. Another part of this identity, or sense of place, comes from British Columbians' pride in their natural surroundings—this feeling is expressed in the provincial licence-plate motto, 'Beautiful British Columbia'.[1]

If Canada is so clearly made up of very different regions that are separate from the whole, how did it ever overcome its geographic and historical differences to evolve from regional clusters into a modern nation-state? Why is Canada more than a collection of economic regions dominated by self-interest? Canada was able to overcome its regional differences through a willingness to share power and thereby create a balance between central interests and those of regions and groups. In fact, the sharing of power was embedded in the Canadian Constitution right from the start. With the creation of a federation, the BNA Act divided governing powers between the provinces and the federal government. This initial sharing of powers set the stage for a trend to develop in which different Canadian regions and peoples would share power and economic wealth. In fact, Canadian history illustrates this trend. A number of struggles between those holding power and those wanting a share of that power have occurred. Often, the result has been compromise, with power being shared between parties. Equalization payments (1957), official bilingualism (1969), multiculturalism (1971), and modern land-claim agreements (the first in 1975) are examples of how the federation has been able to share wealth and cultural recognition between various regions and peoples. Spatial and cultural diversity were not barriers to national unity that needed to be obliterated, but rather, served as the very building blocks necessary to construct a humanitarian society.

Canada's strength—you might even say what makes it interesting—is its regional diversity. Physical geography laid the groundwork for Canada's regional structure while Canada's history unfolded within these natural settings. Economic, political, and social change has occurred—some regions have grown swiftly while other have lagged behind and in the process lost some of their best people—but Canada's regions have proven remarkably stable. Furthermore, compromises have been reached to reconcile the inequities between regions as well as the peoples across these regions.

The Core/Periphery Model

Canada's regional geography has significantly affected the country's national and regional economies. Physiography, climate, permafrost, and other natural elements have in many ways determined what economic activities could be pursued in each region, and therefore how regional economies would develop. A way to examine this phenomenon is through the core/periphery model—cores represent the more prosperous and industrialized regions, which draw on the peripheries, the more resource-oriented, less economically diversified regions. Using Friedmann's version of this model (see Chapter 1), Canada's six geographic regions fall into: core regions (Ontario and Québec); upward transitional regions (British Columbia and, to a lesser degree, Western Canada); a downward transitional region (Atlantic Canada); and a resource frontier (the Territorial North). Furthermore, within each of Canada's geographic regions, natural variations have created internal core/periphery structures. For instance, within Ontario, southern Ontario represents an industrial core, while northern Ontario is a resource periphery.

For years, the core/periphery structure of Canada remained relatively stable. Central Canada benefited from geography and favourable federal policies that allowed its industrial sector to grow by guaranteeing a domestic market for its goods in other regions. The rest of Canada's regions acted as peripheries, supplying raw materials and purchasing the core's manufactured items. However, during the latter part of the twentieth century, a new global economic order has emerged that is affecting, and will continue to affect Canada's core/periphery structure.

Since the Second World War, Canada has taken several steps to a new economic goal, integration into larger economies. First came the Auto Pact, an agreement that integrated the North American automobile industry. Then FTA and NAFTA were signed in order to achieve uninhibited access to the American economy. Finally, Canada climbed aboard the GATT in order to participate more fully in the growing global economy. The net effect of trade liberalization has been to strengthen Canada's economic ties with the United States. For example, from 1992 to 1997, the percentage of Canada's exports to the United States increased from 75.5 per cent to 83.7 per cent (Statistics Canada 1998:1).

The effects of trade liberalization on Canada's regional economies have been mixed. All the regions have increased their service sectors but, except for Ontario, exports are predominately primary products, with most destined for the American market. Ontario, thanks to the Auto Pact, saw its automobile industry thrive, almost doubling its share of the North American market. But following the Free Trade Agreement in 1989, the rest of the manufacturing industry in Central Canada was either subject to fierce competition from American firms or to closure of American branch plants. As a result, both Ontario and Québec's manufacturing sectors went through difficult times.

For Canada's peripheries, the process of globalization accomplished two, long-sought-after goals—more local processing of primary products and access to foreign manufactured goods. New trade links with the United States, and to a lesser degree, Japan and Europe, quickly took shape. British Columbia, for example, benefited from the rapid economic expansion of Japan and other Pacific Rim countries. In the early 1990s, BC exported around 37 per cent of its goods to Pacific Rim countries, with Japan accounting for 25 per cent of this trade (IDAC 1998:6). However, by 1997, the economies of Pacific Rim countries faltered, chilling BC's export-dependent economy. Other hinterlands gained greater access to particular parts of the American market through liberalization. For instance, Atlantic Canada has access to the New England market. McCain Foods and the various Irving enterprises have taken advantage of the American market, but smaller producers were less able to achieve economies of scale and therefore have more difficulty selling their products in the American market. Certainly, trade liberalization has brought some economic benefits, especially through greater trade, but it has also come at some costs.

In becoming a member of NAFTA and GATT, Canada surrendered certain controls over its economy. No longer can Ottawa or the provinces impose policies reflecting economic nationalism, such as having one set of rules for foreign companies and another set for Canadian companies, or maintaining regional development programs that lend support to businesses competing with foreign firms. For instance, subsidies to sustain regional industries, such as the Cape Breton Development Corporation (DEVCO), or to reduce grain transportation costs for Prairie farmers by means of the Crow Benefit, are no longer an option for the federal government. In fact, Canada, being the smaller partner, has had to bend on sensitive trade issues that affect American producers. One example is the Canada–US softwood lumber agreement, which limits the amount of lumber Canada can export to the United States without facing an export tax administered by Ottawa.

Given the changes that have already resulted from trade liberalization, what future transformations await Canada in these times of global uncertainty? Three questions arise, all having implications for Canada's core/periphery structure:

- Will the economic regional alignment of Canada shift from an east-west alignment to a north-south one?
- Will the national core/periphery structure be supplanted by a continental or global core/periphery structure?
- As the powers of the federal government wane, is more and more of Canada's economic future in the hands of external market forces led by multinational corporations?

The north-south 'grooves' of geography naturally encourage north-south regional trade flows. For instance, trade moves more naturally between the provinces of Western Canada and the American states to their south than between these three provinces and the more distant provinces. With the erosion of trade barriers, north-south trade has increased rapidly. All along the US–Canada border, ever increasing quantities of products move between Canadian regions and American markets. Ontario trades with New York and the Midwest, Québec with New England and New York, British Columbia with the US Pacific Northwest, Atlantic Canada with New England, and Western Canada with the American Great Plains. The Territorial North is an exception. While its oil and gas flows southwards to American markets, mineral production has a more widespread distribution, including Europe and the Far East.

However, east-west trade, while growing more slowly, remains an essential element of Canada's economic structure. Recent economic studies by McCallum (1995), Helliwell (1996), and Anderson and Smith (1999) support the notion that the flow of trade between Canadian regions and provinces remains much more powerful than north-south trade between Canada and the United States, perhaps by a factor of 10 to 20 times. The explanation lies in the 'national border effect'. This effect is likely caused by a number of factors: the east-west Canadian transportation and communications system that encourages interprovincial trade; trade inertia based on pre-liberalization trading patterns between Canadian firms that remained strong in spite of slightly lower US prices; surviving trade barriers such as the milk marketing boards that shut out American milk imports and which, through the assignment of a large milk quota to Québec, ensures a high level of Québec milk sales to Ontario customers; and finally, Canadian taste preferences, such as Canadians consuming more domestic beer rather than lower-priced American beer.

Still, Canada's integration into the North

American economy has realigned Canada's industrial core and resource peripheries from within a national structure to within continental and global structures. The Canadian core serves both the North American market and the domestic one while the Canadian peripheries are receiving more of their manufactured goods from adjacent parts of the United States and from other industrial cores, and shipping more of their primary products to the United States and world markets.

However, the core/periphery structure of Canada's region's has so far remained relatively intact. For instance, Central Canada is still Canada's core. The resilience of the Canadian core/periphery spatial structure under trade liberalization is remarkable, though north-south trade flows have noticeably increased. This persistency represents a paradox: as regional economic growth occurs, the regional economic structure remains relatively stable. Geography, by limiting the economic opportunities available to each region, has promoted a particular type of economic growth in each region, even while foreign countries have provided new markets for this surplus production. At the same time, regional economies have undergone major modifications in response to trade liberalization through finding new markets and processing more primary products. For instance, livestock production and alternative crops in Western Canada have expanded because of increased trade with the United States and Japan. But economic integration of Western Canada with the outside world makes this region's prosperity dependent on continued exports to these countries.

In re-examining the core/periphery model in the light of trade liberalization, much change has taken place and more is expected. Canada's core now forms a subset of the American industrial core and, as a result, competes for customers in all regions of North America. Canadian peripheries are changing too. With new trade opportunities, especially in the United States, regional peripheries are realigning their economic ties along continental and global lines. Among the peripheries, however, there are significant differences. Those peripheries that have a rich resource base and a highly skilled labour force are accelerating their economic diversification and are moving toward a more independent economic status. Those peripheries that have a weak economy, high unemployment rates, and an outflow of young people are becoming more and more dependent on Ottawa and on world market prices for their resources.

The global economy has taken Canada and other countries into unexplored territory. Economic structures like trade agreements are in some ways overcoming political structures, like national borders. Furthermore, multinational corporations, growing ever larger and more powerful, hold more and more of the world's investment capital and therefore have a significant impact on global economic structures. Capital, now the key element in the new market economy, is so mobile and multinational firms are so powerful that a new political order will have to emerge to regulate this world economy. Of particular concern is the need to ensure global environmental standards and labour practices, as well as maintain employment levels in Western economies, while bringing greater development to global peripheries. Varying opinions exist on what form a new political order should take, but most agree that political globalization, which can match the economic globalization already underway, is greatly needed (Wallerstein 1997:141 and 158, Dicken 1998:462–3).

What will be the consequences for Canada if an efficient political global order is not established? Stripped of more and more of its economic powers, Canada will, at best, retain control over its domestic affairs. Even domestically, however, international market forces, unless kept in check by a new political order, can exert exceptional pressure on governments. Take Canada's tax system, for instance. Ottawa is under pressure to lower Canadian taxes to the level of US taxes. While such a policy would keep more money in the pockets of Canadians, it could lead to the underfunding our public health-care system, the slashing of social programs, and the widening of the income gap between Canada's regions.

Canada's Faultlines

Canada is a complex and dynamic country. Part of its complexity is rooted in the three faultlines that illustrate the main tensions within Canada and its regions. The first faultline, between non-Aboriginals and Aboriginal peoples, has existed since well before Confederation. Since Europeans first arrived on what is now Canadian soil, Aboriginal Canadians have struggled with non-Aboriginal Canadians over such issues as land, rights, the environment. The second faultline, between English-speaking and French-speaking Canada, grew out of the two founding European nations that first settled Canada. The third faultline, between centralists and decentralists, arose once governing and economic structures began to create have and have-not regions. While Canada's early history is characterized by imposed solutions to these faultlines, the country's wisest leaders, acknowledging the strength of Canada's complexity, sought compromises instead. Compromise usually took the form of shared power rather than power enforced by the dominant party. This trend of sharing power allowed Canada to evolve into a pluralistic society where both individual and collective differences are accepted and respected.

The sharing of power, however, has not been an easy accomplishment. For instance, in the past, Aboriginal peoples felt the sting of 'imposed solutions' that led to both their geographic and social marginalization within Canada—through such mechanisms as reserves and residential schools. Today the place of Aboriginal peoples within Canada has changed. A shift in power is embodied in modern land-claim agreements (Sparke 1998:463–95). The Nisga'a, for example, are close to reaching a compromise through negotiations with the rest of Canada (Joyce 1999:B8). For the Nisga'a, a land-claim agreement with British Columbia and the federal government will secure territory, guarantee a share of resources, and establish a locally designed form of self-government. The Nisga'a will regain their lands in northern British Columbia and then begin the process of re-establishing their place within Canadian society. This journey of reconciliation between non-Aboriginals and Aboriginal peoples through land-claim agreements began in the 1970s and will continue well into the twenty-first century. For some, like the Métis and non-status Indians, the process may be more difficult.[2]

Canada's French/English faultline is perhaps the one that is the most deeply felt. Since Confederation, these two linguistic groups have struggled for power. Over time, French Canadians have won compromises from the more dominant power of English Canada through such policies as bilingualism. Despite the cultural divide between the two groups, the fact that both societies have contributed to Canada's unique diversity is recognized. However, the rise of Québec nationalism as well as a series of failed attempts at constitutional reform have made the need for further reconciliation even more urgent. Gestures, such as the Calgary Declaration following the 1995 Québec referendum, will need to be backed up by a greater sharing of power and compromise between Canada's two major linguistic groups and visions. The possibility of Québec's separation from the rest of the country demands this issue be addressed.

Canada's third faultline, between centralists and decentralists, stems from one of Canada's most prominent features—regional diversity. Regional diversity has led to regional disparity—some regions are endowed with both political and economic power while other regions lack both. This disparity has generated a tension between those regions having power, as well as the federal government, with those regions struggling for more power. Though federalism already affords a sharing of power between the central government and regional (provincial and territorial) governments, some argue power is not shared equally. Because of this sentiment as well as other issues, like the unilateral cuts in transfer payments to the provinces by Ottawa in the 1990s, provincial governments have been demanding a greater share of power, including more taxing power. With trade liberalization and the federal government's reduced ability to inject money into depressed regions of the country, centralist/decentralist tensions will only intensify. Compromises will have to be reached so that provinces are better equipped to deal with their socio-economic responsibilities, while still maintaining a level of power for the central government that ensures the preservation of national standards among the services available to Canadians in all regions.

Canada's faultlines continually challenge the country and its regions, but these dynamic struggles also strengthen the country. One of Canada's leading thinkers, John Ralston Saul, sees Canada's faultlines as pillars. In his book, *Reflections of a Siamese Twin: Canada at the End of the Twentieth Century*, Saul argues that the Aboriginal, the Francophone, and the Anglophone are dependent on each other. They form Canada's triangular reality. 'Each of their independent beings has been interwoven with the other two for over 450 years of continuous existence on the northern margins of the continent' (1997:81). He further argues that this interdependence has, over time, transformed Canada into a special place, one that has avoided the rigidity of nation-states that compel their citizens to conform. Canada's faultlines, in a sense, rather than dividing the country, have managed to unite its different regions and peoples.

The Last Century: An Overview

Since 1867, Canada's society and economy have undergone a remarkable transformation. At Confederation, Canada was a small rural society of some 3 million people clustered into four regions of Canada (New Brunswick, Nova Scotia, southern Ontario and southern Québec). Canada's population was composed of Aboriginal peoples and settlers from two main ethnic groups—English and French. Eventually, the country's territory expanded and new provinces joined or came into being. The country was rich in resources but small in population and therefore seemed destined to produce primary goods for export markets. Ottawa, recognizing the problem of an economy entirely devoted to primary production, set in motion an economic policy that imposed high tariffs on imported manufactured goods. That policy stimulated industrial growth in southern Ontario and southern Québec and encouraged foreign firms to build branch plants in the emerging industrial core of Canada. The same policy forced the rest of Canada to buy Central Canada's manufactured goods but to sell their primary goods on world markets. The settling of Western Canada added a third dimension to Canada's population—people from northern, central, and eastern Europe added to the diversity of Canada's population and culture. As Canada grew more

independent from Britain, it had to face the harsh realities of federalism, including Aboriginal issues, the two founding peoples concept, and regional self-interests.

At the end of the 20th century, Canada is an independent nation consisting of ten provinces and three territories. Technological advances have transformed Canada. Natural barriers like the Canadian Shield and the Rocky Mountains have been tamed by innovations in transportation; advances in plant and seed genetics have shortened the growing time for crops and provided more crop alternatives in Western Canada; new techniques such as insulated buildings, winter roads, and utilidors have helped overcome the cold environment found in the Territorial North. Political decisions have also transformed Canada. The liberalization of immigration and trade has had a profound impact on Canada's society and economy. Immigration from a broader range of countries is creating a more cosmopolitan society with strong international links. Free trade has moved Canada into a North American regional economy and strengthened its economic links with other countries of the world. Canada is now a more urbane society composed of Canadians whose origins are from all corners of the globe. While progress has been made in building compromise among Canada's three faultlines, more work remains. In fact, Canada's faultlines will continue to demand attention because absolute solutions are impossible within such a diverse but democratic country—but then that combination of democracy and diversity has become Canada's true strength.

The Next Century

First and foremost, the feature that characterizes Canada and makes it strong—diversity—
has evolved through compromise. But compromise is never fully achieved: it takes
ongoing and committed efforts to maintain. In
the next century, Canadians will face several
challenges. The most daunting issue remains
the French/English faultline, one that could
literally split the country. Canada will need to
come to an accord with Québec. The Calgary
Declaration was a very small step in that direction. Except for hard-core separatists, few
Canadians can contemplate their country without Québec and some fear that Québec separation would lead to Canada's disintegration.
The potential outcomes of Québec separating
are sobering to consider. Many argue that once
split, the durability of 'the rest of Canada' is
uncertain. For instance, history and geography
have shown that discontinuous states have
short life-spans.[3]

Canada's regionalism, if not fostered with
ongoing compromise, also has the potential to
fragment the country. Some, like writer Joel
Garreau, argue that Canada's regions could be
'picked off', one by one, and absorbed into the
United States. Robert Kaplan, an American
writer, argues that economic integration with

**Diversity—among its regions, peoples, geography, and history—is what united Canada in the
past. In the future, Canada and its regions will need to continue fostering this unique and essential characteristic. (Courtesy Andréa Vanasse)**

the United States will inevitably lead to regional self-interest fracturing Canada into new political amalgamations. For instance, British Columbia, Kaplan speculates, could become part of the Pacific Northwest, which might be renamed 'Cascadia' (Kaplan 1998:322–4). Economists Courchene and Telmer (1998: 1–3) support this economic integration thesis by declaring that Ontario is no longer Canada's heartland but is now a North American region state. They also base their thesis on 'regional self interest'.

However, geographic regions and their senses of place are enduring. In the next century, the six major regions of Canada will continue to exist. But are Canada's regions likely to become part of a new spatial arrangement (i.e. North America), as predicted by some experts? Or is Canada and its regions an idea driven by more than just economic concerns? Clearly, to ensure that Canada is enduring, internal tensions will have to be addressed. This demands a political will and a recognition that Canada is not a homogenous nation-state but a pluralistic country where regional differences not only exist but are cherished by those living in these regions. Canada's survival will depend on our ability to accept the country's geographic structure, and as a result, to recognize that Canada's identity is a regional matter (Frye 1971:i–iii). In the twenty-first century, Canada will need to build cultural bridges between its regions and, in so doing, strengthen Canada's identity and unity.

Accepting Canada's cultural diversity, then, is the linchpin holding Canada and its six regions together. The future, while full of uncertainly, contains two certainties: Canada will remain a heterogeneous country and, as such, will continue to experience differences. However, by engaging in open and vigorous debate, compromises—not solutions—will follow. Compromise strikes at the heart of the exercise of democratic power in a diverse society by demanding a balance between individual and collective rights, between regional and national interests, and between those with power and those in need of it. The ability to strike this balance will determine whether the next century will bring further compromise among Canada's regions or irreversible divisions.

Notes

1.   Geographers view cultural landscapes, particularly cityscapes, as creating a human imprint that evokes a sense of place. However, this human imprint has been affected in recent years by a sense of urban placelessness wherein all cities seem alike due to a spatial standardization by global-based firms like McDonald's and Shell. Uniformity is erasing the uniqueness of place (Relph 1976:79–80). While the impact of standardization of cultural landscapes is most apparent in urban areas, it does exist elsewhere, especially in international tourist centres. Some cities have opposed such imposed architectural regularity in an effort to conserve their unique cultural landscapes. For instance, Québec City forced McDonald's to construct a fast-food restaurant that matched the architectural style of surrounding buildings in old Québec City.

2.   The ability to negotiate modern land-claim agreements varies among Aboriginal peoples. Those

Aboriginal peoples who did not sign treaties are best positioned to negotiate *comprehensive* land-claim agreements, which generally include rights to land, resources, and self-government. First Nations in British Columbia, the Territorial North, and Labrador fall into this category. Aboriginal peoples who have signed treaties but have not received all the benefits entitled to them under their treaties can seek redress under *specific* land-claim agreements. Aboriginals who are classified as non-status Indians by the federal government have lost their legal right to land claims.

3. In today's world, discontiguous states are rare. Of the more than 200 nations in the world, less than ten states are divided. Most are small enclaves, such as Kaliningrad, which is separated from Russia by Lithuania and Belarus. Others have evolved into separate states, such as Pakistan, which existed as two parts for 25 years, but saw its eastern lands form a new state, Bangladesh, in 1972. If Québec separated, Atlantic Canada would be cut off from the rest of the country, much like Alaska is from the rest of the United States. Alaska, like Atlantic Canada, is a large, remote region that relies on a resource economy. However, several differences exist between Alaska and Atlantic Canada. Alaska has less than 1 per cent of the US population while Atlantic Canada represents 8 per cent of Canada's population. Furthermore, unlike Alaska, Atlantic Canada relies more heavily on road transportation and road links to keep its economy functioning. Finally, while both regions receive large subsidies, the cost of supporting an isolated Atlantic Canada would place a much greater financial burden on the rest of Canada than Alaska now places on the United States.

References

Bibliography

Anderson, Michael A., and Stephen L.S. Smith. 1999. 'Canadian Provinces in World Trade: Engagement and Detachment'. *Canadian Journal of Economics* 32, no. 1:22–38.

Barnes, Trevor J. 1993. 'A Geographical Appreciation of Harold A. Innis'. *The Canadian Geographer* 37, no. 4:352–64.

Clark, Joe. 1994. *A Nation Too Good to Lose: Renewing the Purpose of Canada*. Toronto: Key Porter Books.

Courchene, Thomas J., and Colin R. Telmer. 1998. *From Heartland to North American Region State: The Social, Fiscal and Federal Evolution of Ontario*. Monograph Series on Public Policy, Centre for Public Management. Toronto: University of Toronto Press.

Dicken, Peter. 1998. *Global Shift: Transforming the World Economy*. London: Guilford Press.

Frye, Northrop. 1971. *The Bush Garden: Essays on the Canadian Imagination*. Toronto: Anansi Press.

Garreau, Joel. 1981. *The Nine Nations of North America*. Boston: Houghton Mifflin.

Hare, F. Kenneth. 1968. 'Canada'. In *Canada: A Geographical Interpretation*, edited by John Warkentin. Toronto: Methuen.

Helliwell, John F. 1996. 'Do National Boundaries Matter for Quebec's Trade?'. *Canadian Journal of Economics* 29, no. 3:507–22.

Innis, Harold A. 1930. *The Fur Trade in Canada: An Introduction to Canadian Economic History*. New Haven: Yale University Press.

The Investment Dealers Association of Canada (IDAC). 1998. *British Columbia: Economic & Fiscal Outlook*. Vancouver: The Investment Dealers Association of Canada.

Joyce, Greg. 1999. 'Treaty with the Nisga'a'. *Star Phoenix* (13 April):B8.

Kaplan, Robert D. 1998. *An Empire Wilderness: Travels into America's Future*. New York: Random House.

McCallum, John. 1995. 'National Borders Matter: Canada–US Regional Trade Patterns'. *American Economic Review* 85, no. 3:615–23.

Ormsby, Margaret A. 1958. *British Columbia: A History*. Toronto: The Macmillans in Canada.

Pally, Thomas. 1998. 'Building Prosperity from the Bottom Up'. *Challenge* 51, no. 5:59–71.

Relph, Edward. 1976. *Place and Placelessness*. London: Pion.

Saul, John Raltson. 1997. *Reflections of a Siamese Twin: Canada at the End of the Twentieth Century*. Toronto: Viking.

Sparke, Matthew. 1998. 'A Map that Roared and an Original Atlas: Canada, Cartography, and the Narration of Nation'. *Annuals of the Association of American Geographers* 88, no. 3:463–95.

Statistics Canada. 1997. The Daily—Provincial GDP and Interprovincial Trade 1996, 16 May 1997, [online database], Ottawa. Searched 14 October 1998; <URL:http://www.statcan.ca/Daily/English/970516/d970516.htm>:1–20.

_____. 1998. *Imports and Exports of Goods on a Balance-of-payments Basis*, [online database], Ottawa. Searched 14 October 1998; <URL: http://www.statcan.ca/english/Pgdb/Economy/International/gblec02a.htm>:1.

Wallerstein, Immanuel. 1998. 'Contemporary Capitalist Dilemmas, the Social Sciences, and the Geopolitics of the Twenty-first Century'. *Canadian Journal of Sociology* 23, nos 2/3:141–58.

Further Reading

Britton, John N.H., ed. 1996. *Canada and the Global Economy: the Geography of Structural and Technological Change*. Montréal–Kingston: McGill-Queen's University Press.

Penrose, Jan. 1997. 'Construction, De(con)struction and Reconstruction: The Impact of Globalization and Fragmentation on the Canadian Nation-State'. *International Journal of Canadian Studies* 16:15–50.

Saul, John Raltson. 1997. *Reflections of a Siamese Twin: Canada at the End of the Twentieth Century*. Toronto: Viking.

Index